PELICAN BOOKS

BIRDS, BEASTS, AND MEN

H. R. Hays holds a B.A. from Cornell and a master's degree from Columbia. He is the author of several novels, and more than twenty of his plays have been presented on television. His nonfiction books include the highly acclaimed *From Ape to Angel.* Hays has taught playwriting at the New York Writers' Conference at Wagner College. He has also been head of the Drama Department and assistant professor of English at Fairleigh Dickinson University, and associate professor of English and head of the Drama Department of Southampton College of Long Island University.

H. R. HAYS

BIRDS, BEASTS, and MEN

A Humanist History of Zoology

Penguin Books Inc
Baltimore · Maryland

Penguin Books Inc
7110 Ambassador Road
Baltimore, Maryland 21207, U.S.A.

First published by G. P. Putnam's Sons,
New York 1972
Published in Pelican Books 1973

Printed in the United States of America by
Kingsport Press, Inc., Kingsport, Tennessee 37662

THERE IS A STORY THAT, WHEN SOME STRANGERS WHO WISHED TO MEET HERACLEITUS STOPPED SHORT ON FINDING HIM WARMING HIMSELF AT THE KITCHEN STOVE, HE TOLD THEM TO COME BOLDLY IN, FOR "THERE ALSO THERE WERE GODS." IN THE SAME SPIRIT WE SHOULD APPROACH THE STUDY OF EVERY FORM OF LIFE WITHOUT DISGUST, KNOWING THAT IN EVERY ONE THERE IS SOMETHING OF NATURE AND OF BEAUTY.

—ARISTOTLE, *De partibus animalium*
Book I, 5.

The author wishes to express his gratitude to Dr. Daniel Lehrman, professor of psychology, director, Institute of Animal Behavior, Rutgers University, to Dr. John Raul Welker, assistant professor of marine zoology at Southampton College of LIU, and to Sir Julian Huxley, who read the manuscript before publication, for their valuable comments and criticism.

Contents

Introduction

❧ CONCERN with the animal world takes on a new urgency in the contemporary period. The glittering promises of science have been partially fulfilled, but what no one foresaw was that science could take away with one hand what it gave with the other. The natural balance is gone forever. The bulldozer, insecticides, waste of all kinds, the phenomenon of megacities, automobiles, even the indestructible presence of plastic containers are altering the landscape and, day by day, the very air we breathe.

And what of the other living creatures with which we share the planet? Are we to brush them aside, are we content to exterminate them? Or are they an important factor in keeping us human? Do we wish to preserve the planet in its many-colored variety?

The study of zoology really needs no justification. Any one who has ever kept tadpoles in a jar or owned a pet can testify to the fascination animals hold for us. But in addition to this, although man's attitudes toward other living creatures have varied, from them he has always learned something about himself. And so today, when we are reappraising our position in nature, when we are being forced to decide what we wish to make of this world we live in before it is too late, there seems to be a wave of almost nostalgic interest in the creatures with whom we share it.

The decision to write a history of zoology stems not only from the above facts but also from the discovery that so little has been written about one of the oldest of the natural sciences. Zoology is the study of life, one of the most fascinating subjects of human investigation. After centuries of intellectual contemplation it still remains a mystery and a challenge. Some of the greatest minds of the past have concerned themselves with our study; as the science unrolls, we become more and more awestruck at the complexity of vital processes; and last but not least, many fine prose writers have been moved to eloquence by their enthusiasm for natural history. Even the insects are not without their Homer.

In this book, therefore, the changing points of view toward other forms of life and the great discoveries in the science of zoology will be sketched. The story will be told in terms of the lives and contributions of the most important scholars, and an attempt will be made to relate them to the cultural and intellectual backgrounds of their periods.

If the material is not to sprawl uncontrollably, some limits must be set. Zoology has, during the centuries, become fragmented into many subdivisions: taxonomy (classification); anatomy; physiology; embryology; microbiology (which covers both tissue and cell structure); paleontology (the study of fossil forms of life); zoography; genetics (which is tied up on the one hand with evolutionary theory and on the other with biochemistry); biochemistry itself, which breaks life processes down into chemical reactions; animal psychology and ethology (animal habits); and, finally, ecology, a relatively new area of interest, the relation between living things and their environment, a subject intimately connected with the survival of the human race.

It can be immediately seen that the study of animal life touches on many abstract and technical areas of science. Mathematics enters into biochemistry and genetics. When we dive below the skin and separate living tissue into its components, we are operating in an invisible world. Although in telling this story it is necessary to indicate how these subsciences came about, we shall keep an eye on the beast as a whole, returning to the natural history point of view as often as possible.

In the beginning man was a hunter, and every hunter is in some sense a naturalist. Among preliterate hunting cultures zoological studies tend to be pragmatic. Nevertheless, as Claude Levi-Strauss points out, Indians and other aborigines are not without intellectual interest in the world around them. He extensively quotes various authorities to show that so-called primitive people are careful to name all the animals and plants in their environment, even those not of immediate use.

"The thought we call primitive is founded on this demand for order. This is equally true of all thought, but it is through the properties common to all thought that we can most easily begin to understand the forms of thought which seem very strange to us." Thus the preliterate, in common with European man, is a taxonomist or classifier and since he studies animal habits in order to capture the creature on which he lives, an early natural historian. His achievements do not lead to modern scientific tradition because his nomenclature is always local, his interpretation of animal behavior is often magical and in any case not written down.

It is well to remember, however, the respect aborigines have for other living things. They often claim blood kinship with them and nearly always admit to a spiritual relationship. Dependent as they are on wild game, they very sensibly develop a mythology that threatens with heavy sanctions those who kill wantonly or waste the bodies of wild creatures. Thus the preliterate can be said to be an instinctive ecologist.

Keeping this in mind, we can pass over the early urban people whose attitudes were not very different from the preliterates' and plunge into our subject with the Greeks and, above all, that great genius of Athens in the fourth century B.C., Aristotle.

BIRDS, BEASTS, AND MEN

1

The Invention of Science

☙§ SCIENCE had to be invented. It should never be forgotten that, like other human activities and conceptions, it was created by the human mind. We now take the attitude that scientific statements are provisionally true if they seem to work. The scientific point of view is not easy to define in a few words. It is, however, a radically different way of looking at the world from the magico-poetic one that we call religious. In the ancient cities of Babylon and Egypt, a strong self-perpetuating priesthood was interested in upholding the dogmas of religion and the supernatural attitude toward life. Thus the invention of science did not take place. In Greece, however, with its disunited city-states, there was no authoritarian king, no united priesthood and no state ideology to keep individual thinkers in line. Except for a few official holidays, religion was almost an individual affair. Thus there was room for private thinking and outbursts of brilliant originality truly amazing for their variety and number in so short a period of time. And it *was* a short period considering that in a couple of hundred years of Greek thinking the basis was laid for the modern world.

In the sixth century, in the Ionian colonies of Asia Minor, Thales (650-580 B.C.) and his follower, Anaximander, became interested in what we call facts. That is, in data that it is possible to verify through the perceptions of the senses, plus the use of reason. Reason itself is not enough. The idealist Greek philosophers were great reasoners and when they had fitted words together neatly, felt they had proved what they set out to prove. But this is not the method of science or of those tough-minded Ionians. Facts must be *observed* again and again, and if they occur many times in the same relationship, we may provisionally accept them as a basis for further investigation.

At any rate Thales and Anaximander possessed the patient down-to-earth kind of mind. Thales, for instance, picked up the records made by the Babylonian priests of the movements of the planets and was able, in 585, to predict an eclipse. The Babylonians were interested in the stars only to predict future historical or human events and to cast the horoscopes of individuals. Astrologers are careful never to check statistically on the success of their predictions. Thales could point to the eclipse, as solid a fact as one can get.

Myth had operated on the basis of fantasy, association, or analogy. Since thunderbolts were like hurled weapons, it was stated that someone with the name Zeus threw them from the sky. But as we know, scientists were eventually to discover they could manufacture miniature thunderbolts in the laboratory without the aid of the ruler of Olympus.

The scientific type of mind, therefore, is inclined to rule out the supernatural when it cannot be substantiated by fact. In the sixth century, however, it was a bold breakthrough for Xenophanes, a disciple of Anaximander, to point out: "If horses or oxen had hands and could draw or make statues, horses would represent the forms of gods like horses, oxen like oxen." Freedom from superstition of this sort is absolutely necessary for rigorous scientific thinking. Once it is attained, instead of staring at diseased livers in order to prophesy the outcome of battles, men attempt to study the function of the liver as a part of the digestive system. Instead of observing the flight of birds in order to decide whether the gods are pleased or not, men attempt to identify the birds and study their migrations. And as regards our own science, instead of placing imaginary zodiacal animals in the sky, men take a look at real animals and work out a system of classification.

As it happened, those who followed the Ionians were not of the same temperament. The mixture of poetry and magic called religion continued to possess political power for many centuries. It sometimes hindered and never helped the progress of science.

This does not mean that myth and poetry do not shed profound beauty on human life—as long as they are understood as myth and poetry and not taken literally or used for the wrong reasons by priests and politicians.

It so happened that Socrates and Plato succeeded in overshadowing the beginnings of science with their own poetic mythmaking and Aristotle (b. 384 B.C.), who for centuries was to cast a long shadow over the young science of zoology, started out under the influence of Plato. When Plato was sixty and had been head of his academy for about fifteen years, Aristotle came to Athens from the northern Grecian town of Stagira, on the Macedonian coast. His father had been physician to the royal family of Macedonia, and although he died when Aristotle was still young, the youth received some training in what was known of biology and medicine.

Coming from a kind of frontier kingdom, from a court which was barbarous in comparison with luxurious and sophisticated Athens, Aristotle naturally fell completely under the spell of his master. He wrote dialogues in imitation of Plato's and accepted the latter's doctrine of ideas. This doctrine was, of course, one of the most imposing constructs of human reason not based on any factual verification. In it Plato maintained that spiritual (verbal) concepts such as beauty, virtue and love were basically as real as a loaf of bread—in fact, more real because the bread would decay and disappear, while the "ideas" were a reflection of the ultimate eternal idea, which was the deity.

When Plato died, the control of the academy passed to his nephew, a man much inferior to Aristotle. The Macedonian left the academy, and from then on his attitude toward Platonism became critical. He was, he discovered, temperamentally at odds with idealistic philosophy; his bent was toward facts and observation, while Plato distrusted the senses and devoted his thinking to imaginative concepts.

Aristotle, on the whole, was a mixed character. Although his speculation was often daring, apparently his early life had made him politically a conservative and a monarchist. He spent some time in Assos, where he married happily, and in 338 B.C. Philip of Macedon invited him to become the tutor of the then-thirteen-year-old Alexander the Great. It is likely that the offer was made to him because of his father's former position at court. He held the post from 338 to 335. The record of philosophers as the mentors of monarchs is discouraging. Seneca, the tutor of Nero, ended by committing suicide. Although Aristotle seems to have maintained good relations with Alexander, his tutorial period is suspiciously short, and judging from the conqueror's tumultuous and uncontrolled career, very little philosophical balance was inculcated by the master.

When Alexander was in control of the Greek world, Aristotle moved back to Athens, where for twelve years he wrote, studied and taught an ever-increasing number of pupils. For his school he was granted the use of a temple dedicated to Apollo Lycaeus, after whom the institution was called the Lyceum, which accounts for the use of the word in the names of so many scientific and educational institutions over the centuries.

Aristotle, a one-man university, taught practically everything that was known in his time. Philosophy was then considered the master study, including logic, metaphysics, art, literary criticism, politics, psychology and biology. Many students and colleagues collaborated with him.

He is described as small and fat, proud, his manner arrogant and sarcastic, his dress and general life-style elegant and refined.

Fragments of a dialogue, which may have been written when he was living in Assos, indicate that by then he was already attacking the Platonic doctrine of ideas. Aristotle denied that these verbal concepts could have any real existence apart from the objects perceptible to the senses which embody them. In other words, beauty exists in a beautiful woman. He also maintained that mathematics, its problems and its proofs were abstractions created by the human mind. His scientific temperament convinced him that the reality lay in the observed world, not in some higher order of spiritual entities of which sight, hearing and the sense of taste and touch provided only a fleeting reflection.

Yet in many ways he never ceased to be a Platonist. He held to the notions that there was a hierarchy in the universe in which the souls of living things aspired toward more perfect form. Although he considered motion to be the stuff of existence, an idea not foreign to modern physics, the various forms of motion were supposed to grow more perfect in the hierarchy of earth, celestial bodies, and finally God, the supreme intelligence and primal source of motion.

Yet his statement, "Motion cannot come into existence or pass away; in other words, it has always existed," puts him into the scientific camp and is a more factual starting point than Hesiod's poetic account of the wars between the older and younger gods.

Although the atomic theory of Democritus (*circa* 460-370 B.C.), in which all matter is believed to be composed of tiny particles moving in empty space, also approaches modern physics more closely than Aristotle's vitalism, Aristotle's hierarchy of form-perfection seeking entities is the first pattern of evolution, which, while purely imaginative speculation, nevertheless is the germ of later theories. Democritus, in contrast, saw only a casual, almost chance change in forms. Aristotle believed the world had been created as a consistent ladder of development from the simple to the complex.

The Lyceum rented a number of buildings, and turned them into a library and a museum of materials which illustrated the lectures. Aristotle discoursed on more difficult abstract subjects in the morning and gave popular lectures in the afternoon. Students, when not attending lectures and discussion, did research, collecting data to substantiate human and natural history. Thus the biological treatises contain a large account of factual material (not always accurate) concerning the structure and habits of plants and animals. Hunters and particularly the Mediterranean fishermen were perhaps the most valuable sources; indeed Aristotle's writing on marine biology is some of his soundest work.

Although the philosopher did not meddle in politics, after Alexander's death, when Athens rebelled against Macedonian supremacy, he fell into disfavor politically. His association with Alexander was enough to annoy political extremists, and since he had done nothing for which to be prosecuted, charges of "godlessness," like those used against Socrates, began to be leveled against him. To escape prosecution he fled to the island of Euboea, where he died in 322 B.C.

Aristotle worked in so many areas of zoology that his contributions need to be reviewed in various categories. To begin with classification, it is clear that he grasped the basic principles, for he points out that species should have a number of characteristics in common in order to be grouped together in what he calls a genus, one of which would include, for instance, the horse, the ass and the onager. Since he had no organized picture of anatomy, however, he failed to set up enough structural criteria to form the basis of a system. He did, however, distinguish between analogical structures such as bat's and bird's wings and identical structures differing only in matter of degree or type, such as fish scales and scales of reptiles. He divided the animal kingdom into what he vaguely called classes—birds and fish, for instance—and he also divided the whole spectrum into creatures possessing red blood and those without it, now called vertebrates and invertebrates. He did not go much further than this, and in his books on natural history he did not treat one group intensively but ranged over the whole kingdom, first anatomically, again summarizing habits, and still again analyzing reproduction, and again treating movement.

Of course, it must be kept in mind that he probably did not edit the works in the versions we have. Many collaborators helped him. Some of the material is considered spurious, while some books may be simply lecture notes taken down by students and hence garbled.

Nevertheless, the range and amount of detail (accurate and inaccurate) are amazing for a first attempt at a natural history treatise: 520 species which can be identified today were discussed. All these species are found in the Mediterranean area.

When he followed other authors, he was often uncritical, since he had no way of checking their material and even in his own thinking there was a tendency to accept folktales. Many of these, once embalmed in his pages, were passed down through history by subsequent authors. His description of the Indian manticore, which he borrowed from Ctesias, a rather gullible physician and historian, is a case in point. "It has a triple row of teeth in both upper and lower jaw, is as big as a lion and as hairy, resembles a man in the face, it has a sting in the tail and the ability to shoot off spines, attached to the tail like arrows." However, he did preface his account with the remark, "If Ctesias is to be believed. . . ." On the other hand, he was perfectly aware that dolphins and whales were warm-blooded and that they breathed air and suckled their young. He never made the mistake of calling them fish.

He wrote: "Some animals share the properties of man and the quadrupeds, as the ape, the monkey and the baboon," a statement remarkable for its calm objectivity.

It is always pointed out that he was most at home in marine biology. The Greeks ate shellfish, squids, sea urchins, lobsters and crabs, sliced sea anemones, as well as all kinds of fish. Since Aristotle lived well, it is evident that he easily took an interest in the creatures that appeared on his table. He divided his nonblooded animals into mollusks (squids and octopuses which we now call cephalopods), Malacostraca (crabs, lobsters and shrimps, which we now call crustaceans), testaceans (including all the shellfish), and insects. Many of these he actually dissected, for we have his drawings which display knowledge of their anatomy.

In the *Historia animalium,* the early books are a catalog of animals, mostly described in terms of their anatomy. In this area Aristotle inherited certain concepts. Hippocrates, the famous physican (460-377 B.C.), and his school had worked out the system of the four "elements"—fire, air, water, and earth. Attitudes toward these were still slightly magical. In living bodies four juices corresponded—blood, phlegm, yellow bile (from the liver), black bile (from the spleen). Harmony resulted if all were in proper proportion; sickness if one predominated excessively. It was a neat system that was to have a long life in medical practice; its only weakness was that it had no real connection with physiology. The Hippocratic school had no notion of the circulation of the blood or of digestive processes or of the function of the lungs. Since dissection of human beings was frowned upon, these physicians apparently worked exclusively with animals. Actually, the fear of the dead and the belief that they could affect the living, and in particular cause sickness, is a universal primitive belief, which has had a long life as a vague superstition. It was an obstacle to anatomists. The Hippocratic physican knew more about bone structure than any other area of anatomy because skeletons were somehow less surrounded with magical sanctions.

Aristotle was rather subtle in his application of earlier ideas. He made much of heat and cold and, in common with the Hippocratic school and an earlier philosopher, Heraclitus, felt that heat was closely associated with the soul (or rather as we should say, life). The association fire-blood-life or soul caused him to locate all the nobler sense activities in the heart. Since the heart was hot and heat rose from it he reasoned that this accounted for man's upright position. This is a good example of a case in which sound Aristotelian logic did not lead to anatomical progress. Having glorified the heart (as poetry has always glorified it since), he degraded the brain. It was cold, bloodless and fluid, composed of earth and water. About its only function was chilling the blood to cause sleep. He believed he had proved his case because he discovered that the exposed brain, when touched, exhibited no feeling. Hence it had nothing to do with sensation. In this he opposed both the later Hippocratic physicians and the atomic philosopher Democritus. Yet he knew enough about the brain to mention what he called a blood vessel between it and the eye, undoubtedly the first record of the optic nerve.

Most of his physiology is set forth in *De partibus animalium,* a book which includes many repetitions of his anatomical descriptions. He granted viscera only to blooded animals and hence refused to see a stomach or liver in insects. He knew that the lungs introduced air into the blood (for cooling purposes), but he saw only three chambers in the heart and did not distinguish between veins and arteries. In this, of course, he was hampered by lack of injection techniques; in the dead animal the blood ceases to dilate the vessels.

His version of nutrition is interesting. The edible portion of the food was "cooked" in the intestines. Since chemistry had not yet been invented, this was an insight, for his period, reasonably close to actuality. Indeed, his use of the element heat as the active agent for change (which suggests oxidation) becomes almost a poetic metaphor for all biochemical processes. Once cooked, the nutriment became a vapor which passed by a kind of osmosis to the heart where it became a serum, was turned into blood, and was fed to the body organs. Again we feel that Aristotle was like a man rowing without oars yet in some mysterious way making progress. He arrived at a kind of metaphor of metabolism.

He carefully described liver, gall and kidneys and decided that they had to do with the elimination of wastes. He had obviously dissected the genito-urinary systems of various animals.

In most cases when he was out of his depth, it was either because he lacked the technique and tools or because he got information secondhand. For instance, his description of the lion, following Herodotus, states that the lion's bones are so hard that they give off sparks like flint when struck and are said to have no marrow.

His ability to unite observation with categorical thinking is illustrated by the fact that he noted there were no single-hoofed animals with two horns and no animals with both tusks and horns. Another of his observations (which was not elaborated until much later in history) was the fact that lizards can regenerate severed limbs.

Sex and reproduction fascinated him; he not only discussed them all through the *Historia* but then wrote a whole treatise, *De generatione animalium,* which covers the same ground in more theoretical detail. It is easy to see why he was concerned with this problem; an explanation of the creation and development of life forms was needed to round out his synthetic view of the world. His ambition was to include all that was animate and inanimate in a complete and finite system. Reproduction, he pointed out, could be sexual, asexual or parthenogenetic and (unfortunately) by spontaneous generation. The last notion, that such creatures as fleas, mosquitoes or barnacles could arise from mud or filth (and lice from the skin of animals), was a myth that was to be repeated solemnly for hundreds of years.

Concerning sexual reproduction, beginning with man, Aristotle generalized as a true patriarchal thinker. Women, and females in general, possessed less heat and were therefore less vital. They supplied (in terms of menstrual flow) only the material of the new being. Semen from the hotter male was the active element, which worked upon the material like a sculptor and also supplied the sensory contribution. Aristotle combated two ideas which, however, he was not able to destroy. Pangenesis, espoused by Hippocrates and continuing in zoological tradition up to Darwin, was a doctrine that saw the semen as extracting material from all over the parent body in order to form the new body. Since in Aristotle's view the semen possessed soul (or vitality) which could creatively work to *form* the new animal, his instinct was fairly correct, as the discovery of spermatozoa was to prove. Preformation, the notion that all parts of the embryo existed in miniature in semen, he disputed because he had dissected embryos and was aware of the gradual formation of organs. He correctly described the nourishment of the new being through the placenta and depicted the development of the chicken in the hen's egg in considerable detail. As is to be expected, his work on the breeding of squids, crustacea and sharks is remarkably sound. When it came to insects, he was not aware of their eggs and mistook the pupae for the equivalent of eggs. Although he wrote extensively on bees and probably got his material from beekeepers, the result is a muddle. He knew there were drones, which he called kings, and workers, but he never seems to have been aware of queens and their role in the hive.

It is in Aristotle's discussions of animal psychology and habits that he exhibited the quaintest mixture of accurate observation and uncritical acceptance of folktales full of the old magical thinking. These, indeed, were probably collected from hunters, herdsmen and farmers whose folklore often consisted of popular versions of astrological myths and various medical superstitions. At any rate, it is rather surprising to learn that serpents "have an insatiate appetite for wine, consequently at times men hunt for snakes by putting wine into saucers and putting them in the interstices of walls, and the creatures are caught when inebriated. Then, too, a singular phenomenon is observed in pigeons with regard to pairing: that is, they kiss one another just when the male is on the point of mounting the female, and without this preliminary the male would decline to perform his function. With older males the preliminary kiss is only given to begin with and subsequently he mounts without previous kissing; with younger males the preliminary kiss is never omitted." Obviously the older males get bored with this formality. It was also well to remember that goats, if they submitted to the male when the north wind was blowing, bore male offspring; if impregnated during a south wind, they produced females. There was more to be learned about goats. "If you catch a goat's beard in the extremity—the beard is of a substance resembling hair—all the companion goats will stand stock still, staring at this particular goat in dumbfounderment." The European bison, on the other hand, had a curious talent for "projecting its excrement to a distance of eight yards; this device it can easily adopt over and over again and the excrement is so pungent that the hair of hunting dogs is burnt off by it." The bison does use it as a territory marker.

In true patriarchal fashion, Aristotle generalized about females, lumping them all together from women to squids. Females were less compassionate, more jealous, more shrinking, more shameless, less courageous and less sympathetic. By way of proof he maintained that when a female squid was struck by a trident, the male stood by to help, but when the male was struck, the female ran away.

Aristotle listed scores of enmities between animals, and although once in a while this enmity was explained on the basis that one preyed upon another or they were rivals for food, in most cases the pairing was fantastic and often astrological. For instance, "the ass is at enmity with the lizard for the lizard sleeps in his manger, gets into his nostril and prevents his eating."

He tells us that stags "when hunted are caught by singing or pipe playing on the part of the hunters; they are so pleased with this music that they lie down on the grass. If there be two hunters, one before their eyes sings or plays the pipe, the other keeps out of sight and shoots at a signal given by the confederate." On a par with this is the behavior of the Egyptian ichneumon, or mongoose, which "when it sees the serpent called the asp does not attack until it has called in other ichneumons to help; to meet the blows and bites of their enemy the assailants beplaster themselves with mud, by first soaking in the river and then rolling on the ground."

On the credit side, the Greek zoologist had made careful studies of the migrations of fish from cold to warmer waters and of the southern migration of birds. (In this connection he repeated, with the qualification "It is said . . . ," the story that when the cranes flew south, they fought with the Pygmies. "And the story is not fabulous for there is in reality a race of dwarfish men, and the horses are little in proportion and the men live in caves underground.")

He knew something of the importance of territory as it was observed among seals and was aware of the hibernation of such animals as the bear. He unfortunately repeated the story that swallows hibernate in holes in the ground in an unfeathered condition. Another myth accepted by him and which was to go down in history was the statement that camels, lions, lynx and hares copulate back to back—that is, facing in opposite directions.

Little details such as a dog eating grass to make itself vomit and a woodpecker inserting an almond in a crack in order to split it with its beak testify to Aristotle's eye for what went on around him. He also knew that the partridge would limp and pretend to be wounded to lure the hunter away from its young. Especially interesting is his report that dolphins had been seen pressed together, holding up a young dead animal at the surface of the water. Only recently, with the new interest in these marine animals, has it been verified that they will support a wounded fellow near the surface so that it can breathe.

As the inventor of the science of zoology, Aristotle must always be given credit for his emphasis on observation and for the elements of a systematic way of viewing the natural world present in his work. There, however, were some aspects of his thought that did not lead to scholarly progress.

As we have said, he tended to arrange animals in a hierarchy that roughly corresponded to what would someday become an evolutionary ladder. He believed however that the motion which was the stuff of the universe was circular and all creation finite. Thus all species had been formed once and for all by the soul-vitalism which, in turn, was set going by the unmoved mover, the source of all motion. Therefore everything was controlled by fixed laws answerable to a supernatural concept. It was a static view which could fit into a religious authoritarian system, and eventually it did. In other words, his system lacked any sense of the creativity of what seems, in commonsense terms, to be chance. The creation of new species involves the idea of uncertainty of outcome or, at least, of conceivable alternate outcomes as living organisms adapt to the environment. What Democritus called necessity, an impersonal force governing the universe, is closer to the cautious and provisional attitudes of modern science than Aristotle's chain of causation going back to an all-powerful abstraction.

In a sense Aristotle, the logician, became a prisoner of his own logic, and many of those who followed him, being less endowed creatively, submitted to authority and instead of rejecting his mistakes, rejected factual proof and slavishly repeated him.

But all in all, his achievement is impressive. The modern specialist, imprisoned in his small territory, looks back with admiration and envy at the Greek pioneer who knew everything that was to be known in his time and quite a few things that nobody else knew. Darwin, who had the right to make a judgment, wrote enthusiastically: "Linnaeus and Cuvier have been my two gods, though in very different ways, but they were mere schoolboys to old Aristotle."

2

A World of Wonder

WHEN the intellectual center of the world shifted to Alexandria, the capital of Egypt, in the third century B.C., the early Ptolemaic rulers interested themselves in culture and science. The Museum of Alexandria was a parallel to that of Aristotle's Lyceum. Scholars from every country were lodged in it and given facilities for research in every branch of science known to the period. Around the year 300 B.C., Herophilus, whose writings are now lost, was able to profit from the support of the government to the extent that he was allowed to carry out vivisectional studies on the bodies of condemned criminals. The church father Tertullian accused him of having tortured six hundred people to death, no doubt a polemic exaggeration. As a result of his dissections he was able to discover the blood sinuses and structure of parts of the brain, which he regarded as the seat of the soul. On the whole he accepted the system of Hippocrates with its four elements. He knew more about the eye than Aristotle, compared the venous and arterial blood vessels, and distinguished between nerves and tendons. His rival, Eristratus, a practicing physician, studied the heart and named the valves. He knew the difference between motor and sensory nerves and established the connection between the veins and the arteries and even investigated the lymphatic ducts.

As the quality of Alexandrian rulers declined, so did the scientific work of the museum, and finally this institution was destroyed in a riot. In order to follow the checkered history of natural science we turn to Rome, to Gaius Plinius Secundus and Titus Lucretius Carus.

Gaius Plinius Secundus who was born in A.D. 23 in what is now Como in northern Italy, exemplified many traits of the Roman character. Unlike Aristotle, a university professor living a life of contemplation, he was a busy Roman official and a high-ranking officer in the army and navy. Actually he made no specific advances in our science, but his influence was so long-lasting and he set going so many of the tales that were to be repeated for centuries that it is important to sketch his mentality and the zoological world he inhabited.

For what we know about his life we are indebted to his nephew, whom he adopted, an industrious, if slightly pompous letter writer. Pliny's first book was a treatise on *The Art of Using a Javelin on Horseback.* Since he was serving in Germany during the reign of Claudius, he was no doubt at this time commanding a troop of horse. He wrote a life of Pompey and a history of the wars in Germany. (He was urged to undertake this by the ghost of Drusus Nero who appeared in a dream.) In 52 he retired from the army to concern himself with literature and science, but by 67 he was made comptroller of Spain, a post he held under four emperors ending with Vespasian. He had written a book on the life and education of a student, but, his nephew tells us, since under the last years of Nero's reign tyranny was at its harshest, he was obliged to avoid political subjects and it was then that he undertook linguistic studies and the thirty-seven volumes of *Natural History,* the last being his only surviving work. Somewhere along the line he engaged in the profession of advocate and was only fifty-seven years old when he died.

His service in Spain made him eligible for knighthood, a rank of which he was very proud. His efficiency was legendary. He began work at midnight, when the August festival of Vulcan took place. In winter he started at one in the morning, never later than two. He had, it seems, the facility of catching a few winks of sleep here and there, nodding over his book, and then going on with his work.

"Before daybreak," his nephew wrote, "he used to wait upon Vespasian, who likewise chose that season to transact business. When he had finished his affairs which that emperor committed to his charge, he returned home again to his studies. After a short and light repast at noon (agreeable to the good old custom of our ancestors) he would frequently in summer, if he was disengaged from business, repose himself in the sun, during which time some author was read to him from whom he made extracts and observations, as indeed was his constant method whatever book he read; for it was a maxim of his that 'no book was so bad but some profit might be gleaned from it.' When this basking was over, he generally went into the cold bath, and as soon as he was out of it, resumed his studies until dinner time, when a book was again read to him upon which he would make some running notes."

But even bath time was not an idle period, for all the while that he was being dried and massaged, he either listened to more reading or dictated.

Once his reader pronounced a word wrong and someone at the table made him repeat it. "My uncle asked his friend if he understood it. Who, acknowledging that he did: 'Why then,' said he, 'would you make him go back again? We have lost by this interruption of yours above ten lines' so chary was this great man of time!"

Romans have often been compared to Americans, and indeed Pliny puts the most efficiency-conscious American tycoon to shame. When he traveled, he never desisted from his labors. "A shorthand writer with books and tablets accompanied him in his chariot, who, in winter, wore a particular sort of warm gloves, that the sharpness of the weather might not occasion any interruption; and for the same reason, my uncle always used a sedan chair in Rome. I remember he once reproved me for walking. 'You might,' he said, 'not have lost those hours,' for he thought all was time lost that was not given to study." Indeed it is a pity the tape recorder was not invented in the first century; we can imagine what delighted use Pliny would have made of it.

In addition to books already mentioned, his nephew says he left one hundred and sixty volumes of a sort of diary, written in very small handwriting, for which he had once been offered a sum equivalent to almost ten thousand dollars.

Pliny had received a good education from private teachers in Rome. In the degenerate days of the empire, he stands out as a patrician of the old school: brave, honest, loyal to his emperors and incredibly conscientious. Philosophically he was a stoic, a doctrine which laid great emphasis on the ability to face life calmly and involved a certain asceticism and a great devotion to duty.

His great work, *Natural History,* is even more comprehensive than the writings of Aristotle, although much less profound and lacking a detailed sense of organization. It is a cross between an encyclopedia and a notebook recording whatever interested him, snatched from those multitudinous readings which went on during his day and most of his night. Books eight to eleven deal with zoology proper, although a number of notes concerning animals are scattered throughout the work.

His interests were certainly catholic. He started in the usual way, following Aristotle, with a view of the universe not very different from his predecessor's. In Ptolemaic astronomy the earth hangs stationary in space while seven planets and the sun circle it. He repeated Pythagoras' idea that the orbits of the planets are equivalent to the seven notes of the scale and hence are supposed to create music, although Pliny admitted he had never heard it. His notion of the deity controlling all this differed from Aristotle's unmoved mover. After suggesting that the sun was a sort of divine ruler of the planets, he then indicated that the whole world was God, so his overall view was rather pantheistic. The divinity, he maintained, "has no power over what is past except to forget it and (to prove our own relationship to God by frivolous arguments), he cannot cause two tens to cease being twenty, or many similar things, which certainly proves the power of nature and that it, and nothing else, is what we call God."

Now, Pliny had been scolded again and again for being uncritical and gullible. It is true that much of his *Natural History* is a book of wonder, but any sort of careful reading of it shows an odd ambiguity. The Roman naturalist's explicit theory was different from his practice. He continually and loudly attacked magic and superstition. For instance, this is what he said about divination:

> Another sort of people banishes chance altogether and attributes events to a star and to their horoscopes, believing for all men, born and unborn, God's decree has once and for all been enacted so that for the rest of time he may rest and sit still. This belief begins to take root among both the educated and the ignorant mob who march rapidly in that direction. Witness the warnings drawn from lightning, the forecasts made by oracles, the predictions of soothsayers and ridiculous trifles—a sneeze or a stumble counted as omens. . . . Thus these things deceive and entangle mortals so that all that remains certain is that nothing certain exists.

Likewise, although many pages were given over to magical folk remedies for illness, he often expressed extreme disdain for them. For instance, he tells us that people think eating hare's flesh will make them look beautiful for a week afterward. "For my own part I think truly it is but a joke and mere mockery; however, there must be some cause and reason for this settled opinion which has thus generally carried the world away to think so."

For doctors he had no respect at all. He made fun of one who prescribed a cold water cure; he accused them of poisoning their patients, of extracting the last penny from them on their deathbeds. Indeed when he cited remedies, he generally called them the remedies of magicians. And for magic itself he expressed scorn for that practiced by the Britons and the Persians with the conclusion: "The benefit is inestimable that the world has received by the great providence of our Romans, who have abolished these monstrous and abominable arts, which, under the show of religion, murdered men for sacrifices to please the gods; and under the cover of physic prescribed their flesh to be eaten as the most wholesome meat."

Basically a sturdy rationalist, Pliny sometimes makes us feel that he collected many of his oddments and fables simply because he wanted to show the extreme credulity of humanity. And perhaps he, too, even as the modern reader, found them rather entertaining.

Yet he certainly, with a straight face, created a kind of anthropology of the absurd: men with dogs' heads who conversed by barking, men with one leg apiece who got about hopping, men with eyes in their chests, men with no mouths who drank through a straw inserted in their nostrils. Most of these wonders were said to occur in India and Ethiopia, but the climax came when he announced: "Upon the coast of Africa are the Ptonebari and the Ptoemphani, who have a dog for their king, and him they obey according to the signs which he makes by moving the parts of his body, which they take to be his commandments and they religiously observe them."

In less fantastic moments Pliny could list recipes for cooking, give a good account of Greek painting, sound agricultural advice, an exhaustive description of the gold mining industry and textile manufacture, then suddenly swing into a diatribe against the Roman mania for statues in which he tells us that they set up statues of Marius Gratidianus in every street and knocked them all down again when Sulla came to power.

As a zoologist he was not an experimentalist and he does not seem to have practiced anatomical research. His method, as we have already noted, was to cite authorities he had read (two thousand books he tells us). If he seldom made a choice between fabulous and more sober statements, he often cited his authorities, leaving the reader to judge if they were to be believed. All his authors were listed in Book I. In general he followed Aristotle but with less scientific organization. He lumped together the land animals, the birds, the sea animals, and the insects, and cataloged each group in terms of its habits. He followed this with a rundown of anatomy that was basically Aristotelian. We are obliged therefore to appreciate him for his amusing fantasy and also more or less as a naturalist who wrote at times rather lyrically about nature.

Examples of his contribution to the folklore of zoology abound in his study of the elephant. He made perfectly sound remarks about elephant training, but mixed in with these observations were such statements that certain snakes are avid of elephant blood. "They submerge themselves in rivers and when the pachyderms drink, coil 'round their trunks and bite them inside the ear. These snakes are so large that they can contain the whole of an elephant's blood, and thus they drink their prey dry at which time the elephant collapses in a heap and the intoxicated serpents are crushed and also die."

Of lions he tells the Androcles story, although he attributed the adventures to one Mentor of Syracuse. From the fabulist Ctesias he cited not only the manticore but also the yale, an animal with movable horns which could be pointed forward and backward. The basilisk also appears in his pages, although he described it merely as a twelve-inch-long snake with a diadem of bright markings on its head. It killed bushes with its touch, scorched grass, burned rocks, and when speared, its lethal power could run up the spear and kill the huntsman. Another story, which was to be repeated many times, was that of the offspring of the bear, shapeless little lumps of flesh, which had to be "licked into shape" by the mother.

India, that hotbed of miracles, produced turtles with shells so large they could be used to roof houses. Indeed charming was the behavior of the small parasitic crab which lives in the fan-mussel. When the shellfish opened, it presented its dark side to tiny fishes. These darted in and filled up the vacant space. The crab, having watched carefully, gave the body of the shellfish a gentle nip. The shell closed at once, killing the fish upon which it fed, leaving a share for the cooperating crab.

The Plinian version of the phoenix must be cited, for it differs somewhat from later fables. According to Pliny, it lived 540 years, then constructed its nest of wild cinnamon and frankincense, and lay on it until it died. From its bones and marrow a sort of grub was born which grew into a child, who celebrated the funeral rites of its parent by carrying the nest to the city of the sun near Panchaia and placing it on an altar. He was critical enough, however, to reject the story of the death song of the swan.

When it came to the nightingale, Pliny wrote with enthusiasm. He tells us: "The sound is given out with modulations, and now is drawn into a long note with one continuous breath, now varied by managing the breath, now made staccato by checking it, or linked together by prolonging it, or carried on by holding it back; or it is suddenly lowered, and at times sinks to a mere murmur, loud, low, bass, treble, with trills, with long notes, modulated when this seems good—soprano, mezzo, baritone; and briefly all the devices in that tiny throat which human science has created."

Let us listen to Pliny the ecologist in a statement which has greater implications now than in his time. He defended the earth, saying:

> It is true she has brought forth poisons—but who discovered them except man? Birds of the air and wild beasts are content merely to avoid them and know well enough how to watch out for them. . . . It is true that even animals know how to prepare their weapons to inflict injury, yet which of them, except man, dips its weapons in poison? As for us, we even poison arrows and we add to the destructive power of iron itself. It is not unusual for us to poison rivers and the very elements of which the world is made, even the air itself, in which all things live, we corrupt till it injures and destroys.

This sound and prophetic statement balances a good deal of Pliny's spinning of tall tales.

As a conscientious observer of nature, when dealing with things he knew, he could be exact:

> Some people think that butterflies are the most reliable sign of spring, on account of the extremely delicate nature of that insect but in the year in which I am writing, it has been noticed that three flights of them were killed one after another by the cold weather and that migrating birds arriving on January 27 brought hope of spring that was soon dashed to the ground by a spell of very severe weather.

One of his most attractive pieces of writing is his discussion of stars in the heaven and insects in the grass which, he maintained, appear at the same time:

> Nature not only assembles a troop of stars, the pleiades, but she has made other stars in the earth to show him the true seasons. It is as though she cried aloud: why gaze upon the heavens, plowman? Why search among the stars for signs? The nights are shorter now and the slumber your weary work imposes upon you is less. Behold I scatter here and there among the weeds and grass, and display in the evening, special stars when you unyoke and cease from your day's work. I cause you to marvel and gaze upon this wonder, so that you shall not pass them by. Do you see how these fireflies screen their brilliance, which resembles sparks of fire, when they close

their wings and how they carry daylight with them even in the night?

Pliny's death has often been cited as a proof of his interest in science. At the end of his life he had been made an admiral and given the command of a fleet at Misenum, an ancient town in the Bay of Naples. In August, A.D. 79, his sister pointed out a cloud of unusual size and appearance. Pliny was just out of his bath. He climbed a hill and was greeted with an eruption in the shape of a pine tree, which he was later to discover came from Mount Vesuvius. At first, he ordered a light vessel to take him to the scene so that he might observe the phenomenon, but then, hearing that people in nearby villas were in danger, he ordered out a large galley and steered for the coast near Vesuvius. Cinders and blackened pumice stones fell on the deck, but he disembarked at another town which commanded a full view of Vesuvius. Although his friends were alarmed, Pliny calmly took one of his innumerable baths. People were abandoning the villas near the mountain which blazed and shook the earth with violent concussions. Pliny passed a dreadful night amid showers of stone but in the morning went down to the beach to observe the eruption. At this time he was already corpulent and suffering from asthma. He was lying down on a sail when a new outburst of fire and smoke dispersed all his attendants. Apparently the thick sulfurous vapor obstructed his breathing, and at this point he died. Three days later, when the eruption had abated, his body was found, his nephew wrote, "Fully clothed as in life; its posture was that of a sleeping rather than a dead man."

This was the end of Rome's most industrious natural philosopher. His work sums up what was known and believed at the end of the classical world.

Although Pliny continued to be quoted, Lucretius, the distinguished poet-philosopher of Rome, who lived from 99 to about 55 B.C., had to wait for recognition.

Lucretius and Pliny are curiously different in temperament. Lucretius was a hard-thinking, tough-minded materialist, while Pliny could never abandon the marvelous. Although Lucretius did not deal specifically with zoology, he was one of those philosophers who kept alive a tradition created by Democritus and Epicurus. His poetic genius made possible a vivid and lasting expression of the atomic theory and a definitive statement of rationalism which was to survive centuries of regression to magic when it was picked up by Renaissance thinkers and served them as a stimulus.

Lucretius has perhaps been more talked about than read; Rolfe Humphries' beautiful translation, quoted here, should make the Roman poet accessible to moderns and prove once and for all that he is one of us. Indeed he started from the basic position that religion has been the cause of much suffering and evil, both because of the error in attributing causation to the gods and because it encouraged a fear of an afterworld of punishment. The gods, for the Epicureans, were lyric abstract ideals, no more than metaphors for glorified human values. Venus, whom Lucretius invoked, is a combination of the abstract principles of creativeness and fecundity, taken together with the maternal image of the earth, suggesting a certain pantheism.

The theme and motto of the poem, repeated four times, is:

> Our terrors and our darknesses of mind
> Must be dispelled, not by the sunshine's rays,
> Not by those shining arrows of the light,
> But by insight into nature and a scheme
> Of systematic contemplation.

Lucretius posited, as did the Epicureans, atoms as the basic stuff of the world, atoms in motion, infinite in number, indivisible and imperceptible to the senses. The very disassociation of the basic component of reality from the superficially visible is a feat in itself. It allowed him to reject the dogma of the four elements which was to be reasserted by Christianity and to remain unchallenged until the sixteenth century. He wrote:

> If out of four things all things are created,
> If into four things all things are resolved,
> How can these four be called the source? Why not
> The other way around?

Lacking the tools of modern science—the microscope, the telescope, the complicated machinery of measurement and mathematics upon which modern physics and chemistry are built—his proofs of material causation were often naïve and erroneous, but it is amazing how often he intuitively arrived at an image that is suggestive of a later detailed scientific explanation. For instance, his idea that sight could be explained by a film of atoms given off by objects which reached the eyes trembles on the edge of being a metaphor for light waves. And note what he says about color when he is bent on proving that atoms have no color:

Even in bright day
Hues change as lightfall comes direct or slanting.
The plumage of a dove at nape or throat,
Seems in the sunlight sometimes ruby-red
And sometimes emerald-green, suffused with coral.
A peacock's tail, in the full blaze of light,
Changes in color as he moves and turns.
Since the light's impact causes this, we know
Color depends on light.

Like Aristotle he rowed vigorously without oars and made astonishing progress. He absolutely refused to divorce the soul from the body as Plato did but pointed out that when the fleshy frame fell ill, the mind was also affected and that the two died together. He believed that nothing came from nothing and matter was indestructible. Combinations of atoms were broken, new ones formed, consciousness could die and be re-created. He believed in trusting to the senses; reason alone could go astray, but observation plus reason was the way to achieve his goal of "systematic contemplation." Although he had no real conception of evolutionary change, his comments upon the identity of species:

. . . we see all things
Maintain the proper order of their kind,
Same kind of parenthood, same kind of seed
Definite causes, definite effects,
A fixed, assured procedure.

assert the existence of laws which, although he could not go further, were eventually to be investigated by the geneticists. Simply by common sense, by rejection of myth and by hard thinking, he arrived at a version of prehistory, some elements of which have been sustained by archaeology. Man at first was a shaggy hunter who lived in caves and killed animals with stones and clubs. Later he began to evolve social conventions, a truce against aggression, as his language improved. Eventually cities were built. First bronze was used for artifacts, but later iron proved tougher and more practical. The invention of the textile industry he assigned to women, but put it too late in time, during the Iron Age. Nevertheless, in this speculative account there is a glimmering of evolutionary arrangement.

It is no wonder that the poem *De rerum natura* was suppressed during the Middle Ages. It is passionately written; it stresses rejection of superstition and has no room for blind faith or the tyranny of authority. Lucretius, by the exercise of his talent, showed that there was another kind of poetry possible than that of fairy tales created to make man feel important or to frighten him into behaving himself.

In his own time he lived the life of a scholar and does not seem to have been active in public life. All that is known of him is that he came from a famous patrician family and undoubtedly had means allowing him to pursue his interests. He does not seem to have affected Pliny, who is the sole important zoologist of the Roman period.

Something should be said of the public and popular relation to animals, consisting of exhibiting them in zoos and using them in spectacles. Emperors and aristocrats had been keeping small private collections of living beasts since Babylonian and Egyptian times. The Mesopotamian kings kept lions because they were royal symbols and also symbols of power and aggression. In Egypt, of course, a curious religious conservatism preserved the sacred character of animals, a cultural lag which probably pointed back to the hunting phase of society. At any rate, Titus Flavius Clemens, a Roman historian, after returning from Alexandria, wrote of his visit to a temple: "If you penetrate to the back of the edifice and seek the statue, a priest advances gravely singing a hymn in Egyptian and raises a veil as if to show you the god. And what do you see? A cat, a crocodile, a native serpent or some other dangerous animal." At Heliopolis, sacred lions were kept in the temple of Ammon, nourished on special food, their dining period accompanied by sacred melodies. Sometimes the spectators enjoyed the treat of watching them eat live prey. Sacred crocodiles in Lake Moeris wore golden collars and came when called to be fed cakes by their worshipers. Queen Hatshepsut, in the second millennium, collected a real zoo as did the first of the Chou emperors, Wên Wang, in about 1100 B.C. These royal collections (Wên Wang kept deer, goats, birds with resplendent white feathers, and many fish) were apparently half esthetic, since animals were decorative, and half to satisfy a certain curiosity about the world. Wên Wang sent for typical specimens from all parts of his empire; Hatshepsut even sent expeditions beyond her boundaries which brought back leopards and a giraffe. This indicates a certain kind of interest in the animal kingdom that is tangential to zoology. Montezuma also kept an elaborate zoo.

The Romans made a real business of animal catching for two reasons. One was for public display—in menageries called *vivaria,* one of which had even a bear pit with trees in it for the animals to climb. The Emperor Augustus took a mildly scientific interest in foreign animals, collecting seals, eagles, crocodiles, a rhino and a serpent twenty-five feet long. The second demand for animals sprang from what, for us, is the most unsympathetic Roman invention of fights between animals. This emphasizes an aspect of the psychological relation between animals and men which seems to have played an important role throughout history. There is plenty of repressed hostility in the human soul. Often enough it finds a direct outlet in man's inhumanity to man, but at times animals could serve as surrogates for human beings. Lions were pitted against bears, rhinos fought tigers, elephants battled carnivores. Human beings—unarmed Christians and trained fighters—took part in conflicts with animals.

The morbid interest in animal combats was such that jaded voluptuaries went into the arena for thrills. There were also professionals of a sort who specialized in this activity. Some wore armor, others only a tunic; the bravest went into the arena stark-naked and fought the beasts with their bare hands, wrestling, strangling their opponents, or choking them by plunging a hand down their throats. Among these were bullfighters who performed like those in southern France, evading the animal's charges, seizing it by the tail, or vaulting over its back with a pole. There were wooden barriers encircling the ring, behind which the fighters could escape when hard pressed, much as the cuadrilla does in a bullfight today.

Although a depraved taste in such bloodshed forced the authorities to spend large sums on these popular entertainments, there were some civilized Romans who disapproved. Both the dramatist Seneca and the orator Cicero condemned the slaughter of animals on humanitarian grounds. Cicero said: "What pleasure can a cultivated man find in seeing a weak man torn to pieces by the gigantic strength of a wild beast or a splendid animal pierced by a spear?"

Not all Roman interest in animals was bloodthirsty; indeed most of the circus tricks which animals now perform were invented in the time of the empire. Monkeys picked out letters of the alphabet and played the flute or zither. Elephants were trained to do pyrrhic dances to the sound of cymbals, and wrote Latin words with their trunks on slates or walked a tightrope. Eagles carried children up in the air and brought them down again. Bears were trained to dance, and lions, oddly enough, taught to pursue rabbits, catch them, play with them as a cat does a mouse, and finally return them to their trainers.

All this is tangential to science, but even as such activities as hunting and fishing lead to more serious study of the game sought, so zoos were eventually to become a source of valuable data.

Before leaving the Roman scene, we must not omit one other scholarly figure, a man who contributed indirectly to zoology by virtue of his anatomical investigations. Galen, born in A.D. 131 in Pergamum, Asia Minor, was a Greek who wrote in Greek, studied the Platonists, and, above all, was well acquainted with Aristotle's writings. One of the great personalities of the late Roman world, a man of wide culture, whose architect father was told in a dream that his son would become a doctor, Galen accepted this semireligious call and seems to have been a man of great piety. He viewed the world as the product of a divine intelligence which he saw exemplified in the human body. Educated in Alexandria, he soon outdistanced the warring Alexandrian schools in scholarship and intellectual breadth. After practicing for a time in Pergamum, he went to Rome where he developed a huge practice and also gave public lectures. His success aroused the envy of his brother physicians. Hostility toward him grew so intense that he joined the ranks of persecuted men of genius who had to leave their homes to save their lives. After a cooling-off period of a few years in Pergamum he was invited back to Rome to become the personal physician of the Emperor Marcus Aurelius under whom and under whose son, Commodus, he found protection and was able to work peacefully. He is believed to have died in Pergamum in A.D. 210.

All of Galen's anatomical work was done on animals, apes being his substitutes for human bodies. His work on the brain and the nerves represents an advance. He distinguished between sensory and motor nerves, pointing out that the latter which he called "hard" nerves, emanate from the spine. Although he described the brain well, because he clung to the theory of "pneuma," a sort of soul gas which he believed was produced in the brain and circulated through the nerves, he did not increase our knowledge of its functions. As far as the digestion goes, he did not advance beyond Aristotle. Galen's work on the circulatory system was mixed in quality. He incorrectly described the heart and thought that both veins and arteries conveyed blood (which was made in the liver) to the rest of the body. Pneuma was conveyed from the lungs to the left ventricle while the soot of combustion in the heart went to the lungs to be eliminated. The idea of pneuma, a necessary gas for combustion, was leading in the direction of oxygen and carbon dioxide, but like Aristotle, Galen was forced to think in metaphors because chemistry had not been invented. Lacking the concept of the circulation of the blood, Galen's doctrine of the functions of the blood was an obstacle to progress because it was reverently repeated for about fifteen hundred years.

After Galen, the newborn scientific spirit that reached its peak in Aristotle began to wane. Magic made a comeback in the guise of the new mystical religions which began to take on importance as a solace for the troubled spirits of the disillusioned citizens of the Roman Empire. There was Orphism with promises of an afterlife; there were the self-mutilated priests of Cybele who celebrated dark mysteries; there was Mithraism with a doctrine of death and resurrection. From these competing mystical cults, Christianity rose triumphant and was to dominate the life-style of the West for centuries.

3

Emperor and Two Saints

☙ IF we are honest, we admit that the early Middle Ages, the period when Christianity triumphed, is a retrograde one for science. The ecclesiastical turn of mind simply was not oriented toward facts. It took off from a dogma, a static pattern which was to be defended for the good of men's souls. Any attempt to modify it was inspired by the devil, and people yielding to such inspiration had to be punished to maintain the establishment. Thus the medieval establishment burned heretics with a clear conscience. Some recanted and this proved the system was a good one.

In this climate anything that could be exploited to support the basic Christian dogma was seized upon and used. As a result there was no need to make a distinction between observed data and fable. Magical thinking returned in full force and to it was added an allegorical emphasis. Medieval preaching, of which there was a great deal, was vivid and down-to-earth. The speaker liked to illustrate his points with short anecdotes or exempla, which in symbolic form supported his thesis. It was a poetic method on a rather pedestrian level, no doubt inspired by the literary forms of the New Testament. Thus it was that animals, when they were recognized for anything other than utilitarian purposes, now had to bear the weight of the new symbolism.

The most popular book in this period dealing with natural history, originating around A.D. 200 in Alexandria, was the *Physiologus*, as it is called in Latin, or the *Bestiary*. E. P. Evans, who has studied its origin, tells us that it stood next to the Bible, so widely was it diffused among many peoples for many centuries. It appeared in Latin, Ethiopic, Arabic, Armenian, Syriac, Anglo-Saxon, Icelandic, Spanish, Italian, Provençal and all the Romanic and German dialects. The last handwritten manuscript appeared in 1724; the earliest written version was Greek, probably between the first and second centuries. It was a compilation which borrowed from numerous sources. It drew on oral Hellenic, Egyptian and Asiatic tradition; it absorbed Aristotle's wildest stories, everything gullible old Ctesias had to offer, and eventually borrowed from Pliny. In the Greek original, there were only forty-nine beasts, but by the time it became the *Bestiary* and numerous copyings went on in England in Latin, each scribe added to it and often elaborated the stories. In the version which we shall be quoting there are about one hundred and sixty beasts.

In the twelfth and thirteenth centuries, this work became one of the leading picture books because, like all natural histories, it lent itself to illustration. Its influence extended well into the Renaissance and penetrated the best natural histories. Indeed it was not until the later seventeenth century that much of its fantasy was subjected to criticism, although such writers as Konrad Gesner and Ulisse Aldrovandi (whom we shall discuss later) were beginning to reject the wilder stories in the sixteenth century.

All in all the *Bestiary* was Pliny plus Christian moralizing in neat capsule form for the use of the ecclesiastical orator. The phoenix, for instance, was a natural:

> When it notices that it is growing old, it builds itself a funeral pyre, after collecting some spice branches, and on this turning its body toward the rays of the sun and flapping its wings, it sets fire to itself of its own accord until it burns itself up. Then, verily, on the ninth day afterward, it rises from its own ashes!
>
> Now our Lord Jesus Christ exhibits the character of this bird. . . .

The fox, who has not distinguished himself up to now, takes on new significance:

> He is a fraudulent and ingenious animal. When he is hungry and nothing turns up for him to devour, he rolls himself in red mud so that he looks as if he were stained with blood. Then he throws himself on the ground and holds his breath so that he positively does not seem to breathe. The birds, seeing that he is not breathing, and that he looks as if he were covered with blood, with his tongue hanging out, think he is dead and come down to sit on him. Well, thus he grabs them and gobbles them up.
>
> The devil has the nature of the same.
>
> With all those who are living according to the flesh he feigns himself dead until he gets them in his gullet and punishes them.

Although Pliny tells us that if a wolf sees a man first, the human being is paralyzed, the *Bestiary* has a number of new additions. A wolf's eyes shine in the night because the works of the devil seem to foolish human beings bright and beautiful. Likewise the wolf cannot bend its neck backward because the devil never turns back to lay hold on repentance.

The book proceeds in this style, in many cases the allegorical moral being longer than the description of the animal. There is practically nothing in it that evidences real observation, and indeed it was Saint Augustine who said that it did not matter whether certain animals existed; what mattered was what they meant.

If we cannot go to medieval ecclesiastics for trustworthy information about living creatures, there is one great figure of the early thirteenth century, semi-legendary himself, who is significant for his attitude toward animals, an attitude which was in a sense heretical. As the historian Lynn White, Jr., has pointed out, "Christianity is the most anthropocentric religion the world has seen." We have already mentioned the respectful attitude of the primitive toward his fellow living beings. In the pagan world of antiquity, natural beings and objects had their indwelling spirits. There were centaurs, fauns, dryads, mermaids, water nymphs and the like. Monotheism with its jealous insistence on a single anthropomorphic god could not tolerate a situation in which semi-divinity was all around. "By destroying pagan animism, Christianity made it possible to exploit nature in a mood of indifference to the feelings of natural objects." In the early Middle Ages, this exploitation was still in a germinal stage. The great sources of power were not being used, men relied on animal and manual labor and hand tools. Ideologically, however, the break had been made with the past. Man's conscience was cleared and he was free to do what he liked. "Christianity, in absolute contrast to ancient paganism and Asia's religions (except perhaps Zoroastrianism) not only established the dualism of man and nature but also insisted it is God's will that man exploit nature for his proper ends."

The single exception was Saint Francis of Assisi, a man so radical, so close a parallel to the modern hippie, that in his own time he embarrassed Popes and a general of his own order, Saint Bonaventure, tried to suppress the early accounts of Franciscanism.

To understand his position, we must be aware of his story. If the charming fables embedded in it seem far from serious zoology, we must remember that the crisis in ecology is something new and only now are we in a position to realize the historical significance of Saint Francis and able to learn something from him.

He was born about 1181 to a prosperous dry-goods merchant, Pieter Bernardone. His mother had aristocratic connections and tended to spoil him. She thought of him as a poet and a knight. He received a good education. His father put him behind the counter to train him so that he might become a branch manager of the business. His mother, a provençale, brought him up in the tradition of the trouvères. When not at work, he was a leader of a group of gilded youth, who danced, drank, sang and recited poetry.

In the year 1205, he had a vision and felt himself instructed to embrace holy poverty and live for spiritual aims. Needless to say, his father was furious. There were clashes, he was beaten, but he stuck to his guns. Like our modern youth, he rejected the materialism of the period, which merely paid lip service to the church and went its own selfish way. The group of young men around him were dedicated to poverty and love. Although he was not a priest, he succeeded in obtaining permission to preach by quoting from the Bible to a materialist-minded Pope. Saint Francis preached what he considered the true way of salvation. He was not a meliorist who felt that the condition of the poor should be improved. Poverty, for him, was a positive value; his approach was parallel to that of the Eastern sage who renounced the world in order to concentrate on the inner world of the spirit.

It was his pantheism, however, which led him to take a different attitude toward other forms of life, again a parallel with Buddhism, for the latter philosophy treats all living things as sacred and forbids killing.

To quote Lynn White, Jr., once more:

> The key to an understanding of Francis is his belief in the virtue of humility—not merely for the individual but for man as a species. Francis tried to depose man from his monarchy over creation and set up a democracy of all God's creatures.

The two stories connected with Franciscanism which illustrate this attitude are in the nature of poetic metaphors. The first concerning his preaching to the birds goes as follows: Seeing a great number of birds in a field, he said to his friends, "Wait for me here by the way while I go to preach to my little sisters, the birds."

Not only did the birds listen, but those in the trees came down to join those in the field. He told them they should praise God who had given them the trees in which to build their nests. They, in turn, opened their beaks, stretched their necks, flapped their wings, and bowed their heads. They also sang loudly in praise of the Lord.

The second episode involves a wolf which had been terrorizing the inhabitants of the town of Gubbio. Francis went out to meet it, made the sign of the cross, and said, "Come hither, brother wolf, I command thee in the name of Christ, neither to harm me or anybody else!"

The wolf came and lay down at his feet. After a short sermon in which the wolf was admonished that eating human beings was wrong, Francis demanded that it promise never to sin again. The wolf gave him its paw as a sign it would obey. From then on, it became a great favorite in the town of Gubbio.

These amiable conceits indicate an attitude toward animals different from the one current in St. Francis' time. Instead of making them into abstractions which could be manipulated to point a moral, animals were recognized as fellow inhabitants of the earth and even anthropomorphized. The doctrine of love, broad enough to cover everything, is best stated in the charming "Canticle of All Creation," which includes even inanimate things:

> My Lord be praised with all your creatures
> Most of all brother sun;
> It is your day and light he brings us.
> For he is lovely, shining with great splendor
> And of you he brings us tidings.
>
> My Lord be praised by sister moon and stars,
> Lovely, clear and precious, you shaped them in the sky.
>
> My Lord be praised by brother wind,
> By air in every weather, bad or good,
> It is from them all your creatures gain their food.

As we have said before, Saint Francis was a radical. His thinking pointed the way toward a tolerant philosophy of man's place in nature, but it went counter to the general selfishness and greed that has pervaded Western Christian tradition.

Interestingly enough, the greatest zoologist of the Middle Ages was a contemporary and, in some way, the antithesis of Saint Francis. The most luxurious, brilliant and despotic monarch of the period, Friedrich II of Hohenstaufen, was born twelve years after the saint, in 1194.

In a period in which feudal barons jockeyed with Popes for the political power, the concept of the Holy Roman Empire still existed. Friedrich's father, Henry VI, was emperor for six years, but the area he controlled had little unity. Germany was an elective monarchy. Sicily (which comprised not only the island but half of Italy) was Henry's hereditary realm. Various Teutonic and more or less Romanic principalities and duchies disposed their influence in different ways. Some, like the Lombard cities, were rebelliously independent, some played politics with the Pope, some were fairly loyal to the empire. When Henry died (some said as a result of a plot between the Pope and his wife), the empire became a political plum. Innocent III, a thoroughgoing intriguer, became Pope. The Hohenstaufen were Waiblingers (Ghibellines), but the Welfs (Guelphs) were more acceptable to the papacy. Hence, although the young Friedrich had been elected King of Germany, Innocent supported a Welf pretender, Otto of Brunswick. The Pope wanted above all things to keep the kingdom of Sicily and the empire from uniting, for he feared the temporal power would overshadow the supreme spiritual dominion which the head of the church had always claimed. Although young Friedrich's uncle, Philip of Swabia, opposed Otto, the Pope managed to deprive the young heir of the German crown. Left without parents at the age of four, Friedrich was bandied about between regents who were more concerned with exploiting Sicily than educating their ward. The boy was actually left without food until various citizens of Palermo took turns in supporting him. From ages eight to fourteen he ran wild, wandering about the streets, markets and gardens of the semi-Oriental Palermo, which combined the palm trees of the tropics with Saracen mosques and Norman cathedrals. The people with whom he mixed were a symbolic cross section of the empire he was eventually to control: Normans, Saracens, Jews, Greeks, and Italians. No one knows what individual influences molded him, but it is certainly clear that he was untouched by the scholastic minds of his time. Somewhere along the line he learned languages and mathematics, but above all, his childhood in the gutter made him a supreme realist. Undoubtedly his awareness of at least three different faiths made him a cynic and, eventually, pragmatically tolerant of non-Christian belief. Judging from his later behavior, he must have come in contact with Saracen wisdom and learned to respect it. Living by his wits, as no royal youth had ever done before, he must have realized through bitter experience the motto he was later to promulgate: "No certainty comes by hearsay."

Yet despite the neglect and rejection and the frustration of quarrels between regents who sought to control his person as a pawn in their struggles, the prince never wavered in his belief in himself, in his almost demonic strength of character. When he reached the age of fifteen, the Pope gave up his technical guardianship and Friedrich took over the control of Sicily. He immediately quelled some disloyal barons. He had been married to a Spanish princess while still a minor and hoped for the support of her adherents. Although these failed him, he ruthlessly put down disorder in Sicily and then looked for new worlds to conquer. In 1209 his uncle, Philip, the imperial Hohenstaufen pretender, was murdered. This left Friedrich the last Hohenstaufen. The Pope, who was in the process of transferring his support to Philip, now switched back to Otto of Brunswick, who was elected King of Germany. In 1209 Innocent crowned him emperor. Aside from the traditional papal ties with the Welfs, Otto IV was supposed to be an unintellectual tool of Innocent's strategy. Unfortunately he showed himself capable of ambition and invaded Sicily. Innocent now tried to play both ends against the middle. He wanted Otto to fail but did not wish to strengthen Friedrich. He therefore excommunicated Otto. Popes used this ban as a temporary harassment which, after negotiations, could be lifted when the papacy gained its ends. Friedrich was in great danger, almost ready to flee from Sicily, when the Pope's secret machinations bore fruit. Otto's German princes turned against him. Pressured by the King of France, the Germans deposed Otto and elected Friedrich emperor. Otto fled north; the Pope at once joined the winning side and acquiesced.

At seventeen, as a result of this twist of fate (which he was to consider divine intervention), Friedrich, with no experience and no money, no strong ties with powerful princes, with only the lukewarm support of the Pope, was emperor of a vast area split into hostile nationalities and plagued by deceitful and conniving feudal lords.

Otto rallied his followers and showed every sign of intending to depose the as-yet-uncrowned emperor. But a combination of daring and luck allowed Friedrich to fight his way through hostile towns, gaining many which belonged to his adversary by the charm of his personality. A kind of myth began to develop; the people of Apulia called him Our Boy when he had himself crowned at Mainz. Judging from reconstructed portraits, he was a blond with regular features—Teutonically good-looking. It was David and Goliath, for Friedrich was slender and attractive and Otto huge and lionlike, stupidly brave and above all no great intellect. Otto allied himself with the English against France, and once more Friedrich profited from fortunate circumstances, for his rival was roundly defeated in 1214 at Bouvines along with the unfortunate King John of England. Otto never recovered, and Friedrich was again crowned, this time at the traditional seat of imperial power, Aix-la-Chapelle, while his grudging supporter, Pope Innocent, died the next year, 1216.

From then on the story of Friedrich is that of a man who trusted in his lucky star, which never failed him. Always an enemy of the papacy, for he gradually assimilated much of the loyalty claimed by that institution to his own person, he was basically as much a heretic and precursor of Protestantism as Saint Francis, for he, too, was inclined to believe that he was in closer touch with the divine will than the church establishment.

The story of his bloodless conquest of Jerusalem as a crusader and his long friendship in the Near East with Malik al-Kamil, Sultan of Egypt, the struggles of Popes Gregory IX and Innocent IV to destroy him with murder plots, excommunications and even armies, his record of statesmanship and the growth of his myth need not concern us here. His importance for science lies in his attitude toward life, his freedom from superstition which made him the first Renaissance figure long before the actual Renaissance. All the evidence indicates that he was without traditional religious belief. He mocked the church dogmas to his intimates and was capable of sitting through a Muslim service out of courtesy to his Saracen friends. (For this the papists called him the Antichrist.) Since he also openly admitted his lack of respect for the doctrine of an afterlife, he was free of the blackmail of heaven and hell. Clearly enough, his official Christianity was a political affair and the will of God was the will of Friedrich II. In his writings on monarchy he developed the theory of *necessity* (reminiscent of Democritus). The imperial institution rested upon divine foundations; it was necessary for the safety and the advantage of the world. If his divine foundations are, in turn, examined, they turn out to be the laws of nature. Thus the divine will, with which Friedrich was in such close contact without benefit of the church, was a mixture of fate, the law of cause and effect, the laws of nature as he interpreted them by observation. It was at this point that his slogan, "No certainty comes by hearsay," also applied. This revolutionary doctrine, barely covered by a few Christian phrases, ignored the prevailing mentality of miracle and magic. It will be seen that he bypassed sin and raised himself to the heights of a kind of scientific divinity since he, himself, was the arbiter of everything. It was no wonder that Popes recognized him as their mortal enemy.

And, indeed, his way of life was in accord with his philosophy. He surrounded himself with faithful Saracen servants, often elevated countrymen of low birth to important positions, and created in his court a kind of lyceum in which poets (who for the first time wrote in the vernacular), Muslim mathematicians, Jewish philosophers and doctors, even a Scottish astrologer, Michael Scot, combined to create an institution of learning and culture. Over this, first among peers, the gifted and learned emperor, who spoke nine languages and wrote seven and somehow during his busy and active life had managed to acquaint himself with most of the learning of his era, presided, himself a poet, himself a practical philosopher and scientist. Owing to his mixed culture and his residence in the East and friendship with Muslims, he became a kind of link between Eastern and Western thought. To round out the picture, the bevy of beautiful Saracen maidens whom he kept in his court, spinning and weaving, were regarded by contemporaries as an Oriental harem, and indeed some were probably his mistresses, for wives, once married and bedded, took no further part in the emperor's daily life.

A rather ruthless example of his experimental attitude is the case of his investigation into digestion. He wanted to know whether digestion proceeded better in one who rested after a meal or in one who exercised. He fed two men, made one exercise, and cut them both open to scrutinize the results. In order to study the longevity of fish, he put a copper ring in a carp's tail and set it free in his fishpond. He sent divers into the water to learn about sea animals and plants. His menagerie, which often traveled with him, tended by Moorish and Ethiopian keepers, included camels, dromedaries, apes and leopards. Close to Foggia, he had a large marsh laid out with ponds and walled water conduits in which, next to his luxurious palace with marble columns and bronze and marble statues, he raised and studied pelicans, cranes, herons, wild geese and exotic marsh birds.

We have suggested earlier that all hunters are practical zoologists. It was Friedrich's passion for falconry which made him an ornithologist. Hunting he considered mere physical exercise, but falconry required skill, patience, and an artful training of wild birds, which was undertaken with success mostly by the nobility, and this accorded with his aristocratic outlook.

His book, *The Art of Hunting with Birds,* for which he collected material for years, was written on the instigation of his natural son, Manfred, sometime between 1244 and 1250. A two-book version was published, although the manuscript versions contain six books. The published version was edited with some minor editions by Manfred. The manuscripts are illustrated on the margins with nine thousand colored miniatures which are a notably accurate first contribution to zoological illustration. Some are probably by Friedrich's own hand. What makes the book important is that aside from being a detailed, well-organized, eminently practical and still usable treatise on falconry, the first hundred pages constitute a general work on birds, a compendium of everything Friedrich had observed in his short but active life.

He had seen to it that his philosopher, Michael Scot, translated Aristotle's natural history into Latin, in itself a service to science; but although he respected the Greek, he was too independent to copy authority slavishly:

> The deductions of Aristotle, whom we followed when they appealed to our reason, were not entirely to be relied upon, more particularly in his descriptions of the character of certain birds. . . . In his work, the *Liber animalium* [*sic*], we find many quotations from other authors whose statements he did not verify and who, in their turn, were not speaking from experience. Entire conviction of the truth never follows hearsay.

The amazing quality of Friedrich's book is its completely factual style, its freedom from digressions, magical and miraculous tales, and empty shows of erudition based on encyclopedic quoting—in short, its rejection of the whole medieval scholarly apparatus. In reading it, one's conviction grows that Friedrich always applied his motto and spoke either from experience or from a carefully verified report. Indeed, in the footnotes of the contemporary English translation, scarcely more than a half dozen minor errors of fact are pointed out.

A good example of his skepticism is his treatment of the so-called barnacle goose:

> It is said that in the far north old ships are to be found in whose rotting hulls a worm is born that develops into a barnacle goose. This goose hangs from the dead wood by its beak until it is old and strong enough to fly.

He tells us he sent to the north and obtained specimens of such old and rotten wood with shells on it but no geese developed:

> We therefore doubt the truth of this legend in the absence of corroborating evidence. In our opinion this superstition arose from the fact that barnacle geese breed in such remote latitudes that men, in ignorance of their real nesting places, invented this explanation.

Friedrich's comments on the phoenix are equally in character: "It is said that there are certain harmless species that consist entirely of a single male and a unique female, and that the phoenix belongs to such a species. This, however, we do not believe."

In the matter of taxonomy, Friedrich made no advance because he classified his birds as water dwellers, land dwellers and neutral birds. This results in rather vague divisions. A better classification is raptores, or birds of prey, and noncarnivorous birds. His description of the former as possessing hooked beaks, bent claws, short necks and legs, with a strong back toe, observing that they abandoned their young after they have pushed them out of the nest, begins to be more practical.

He tells us that he had planned to write on genera and species, but this was never done. In the falconry book, however, he discussed the habits of the different groups (feeding, nesting, migrating, with occasional detailed reference to individual species), following this with a generalized rundown on avian anatomy.

The amount of detail which the busy emperor took the trouble to note down is amazing, especially since it includes precisely the kind of material that ecclesiastics would have considered of no significance. He tells us that swimmers, such as ducks, sleep with one foot touching the bottom so that they may be able to push off fast if an enemy approaches. He further recorded that they turn their heads backward and bury their cold beaks in their feathers. Since they are prone to respiratory diseases, Friedrich considered this a wise precaution. They also sleep facing the wind so that their feathers will remain unruffled. (Manfred added that they alternate the foot kept on the bottom in order to warm their appendages in turn.) Swallows which walk badly and have difficulty taking off from the ground often perch on a cliff from which they can launch themselves with ease. On the whole, birds which nest in trees sleep there at night, bush nesters sleep in bushes, but, he added acutely, pheasants, which nest on the ground, fly up in the trees at night for safety. Vultures, he discovered by the experiment of sealing their eyes, are attracted to their prey by sight, not by smell. He also pointed out that there is a reason for their naked heads, for when they feed on a corpse, they insert their heads into openings in the body. In a careful study of the raptores, his specialty, he distinguished the prey of each type of falcon and tells us that they perch to eat, since they pluck and tear apart the animal or bird caught for food.

Migration interested him intensely. Birds which fed on worms had to migrate to find unfrozen ground; those which fed in water sought areas where ice did not form; those which nested in the far north did not fly so far south. Migrating birds instinctively waited until they could fly before the wind. Land birds, such as starlings, flew in disorderly flocks, but the water birds grouped themselves in a V with a leader at the apex (and in contradiction to Aristotle's opinion, this leader changed during the migration). On the whole, the return flight was less organized. Most birds came back in short stages, following the warm weather.

When nesting began, the males attracted the females by song. He pointed out that in the case of raptores, either the male or the female may return to the same nesting place and wait for some time until the mate arrives. In an age in which sex and sin were fairly identical, Friedrich's comment on procreation once more underlines his independent spirit:

> Nature, in her endeavor to preserve the race by the continual multiplication of individuals, has decreed that every species of the animal kingdom, whether it progresses by the use of wings or walks on the ground, shall take pleasure in sexual union so that they seek instinctively to bring about such enjoyment.

With his usual perseverance, Friedrich checked on the habits of the cuckoo. Having been brought a nest, in which one young bird was scrawny and obviously of a different type from the rest of the chicks, he fed the alien and found that it grew into a cuckoo, proving that "these birds do not build their nests of their own but make use of other birds' nests in which to lay their eggs." Aristotle had never seen a vulture's nest and accepted the notion that they reproduced after impregnation by the wind. Friedrich had seen such nests containing but one egg. Another shrewd observation was the fact that graminiferous birds, which would find seed feeding too slow for their young, substituted locusts, crickets and worms. The raptores brought a partly plucked prey to their young upon which, since it was disabled, the young carnivores could practice.

Friedrich's philosophy of anatomy ascribes form and function to nature and makes no mention of God. He tells us that the characteristics of organs are manifest through their action and their use. Each is suited to its purpose. Yet nature can be both helpful and hostile because even though it creates effective animal forms, it also creates other forms which prey upon them.

Although in life Saint Francis and Friedrich were opposed, there were certain parallels between them. (There was a legend that the emperor had met Saint Francis, tried unsuccessfully to tempt him with a woman, and had long conversations with him.) Although the emperor's lack of humility, Epicureanism, acceptance of sensual pleasure and general heedlessness of church authority were foreign to Francis, both were Protestant in their emphasis on personal relationship to their sense of divinity. Both expressed an almost pantheistic feeling about nature, Saint Francis on an emotional, the emperor on an intellectual, level.

Interestingly enough, in the emperor's later struggles with the church he called for reform and a rejection of worldly goods, which he did not, of course, apply to himself.

To return to Friedrich's anatomical writing, he was forthright in ascribing to the head and brain the functions of the senses, hence they are the noblest part of the body. His descriptions of the skeleton are exact and include the observation that the sternum of the crane is exceptionally light and contains a hollow which holds the trachea. Likewise he correctly stated that birds possess kidneys, something which Aristotle denied. His observations on the oil gland and its use to oil down the feathers are detailed.

Wings and the mechanism of flight fascinated him. He had counted the feathers in the wings of raptores and pointed out that birds with short soft wings had to flap fast to stay in the air while the long-winged birds could soar. He does not seem, however, to have understood their ability to use air currents to stay aloft without wing movement. Investigating the movement of the wing, he pointed out that these feathers at the end described the largest arcs in flight, had the greatest lifting power, and hence had to be the longest and strongest.

Finally, his discussion of self-defense by birds makes it clear that long-billed creatures stab, others with blunt bills bite, waterfowl strike with their wings, while the domestic chicken and its wild counterparts such as pheasants and peacocks use their spurs. Brownish birds such as partridges and larks sometimes depended on their protective coloration and lay still, hoping to avoid the notice of enemies.

In the emperor's discussion of birds of prey, he remarked that the young birds are driven away by parents after they are out of the nest and that this must be done in order to make them seek a different area in which to live. This is perhaps the first mention of the significance of territory in zoological literature.

Judging by his one contribution to zoology, it is a pity that the emperor's keen mind was occupied so much of the time with politics and warfare. His book on falconry inspired imitations, the most notable being Gaston de Foix' *Miroir de Phebus des deduits de la chasse des bêtes sauvages et des oiseaux de proie,* which appeared in 1387.

By the twelfth and thirteenth centuries, however, some scientific activities were taking place in university circles. The stimulus of Greek mathematics and philosophy transmitted by the Arabs produced some tentative original thinking. Most of it dealt with individual problems. Robert Grosseteste, for instance, because he felt that light, according to Neoplatonist ideas, was the fundamental form of the universe, studied optics seriously and explained geometrically the laws of reflection.

Of the scholastic scientists one of those whose attainments rank highest is Albertus Magnus (Saint Albert), a thinker who worked in many fields and actually wrote a *De animalibus* in twenty-six books. He is credited with making some progress in the description of classification of plants, and he also wrote on theology and philosophy. Born in 1193, he studied at Padua, a university destined to be important for our science, became a Dominican in 1223, and after becoming a professor of medicine taught in Cologne, Hildesheim and Freiburg. For a time he was bishop of Regensburg, but he apparently preferred to teach and study, for he resigned his post and traveled around to various universities. He died some time in the 1280's.

The 2,600 pages of Albert's treatise become somewhat less impressive when it is pointed out that nineteen of the twenty-six books are no more than a rehash of Aristotle, Pliny and a few church fathers. This leaves seven books which deal specifically with animals, but these, in turn, are less impressive when the fact is taken into account that he was indebted for his description of 400 of his 476 species to Thomas of Cantimpré, a cleric, who wrote his *De natura rerum* about 1250. Studies show that he repeated most of Cantimpré's fables and errors, although he omitted the earlier author's moralizing. Albert is credited with being the first to mention the weasel, two kinds of martens and an arctic bear. Although he accepted the story (from Pliny) of blue worms of the Ganges which can pull off an elephant's trunk, he rejected the *Physiologus* story of the pelican which feeds its young with blood from its own breast. If we compare Cantimpré on the lion, we learn that the king of beasts is divided into two types, long-maned brave and short-maned cowardly. Those which are the offspring of sex with "pards" are weak. Lions are highly sexed. They will not touch old meat. They fear scorpions and fire. When sick, they eat the flesh of an ape. They have no joint in their necks and cannot turn their heads. Eating their flesh or hearts will cure colds. Their fat is an antidote for poisons.

All this traditional folklore was repeated by Albert, except he maintained that lions are also afraid of cocks, and when the female commits adultery with a leopard, she washes to rid herself of the alien animal's smell before returning to her mate. There he added: "But I think it is a fable." He also made a few additions to the lion as a source of materia medica. A distillation from its ears cures deafness, its testicles ground and fed to the victim cause sterility, while its dung, if drunk with wine, will cause a hatred of wine.

The strongest claim made for Albert's originality is based on his section on whales, for he had firsthand knowledge of whaling in Friesland. In the first place, when he wrote of various types, mention of long canines makes it clear that he classified walruses as whales. He pointed out that whales do not have gills but breathe through a tube like dolphins. He granted the males a penis and testicles and the females a womb like a woman's. When they breathe, they expel so much water that they swamp ships. He also stated that their tails are twenty-five feet wide. The manner of capture with a harpoon is exactly like that of more recent times. If the whale sounds, he tells us, it is badly hurt and they eventually capture it, but if it withstands the blow and makes for the sea, they are obliged to cut the rope and lose all their labor. Many barrels of oil were obtained when the head was opened. The sadistic capture of another kind of whale, hairy and with claws by which it clung to the rocks when it slept (evidently a seal or sea lion), makes it clear that Albert was no Saint Francis. He maintained that the hunters slit the skin near the tail when it is asleep and fasten a rope to the skin. They then arouse the animal which, in its effort to escape, divests itself of its skin, for the other end of the rope is fastened to a post.

All in all, the zoological writing of Albertus Magnus is no more than transitional. If he showed some skepticism toward his sources and rejected the moralizing attitude of the *Physiologus* for a purely descriptive approach, his original contributions were merely numerical. The scientific revolution was not yet really under way, even though a few minds had begun to view the world more objectively. Actually, as Lynn White, Jr., points out, the intellectual did not play the leading role in change. "Praying and fighting might be all very well for clerics and nobles but the new third estate was interested in concrete tangible goods. . . . In the new social environment, men of learning and leisure came to regard the craftsmen and the technician with new respect. Manual operation gradually became fashionable and theory and practice combined to produce modern science." Friedrich and Saint Francis were men of genius with insights transcending their time. Immediately after their deaths, they were ignored. The real revolutionaries were such humble artisans as glassmakers and printers.

4

The Assault on the Establishment

☙ THE early Renaissance inherited a picture of the world from scholasticism, or Christian philosophy (which it modified somewhat) called the Great Chain of Being. It is well to pause and consider this concept, for it was important to zoology and was not wholly abandoned until the nineteenth century. It was derived from Aristotle's hierarchies, plus some Neoplatonist thinking, and welded to Christian concepts until it became an accepted commonplace of the cultural scene. Poets such as John Donne were familiar with it; even Alexander Pope in the eighteenth century was still convinced of its validity. John Fortescue, a fifteenth-century lawyer-philosopher, pointed out:

> God created as many different kinds of things as he did creatures, so that there is no creature which does not differ from all other creatures and by which it is in some respect superior or inferior to all the rest. So that from the highest angel down to the lowest of his kind there is absolutely not found an angel which had not a superior and inferior; nor from man down to the meanest worm is there any creature which is not in some respect superior to one creature and inferior to another. So that there is nothing that the bond of order does not embrace.

Man, occupying the central point, shared the nature of the animals below him and the angels above (Plato's flesh and spirit). There were three main orders of angels; highest were the seraphs and cherubs, next archangels (sometimes this order was reversed), and below, the rank-and-file ordinary angels. Some angels attended God, who was, of course, the peak of everything in the empyrean. Below the latter was the created universe, which in Ptolemaic astronomy consisted of nine spheres carrying the planets around the earth with fixed stars outside them. The movement of the spheres was said to create music and was controlled by angels, who were sometimes depicted as turning cranks. The four elements were a sort of parallel chain, also arranged in an order, fire at the top, earth on the bottom. There was a subsidiary theory which stated that below the moon the air was thick and dirty, while above, it was known as ether and was pure.

Working downward from man, there were the animals, the plants (possessing vegetable souls) and the minerals. Each of these subdivisions of the chain had a divinely appointed head: diamond and gold were the leading stones and metals, the oak led the vegetable kingdom (the rose ruling over the flowers). There was some vagueness as to whether the dolphin or the whale was head of the fish and the lion or the elephant leader of the animals. Men were divided into hierarchies with the serfs on the bottom, the secular nobility leading up to the king or emperor, and the church ruled by its dignitaries with the Pope supreme. Unfortunately, whether the emperor or Pope was the final authority remained unsettled as the quarrels already described between Friedrich II and various pontiffs indicate. However, ideally, everything in the great chain had its permanent place, and since it was divinely ordered, any insubordination was not to be tolerated.

Objectively this concept can be seen as a neat rationalization of the feudal system. Like so many ideologies, this poetic metaphor seemed to have evolved after the social and material facts. And like so many products of human culture, it persisted long after it was moribund and a social liability.

The Great Chain of Being in its static character is linked with the ecclesiastical establishment's monopoly of information and its intellectual dictatorship. We have already mentioned that a strong clerical establishment in Babylonia and Egypt stood in the way of individual intellectual progress. In these ancient worlds religious and civil governments were one. There was no evolution from within; the old civilizations fell apart and were conquered, but in the Western world a gradual cultural change took place which was unique. Out of this change modern science arose as the dominant philosophy and strategy for controlling man's environment.

What motivated innovation? The many variables involved make it difficult to analyze how change comes about and what comes first, the hen or the egg. For the emergence of a new point of view there had to be a rejection of authority, and for the rejection of authority there had to be a type of mind that rejected authority.

It has been customary to say that with the fall of Constantinople, which released Greek manuscripts for Western study, and with the transmission of Greek philosophy through Arabian intermediaries men's minds were liberated. But in the twelfth century Plato and Aristotle were already being absorbed into the Christian synthesis. For our science of zoology Aristotle was the most important of the classical thinkers. He was read and copied, but the typically monkish device of quoting authorities slavishly, reinforced by Arab veneration for the Greek scholar, meant that his mistakes and less constructive speculations were treated respectfully, while his drive toward accurate observation was passed over.

Another approach to change lays stress on invention. There is of course intellectual invention exemplified by what we call Arabic numerals. These were really invented by the Hindus, and transmitted westward through Sicily and Spain by the Arabs (together with the concept of zero), they made advances possible in mathematics, a science which was eventually to fertilize other fields.

Material inventions that fulfill practical needs, however, begin to loom larger on the historic scene. As Marshall McLuhan has been pointing out with respect to communication, the side effects of mechanical changes in man's environment are often more profound than, and perhaps precede, the intellectual rationalization which accompanies them.

That significant innovation of the mechanical sort was taking place all during the Middle Ages is not generally appreciated. Lynn White, Jr., points out that by the year 1000 the Western world had begun the process of saving human labor by applying natural power to industry, which eventually so differentiates European from Oriental culture. By 1083 there had been a fulling mill on the banks of the Serchio in Tuscany. Before the end of the eleventh century, water power was used in the metal and textiles industries from the Pyrénées to Britain. In 1185, the first horizontal-axle windmill was invented. By the thirteenth century one hundred and twenty windmills were built in the neighborhood of Ypres. By the fourteenth century there were mills for tanning, sawing, crushing, operating the bellows of blast furnaces and the hammers of forges, mills for grinding pulp for paper and pigment and for polishing tools. Thus a primitive industrial revolution (based on water power) was in progress, which reached such refinements that by 1534 at Paris a mill was set up to polish diamonds.

Lynn White, Jr., as we have already noted in connection with Saint Francis, stresses the fact that Christianity, by driving out the animism of pagan belief and creating a despiritualized natural world at the service of man, encouraged the kind of technological progress that would develop more and more machines to exploit the terrestrial environment.

Meanwhile other inventive innovations were taking place which had important social repercussions. The device of movable type (Johannes Gutenberg's Bible, 1456) made possible the duplication of the written word, broke the monopoly of information held by the monasteries, and created a new kind of intellectual democracy. It led to the dissemination of the Bible in translation into the vulgar languages, a process opposed by the Church, which proves that the establishment felt itself threatened. This ability to duplicate and disseminate information was evidently a need felt by the rising merchant class for which we use the rather emotionally colored word "bourgeoisie." The same bourgeoisie can probably be held responsible for the widespread introduction of mechanical clocks in the thirteenth and fourteenth centuries. With a new emphasis on getting work done, time had to be measured accurately.

At this point we can begin to see that the Great Chain of Being metaphor was only suitable to the feudal picture of isolated castles, rather independent baronies and small peasant villages, firmly founded on a lack of class mobility. The merchants were city dwellers and the Renaissance city was undergoing a revolution of luxury, for, as the Marxists never tire of pointing out, the profit incentive was coming into its own. The growing use of credit (and interest in spite of clerical prohibitions) made possible the growth of large mercantile houses such as the Fuggers of Germany and the Strozzi of Italy. This amounted to a rejection of the ecclesiastical establishment, for the Church had always been hostile toward money made with money and currency in general. The Church proclaimed all human motivation should be based on faith; it was only too evident that a desire for profit was destructive of piety. Insofar as Saint Francis was trying to return to the primitive church, his fierce insistence on poverty was a reaction to social change he saw beginning as early as the thirteenth century.

A secular revolution was taking place. Increasingly, the philosophy of postponement of rewards held less and less appeal for a larger and larger group of citizens. Despite Saint Augustine's teachings, the City of Men was more alluring than the problematic City of God. Again in this period, the phenomenon of class mobility is striking. People of low birth could rise to be kings' counselors; Thomas Cardinal Wolsey is a good example. If a life of comfort and elegance could be achieved by a man's own efforts, without benefit of lineage, what became of the unalterable hierarchy of the Great Chain of Being? Material comforts made men aware of the potential of their environment; they were learning how to exploit it; and hence the urge to explore it also grew.

All this is parallel to the rise of scientific observation. As the material world became a greater source of enjoyment, interest in it increased. Some of the intellectual satisfaction formerly found in abstract argument could now be felt in accumulating the data of natural science. To solve the problem of how the blood circulates could be more rewarding than deciding how many angels could dance on the head of a pin. There was a new interest in materialistic theory.

Although the danger of heresy was always present, Alistair C. Crombie, the cultural historian, points out that already by the twelfth century Adelard of Bath, a student of Arab science, was rationalizing a departure from the moralizing point of view:

> I do not detract from God. Everything that is, is from Him and because of Him. But Nature is not confused and without system and so far as human knowledge has progressed, it should be given a hearing. Only if it fails utterly should there be a recourse to God.

While such changes were taking place in the Renaissance soul, Martin Luther's rebellion must not be forgotten. The nailing of the Ninety-five Theses on the Wittenberg church door in 1517 signified that the attitudes foreshadowed by Saint Francis and Friedrich II had become respectable by gaining political power. Wave after wave of heresy had been put down with blood and fire. Now the Protestant revolt split the establishment and intensified the feeling that the individual had a right to think independently. All the forces that were combining to create a new intellectual climate came to a head in the sixteenth century.

It must be remembered that this was an age of exploration and colonization, and here again invention plays an important role, for the mariner's compass made long voyages possible. As early as 1490, the Portuguese had taken over a large chunk of the Congo. Later they were to acquire Brazil. In 1521 came Cortes' amazing coup which garnered the Mexican Empire for Spain. Central America was to follow; the Incas of Peru succumbed between 1526 and 1532. North America was claimed mostly by the British and French; by 1520 Magellan had reached the straits which were to bear his name. All in all, the seafaring pioneers and gentlemen adventurers between 1420 and 1566 nearly quadrupled the known surface of the earth. The effect of reports of countless new animals and plants was bound to be a tremendous stimulus to natural science.

Astronomy, however, led the scientific revolution with Copernicus' *De revolutionibus orbium coelestium* (Of the Revolution of Heavenly Bodies) in 1543, in which he suggested that the earth and planets moved around the sun, even though he still considered the sun static. Since he presented this as a hypothesis and dedicated his book to Pope Paul III, there were no official sanctions, although the book was sometimes censored. It had the effect, of course, of destroying the symmetry of the top half of the Great Chain of Being, for the earth was no longer the center of everything and the scholastic philosophy that all was created for man's use was also weakened. It remained for Giordano Bruno, a lyrical philosopher and theologian, to accept Copernicus' doctrine and to go on to the concept of an infinite universe inhabited by other peoples. He also managed to work the Democritus-Lucretius atom theory into his cosmology. Bruno, naïvely unable to see the implications of his own ideas, went to the stake still clamoring to be heard and protesting that he was an orthodox Catholic. Galileo, more easily quelled for reaffirming and correcting the theories of Copernicus, as a result of direct observation through a telescope was muzzled and silenced in 1633.

All through the fifteenth and sixteenth centuries new types of thought struggled with the old, and various compromises and adjustments were made by those who were, on the one hand, drawn toward speculation and experiment but, on the other hand, clung emotionally (and perhaps from fear of sanctions) to some remnants of traditional theology.

A most important figure, although not a full-time scientist, was Francis Bacon. More clearly than anyone else in his time, he grasped the potential of science and its possible impact on human society. Beyond this, he was concerned with a constructive relationship between government and science (a problem we have not yet solved). Because of his importance as a link with the future, his significant contribution to a philosophy of science, and the attitude he typifies, it is worthwhile spending some time on the man and his work, even though he made no specific contributions to zoology.

Bacon, born in 1561, came from more or less bourgeois stock. His father, Nicolas, because of his intellectual ability, was sent to Cambridge. A self-made man who became Elizabeth's chief counselor, he was knighted for his services. Francis' mother came of a well-connected family, her sister being the wife of William Cecil, Lord Treasurer. Although Francis Bacon laid much emphasis on his aristocratic connections and the court circles in which he moved, his outlook was that of the rising merchant class, with all its pragmatism and concern with material progress. He was well educated during three years at Cambridge and spent some time in France on the staff of the British ambassador, at which point he read Montaigne's essays. His acquaintance with the French thinker probably influenced his choice of the essay as his favorite form, and certainly the content of the Frenchman's work must have taught him tolerance and inculcated an alert attitude toward intellectual problems. Unfortunately, when he returned home at his father's death, he discovered that Nicolas had not succeeded in providing for him as adequately as had been expected. Bacon, with his liking for aristocratic elegance, spent most of his life struggling with debts and conniving for positions that would both increase his political importance and improve his income.

He studied law and worked as a lawyer, but since he had been brought up with the expectation of being a statesman, his ambitions drove him to attach himself to persons in high places who might further him in politics. Elizabeth had a certain respect for him and made use of him in minor matters, but positions such as that of Attorney General were dangled before him like a carrot and then withdrawn. During his early years, when he was a member of the House of Commons, he supported that body's right to decide how much the subsidies for the queen's expenses should be and in what manner they should be collected. Elizabeth never seemed to have forgiven him for this. All during her reign, he was the victim of her feminine caprices. He tried through Robert Devereux, Second Earl of Essex, to better his fortunes, but that willful adventurer was no help to him, and after his rebellion Bacon found himself in the unhappy position of having to take part in the prosecution of his patron.

Bacon is described as of average height, slender, with dark hair and greenish eyes. Portraits show him wearing a mustache and a small pointed beard, a lingering melancholy expression about the eyes. He seems to have been of a nervous disposition, probably a prey to psychosomatic illness when his hopes were dashed, which was often enough. Some of his contemporaries have suggested that he had homosexual tastes. The fact that the shadow of a fanatical puritan mother hung over him much of his life, in the form of letters of exhortation and advice, is a bit of circumstantial supporting evidence when taken together with the fact that he did not marry until he was forty-five and had no children. All that seems to be known of his marital relationship is that his wife, Alice, talked too much. She was, after all, a girl of less than twenty. His views of marriage, expressed in his essays, show no enthusiasm for the institution: ". . . certainly the best works and of the greatest merit for the public have proceeded from the unmarried or childless men. . . ."

Bacon approached every problem intellectually; tact and the ability to flatter, so necessary for advancement in court circles, did not come naturally. He was well aware of his own exceptional gifts and also of the fact that his ideas were too unconventional to gain much approbation from the establishment. Thus it was that Bacon, the scientific philosopher, and Bacon, the statesman, were somewhat in conflict; one did not coalesce with the other. Yet he could never make up his mind on which career he should concentrate. As a result, his literary projects were seldom completed and his public career had a bitter ending.

When James I followed Elizabeth, Bacon's prospects brightened. Sir Robert Cecil, his cousin on his mother's side, had helped bring the new sovereign to the throne. The Cecils, however, never seemed to have backed poor Francis with much enthusiasm. Bacon showered Sir Robert with letters of advice and useful suggestions. It was a practice which had become a habit with the philosopher-statesman, rather like writing letters to the New York *Times*. Bacon never seemed to realize that it could have a negative effect. He was still anxious to obtain the position of Attorney General, and it was finally granted him. Except for the minor office of king's counsel, it was the first recognition he had received in a lifetime of proffering his services to the nation.

By this time Bacon's scientific ideas, on which he had also been working all his life, had coalesced into a work which he called *Instauratio Magna* (Great Restoration), a comprehensive project which would summarize all the sciences and indicate how they could be utilized for the benefit of mankind in order to restore the world to the happy state before the Fall. The plan was to involve a kind of technocracy of leading scientists who should revolutionize society and to whose advice statesmen must listen. Obviously, Bacon dreamed of himself as the link between this elite of intellect and the pragmatic world of political administration. Communication between intellectuals and politicians, however, has always been a prey to what in communication theory is called interference. When Bacon showed a draft of some of his ideas to Sir Edward Coke (who had beaten him in the competition for the Attorney Generalship under Elizabeth), Coke wrote on his copy:

> It deserveth not to be read in schools
> But to be freighted in the ship of Fooles.

On the other hand, the second-rate academic mind found Bacon slightly crazy. At various points in his life, he toyed with the idea of trying for the rectorship of some university. It was a vain dream.

Bacon worked his way up under the eccentric, conceited, boorish king by sheer perseverance. In addition, Cecil's death in 1610 left a gap in administrative circles. Bacon was now approaching the highest office, the Lord Chancellorship. By now, however, George Villiers, who had become the king's favorite, was made Duke of Buckingham, and advancement could be obtained only through his support. Bacon, fifty-five and at the height of his powers, had to swallow his pride and flatter the effeminate twenty-three-year-old. Eventually Bacon became Lord Chancellor, but in 1618, as head of the Court of Chancery, he had to listen to hints and even more than hints from Buckingham concerning cases in which the favorite was interested; in one instance he was pressured into reversing a decision.

In 1620 Bacon published a final version of his theories on applied science under the name *Novum organum,* or New Method. He sent it to King James, who wrote a kind letter four days later saying he had every intention of reading it through and would even steal some hours from his sleep to do it. Alas, poor Bacon! It appeared that far from stealing hours from King James' sleep, the book increased the length of the sovereign's slumbers. The king was reported to have said the book "was like the peace of God that passeth all understanding."

During the following year Bacon came into conflict with the House of Commons. In order to discredit him, witnesses were brought forth who testified that he had received numerous sums of money from litigants in the Court of Chancery. Bacon indeed had received money, but there was a tradition behind giving presents to government officers which in the past had not been thought of as bribery. Bacon insisted that he had never been influenced in his decisions, and no evidence was ever discovered to disprove his defense. The scandal ruined him, however. He spent some time in the tower and passed his last years in seclusion.

What, then, were Bacon's major contributions to scientific thinking?

First, he had the courage to attack Aristotle and Plato for whose idealism he had no use. Bacon's position was that the philosophy of science must be materialistic, that theology was a separate study which was not affected by thought in the scientific field. This was, of course, having your cake and eating it too. Some critics have suggested that his professions of orthodoxy were merely a cover to avoid persecution. This may or may not be true; plenty of scientists in our own era have cultivated the divided mind, materialists in relation to their profession, and religionists in their social lives. Bacon severely criticized the subjective models of the universe to which the classical philosophers subscribed. Nature was extremely complicated and knowledge of its laws could only be obtained by careful observation and classification. Moreover, he went back to the ideas of Democritus and Lucretius, saying that the atomic theory, if not wholly adequate, was a most useful explanation of the world. His attack on Aristotle's cosmology was uncompromising. He said the four elements were arbitrarily chosen and earth in this formulation had nothing in common with the perceived earth. Since he felt that knowledge of God came through revelation, he rejected Aristotle's progression from the theory of the moveless mover, to first cause, to God, as unjustified, seeing it for what it was, a muddling of theology and physics. Although Bacon never accepted Copernicus' revision of astronomy, nevertheless his rejection of Aristotle's cosmology and his separation of physics and astronomy from theology weakened the metaphor of the Great Chain of Being, which was a compound of all three.

As a result of his rejection of the four elements, he also expressed skepticism of the humors, remarking that not enough was known about them, that one could not be sure in what cavities or receptacles they were to be found, he himself "not relying herein too much upon the accepted divisions of them." Indeed experimental medicine was sadly lacking. Too many diseases were pronounced incurable, too few diseases had been carefully observed, and too few cases fully recorded.

His critique of the universities was also severe. He was fully aware of their medieval tradition of oratory and disputation which easily degenerated into "childish sophistry and ridiculous affection." He even suggested that a special university be set up to concern itself with invention.

The "fundamental sciences have been studied but only in passage and deeper draughts have not been taken of them." He bewailed small salaries and lack of funds for research. "There can hardly be made any great proficiency in the developing and unlocking of the secrets of nature, unless there be a plentiful allowance for experiments whether of Vulcan or Daedalus [furnace or engine] or any other kind whatsoever." He also called for learned societies in all parts of Europe. The British Royal Society was thus his spiritual brainchild, as indeed were all scientific associations which were subsequently to arise. He foresaw clearly that there was a need for an organization run by scientists, since universities would always be subservient to the establishment.

The conception of Bacon's *Instauratio magna* and its further implementation in the *Novum organum* is truly impressive. He began by pointing out that science had only about two hundred years of history. The reasons it had not progressed faster were that its history had not been studied or its aims defined. (The idea of progress which he always stressed was typically a point of view of the rising bourgeoisie.) "Now the true and lawful goal of the sciences is none other than this: that human life be endowed with new discoveries and powers." Interestingly enough, he drew the parallel with the progress of exploration. "It would be disgraceful if: while the regions of the material globe—that is of the earth and of the sea and of the stars—have been in our times laid widely open and revealed, the intellectual globe should remain shut up within the limits of old discoveries."

Bacon made it clear that the social position of science must be changed, once adequate salaries and funds were forthcoming. His utopian vision was of an international brotherhood of experts to be advised by a kind of supranational body which should continually survey the field and point out those areas which were backward or neglected. As a beginning, his *Novum organum* was to be a survey of what was to be done in all areas and a setting up of guidelines for the future. This of course called for a book, as he himself said, six times as long as Pliny's. Ideally he should have had a staff of experts whose work he could have coordinated and edited. Indeed in 1608 he spent six days jotting down notes on his aims and the possible methods of achieving them. He noted all the people who might be persuaded to collaborate on his plans for research, and these included his cousin, Edmund, who was studying medicine and natural history; Walter Raleigh; Bishop Lancelot Andrewes (James's favorite preacher), who was rich and had done some experiments; Thomas Harriot, an eminent astronomer; the king's doctors; plus "learned men beyond the seas." Unfortunately, Bacon had no ability for binding men together, not even in a political faction. The idea of scientific teamwork was far in the future. Bacon remained a loner and did what he could by himself, producing only rough notes and sketches of the great plan.

Most of what he set down on paper had to do with physics and chemistry. For instance, not enough was known about the winds. He demanded detailed notes of directions, heights of currents in terms of cloud drift, the relation of their direction to the sun. Were there weather cycles? Could continual discharges of cannon produce wind? If forests were cleared and lands tilled, would the winds be affected? In other words, he had a clear view of how a science of meteorology could be developed. He went so far as to try a few primitive experiments. He made model windmills with paper sails which he drove with a bellows in order to experiment with the design of the sails. These he made of different shapes and numbers, added vanes, and studied their rotary force. From this it can be seen that he was pioneering in the sorts of experiments which now are done in order to develop new airplane design. Without knowing the word, he was working in aerodynamics.

He noted that when iron was placed in nitric acid, it dissolved, emitting saffron fumes and great heat. He compared mass and weight, thus working out the density of seventy-eight substances, ranging from iron and gold to beer and petroleum. His method was to place his material in a cube-shaped container and then weigh it. He investigated gas by weighing a known amount of volatile liquid, heating it, and finding out how large a space the gas occupied at about atmospheric pressure.

Although Bacon made no great discoveries and was no great mathematician, his methods were sound and his results were written up with admirable simplicity and clarity. A map maker of the new learning and a prophet without honor in his own time, his exposition of scientific method is as valid today as it was in the sixteenth century.

Bacon's end is ironically in keeping with his life. On a cold day in March, 1626, he was driving through the snow, meditating on heat which he had suggested was a type of motion. He had also begun to think about heat and cold and the use of cold for refrigeration. He stopped his coach, bought a chicken, and proceeded to stuff it with snow to see if it would keep fresh.

He was taken with a chill, stopped at Lord Arundel's house, and went to bed. From his sickbed he wrote to his host, who was in London at the time: "I was likely to have had the fortune of Gaius Plinius, the elder, who lost his life by trying an experiment about the burning of the mountain Vesuvius. . . ." He explained his experiment, which he said had succeeded. He also apologized for being obliged to dictate the letter. The unfortunate philosopher-scientist had prophesied more accurately than he knew. The infection turned into bronchial pneumonia of which he died.

5

Doctors and Picture Books

⁂ ALTHOUGH zoology began at a Greek university, there was a gap of many centuries before it returned to the academic fold. On the whole, the medieval university made progress in mathematics, optics and mechanics. It is interesting to note that science often forges ahead on the practical side before a real body of theory develops. The primitive industrial revolution which took place in the twelfth and thirteenth centuries gave rise to such devices as spectacles, which were, of course, the forerunners of microscopic and telescopic lenses, but for these scholars had to wait until the quality of glass improved and also the technique of grinding lenses. Chemistry was both begotten and held back by alchemy in the later Middle Ages. By the thirteenth century alchemists knew how to distill and sublimate and had invented various types of basic apparatus, but their preoccupation with changes in color and appearance rather than exact analysis and measurement led them in false directions. Above all, their attitude toward gold is significant of the old mentality. Aside from the desire to make gold in order to acquire riches, theoretical considerations which lead us back to the Great Chain of Being were attached to this metal. Since, it will be remembered, it was hieratic leader in its domain, gold was an example of perfect natural harmony and proportion of combination. Other metals were formed as a result of defects in purity or proportion of ingredients. Hence the aim of the alchemist was to remove such defects and turn "base" or inferior metals into gold.

Industrial chemistry, however, improved the metals used for casting cannon and bells, created pewter, invented plate glass, and made many pragmatic contributions to such homely arts as pottery making, tanning and the production of soap.

Similarly, the science of zoology for centuries was no more than a practical appendage to medicine as far as the university was concerned. Since medicine involved drugs, such a many-faceted scholar as Albertus Magnus carried the pharmacology into a real interest in botany, while animals ran a poor second in his writings. On the whole, animals were chiefly used for the purpose of dissection by medical scholars.

Nevertheless, when zoology began to emerge as a separate discipline in the early Renaissance, the great names were those of doctors or at least of those medically trained. Konrad Gesner and Guillaume Rondelet were professors of medicine and practicing doctors. Pierre Belon was educated as a doctor but fortunately encountered patrons who supported his labors as a natural scientist. Ulisse Aldrovandi had a medical degree but taught botany and pharmacology.

The universities reflected the split in the establishment created by the Reformation. In the Protestant areas of northern Germany and Switzerland, scholasticism was attacked and for a time there was considerable confusion. Then there was a gradual process of rebuilding. Orthodoxy was still an authoritarian force. Protestants could not teach in Catholic institutions, and many of these did not even accept Protestant students. Calvinists were barred from universities dominated by Lutherans. After Henry VIII's break with Rome, a law was passed in 1533 obliging all candidates for degrees in British universities to subscribe to the Anglican faith.

In France, Spain and Italy, the Catholic universities flourished as far as endowments and numbers of students went, but many were far from open to new ideas. Francis I, however, founded what was to be the Collège de France on a more progressive basis than the Sorbonne. Neither religious camp, however, stressed science. Mathematics, classical languages and thought and Aristotelian natural science were taught. Professors were mostly poorly paid. Students lived miserably, brawled, whored, starved and wandered from one university to another, for life was often disturbed by imperialist wars.

The life of Konrad Gesner, born in 1516 in Zurich, reflects the problems of the Protestant scholar. His father, a poor artisan, was killed in the battle of Kappel in which the troops of the reformer Ulrich Zwingli clashed with the Catholics in 1531. Switzerland was split with Zwingli, theocratic dictator of Zurich, leading the Protestant cantons. The Protestants lost this particular battle in which Zwingli was killed. He was succeeded by Heinrich Bullinger, later to become Gesner's support.

The young Gesner was sent to school by an uncle on his mother's side of the family. Konrad distinguished himself sufficiently as a scholar under Johannes Frick, who taught him botany, so that he was eventually sent to Oswald Mycenius, the Protestant leader of Basel, under whom he studied the classical authors. From then on, he led a precarious existence on scholarships, which barely supported him. Thanks to the patronage of Mycenius, money was raised to send him to France, where he studied theology and medicine at Bourges. He felt that the level of instruction was low. By 1534 he was in Paris, where he was reading philosophy, rhetoric and medicine. He learned classical and Oriental languages, and picked up natural science for himself—still complaining of his economic difficulties. Nevertheless, by the time he was ready to return to Basel to finish his medical studies he had decided to marry. With Protestant Germanic sobriety he tells us that he determined to marry a girl not rich but of good character. Unfortunately, Barbara Singerin was rather sickly and not even a good housewife. For a time he moved to Lausanne where he was professor of Greek. Since he had few students he made use of his time to continue his medical studies and to enter the field of botany. He made or collected 1,500 drawings to illustrate a *Historia plantarum* in which he classified plants by their reproductive structures, thus anticipating Linnaeus, but the work was so expensive to print that it did not see the light of day until long after his death, in 1751. For a time he then taught in Montpellier and studied with Guillaume Rondelet, but by 1541 he was back in Zurich, where he got the moderately salaried post of town physician. Although he also taught physics and psychology in the Collegium Carolinum and did an immense amount of hack work, including a *Bibliotheca universalis* (an encyclopedia of writings in Greek, Latin and Hebrew), letters to Parson Bullinger are on record in which he begged for financial assistance. He complained that teaching and his medical practice left him little time for writing, that he worked far into the night, that his family and relatives were all dependent on him.

Actually he was offered a post by the commercial house of the Fuggers and he could also have taught in England, but a patriotic love of his hometown, Zurich, caused him to reject these offers. Indeed, he took time off to write a treatise on the beauty of the Swiss mountains.

The renown of his many-sided scholarship (he knew Latin, Greek and Hebrew and spoke German, French, Italian, some English and was learning Arabic) penetrated far and wide. The Emperor Ferdinand granted him a coat of arms with the imperial seal for his work and was quoted as saying, "Gesner is learning itself." In 1565 he died fighting a plague ravaging the town of Zurich.

The important work which makes him the leading Renaissance zoologist is his *Historia animalium,* published in four volumes between 1551 and 1558. The fifth volume on snakes and insects was still in manuscript form at his death and was not printed until 1587.

Ulisse Aldrovandi, Gesner's parallel in Italy, was born in Bologna in 1522. His father was a notary, his mother an aristocrat. From his parents he was to inherit a fortune which freed him from the problems which Gesner had to face. As a child he was educated by a tutor but suddenly, at the age of twelve, set off for Rome without any money and without a word to his relatives. He tells us that everyone he met was happy to give him lodging. His mother had a brother in Rome to whom she wrote to look after her son. Ulisse was steered into a position in the household of the Bishop of Sardinia, but the life was not to his liking, for he was back home four months later. His next position, at the age of fourteen, was that of an accountant in Brescia. A couple of years later, this impetuous young man was off on a pilgrimage to Compostela, in Spain. He was always a pious Catholic, and a chance meeting with a pilgrim resulted in his joining his new friend. Unfortunately, the Emperor Charles V and Francis I of France were at war, a situation which did not make western Europe a good place for traveling. Ulisse and his friend got as far as Riez under the protection of Francis' commander in chief, but in the Pyrénées they were robbed of everything, except for their shoes and hats, by bandits. Eventually they reached Compostela. On the way home they nearly starved, going for two days without eating. They were then shipwrecked, took ship again, were pursued by pirates, and in the skirmish that followed three members of the crew were killed. Nevertheless, no sooner did Aldrovandi get home than he had an impulse to set off on another pilgrimage, this time to Jerusalem. However, since his elder brother had died, he was persuaded to stay home. Curiosity about the world was as strong a motive as piety in the case of Aldrovandi, for he made notes about plants and animals. He wrote that he liked the roving life "because of the many facts of nature that, at that age, he was anxious to observe and because of what he could learn of natural history." Moreover "he had collected much and been given so many things by important people on his travels that he made a museum in his own house." From this we can see that his travels did much to make him a natural scientist.

He went on to study belles-lettres, logic and law, earning a doctorate in law and later obtaining a degree in medicine, which he never practiced. He became a botanist and a teacher.

Strangely enough, in 1549 he was called to Rome for questioning concerning heretical opinions. Since he never entertained unorthodox ideas, the affair is mysterious. Perhaps he had enemies. He was, however, anxious to clear himself and did. Rondelet, already working on fish, was in Rome at the time. Since he and Aldrovandi went on collecting expeditions together, the meeting with the older naturalist must have been a further stimulus to his own penchant for natural sciences. It is probable that Aldrovandi at some time made the acquaintance of Gesner.

The senate of Bologna next elected Aldrovandi professor of philosophy. He also lectured on natural history, keeping his position in the University of Bologna for the rest of his life.

Aldrovandi married twice. His first wife died after they had been together only nineteen months, in 1565. The second bore him two children, who died young, and helped him with his notes. A natural son fell from a balcony and died at age eighteen.

Positions of importance continued to be given him, for he was made inspector of pharmacy in Bologna, responsible for standards in the drug industry. His fame spread to such an extent that he was invited to travel around Italy lecturing on natural science. In 1579 he suffered from kidney stones. A comment upon the medical practice of the period is his belief that he was cured by drinking turpentine. In 1564 he founded a botanical garden, and by 1592 he acquired a disciple in that study, Johanis Cornelis Uterverius of Delft, who succeeded him in Bologna as professor of botany. After forty-six years of teaching he retired in 1600 and died five years later. Short, stocky, bearded like a sea captain, there was something cherubic in his round innocent face. In contrast, Gesner whose beard flowed down onto his chest is solemn and serious, with stern, regular features.

The work of these two men is literary and looks toward the past, for both indulged in the kind of scholarship which consisted of citing, uncritically, every reference to an animal which they could find by a diligent search of the works of Aristotle, Pliny and as many Greek, Latin, Hebrew or Oriental writers as they were acquainted with. Gesner arranged his beasts alphabetically. Aldrovandi accumulated enormous amounts of material but published only four volumes on birds, the first of which did not appear until he was old and his sight impaired, in 1577. After his death, ten more volumes on animals, plants and stones were published, with much revision, by his students. Many volumes of manuscript and many color plates still remain unpublished in the Bologna library.

Both Gesner and Aldrovandi demonstrate the handicap of lack of terminology, for in every case they felt it necessary to bring together all the names of each animal in as many languages as it was mentioned. Gesner was the better stylist and more down-to-earth. On the whole, the contributions of these literary naturalists (aside from pictures, to be discussed later) were fairly accurate verbal descriptions of the animals about which they had firsthand knowledge or fairly reliable reports. They had not, however, freed themselves fully from the *Physiologus,* for the satyr, a shaggy man with horns, and the unicorn still appear. It is true, however, that Gesner was a bit cautious about the latter, which he called a monoceros. "There are perhaps more than one kind of unicorns because there are so many descriptions by different people. And it is surprising, since it seems to range all over the earth, that it never arrived in Europe. Therefore, I will bring together the opinions of ancient and modern writers as each one describes these animals. . . ." So he tells us it has "a stag's head, feet of an elephant, tail of a wild bear, a loud bellow and a horn in the middle of its forehead." He also repeated the medieval legend that it can be caught by using a young virgin as a decoy, the animal being so susceptible that it would come and lay its head in her lap. Gesner explained that the hunters used a boy dressed as a girl, which worked just as well.

Quite interesting is the discussion of the fox-ape, which at first glance appears to be a fabulous animal. Here is the discussion of the beast: "Only fifty years ago, Christopher Columbus, the Genoese, at the service of the Queen of Spain, set out with other brave men to seek new islands and Vicente Pinzón found himself near Payra [*sic*] and discovered a tree so thick that sixteen men could not reach around it. At the same place among these trees they found an animal with the front half of a fox, ears like a bat, forelimbs like human hands but the breast like an ape. And on the belly it had a bag like a great leather sack in which it hid its young and carried them there after they were born, here and there, still in the same bag. They only came out to nurse or when strong enough to help the mother find food. Thus has nature taken care that these young animals have a refuge from the hunter or other wild animals. Pinzón captured the animal with young and brought it into the ship. The young died immediately; the adult was brought to Seville, in Spain, and lived several months but from change of air and food it then died." The picture, obviously drawn by an artist who had never seen the creature, looks like nothing recognizable, but the description convinces us that Gesner had recorded the discovery of the opossum.

A description of the uro or urochs (*Bos primogenius*), which was last observed in 1627 in Poland, is of historical interest: "In Hercynia an animal, called the Uro, flourishes. In size it is little smaller than an elephant, in aspect color and shape it is like a bull; it has great strength and speed and it spares neither man nor beast it sees." (In the German version its habitat is given as the Pyrénées between France and Spain.) In the woodcut the beast is shown charging a tree behind which the hunter hides to kill the animal with a spear as it rams its horns into the trunk.

Gesner followed Aristotle's classification in his thirty-five-hundred-page treatise—in his large subdivisions of viviparous and oviparous, quadrupeds, birds and fish, reptiles and insects. The alphabetical arrangement is not entirely helter-skelter since it contains generic grouping—all the cattle under *bos* and all the apes under *simia,* for instance. He then discussed each beast in terms of habitat, origin, physical description, natural functions of the body, qualities of the soul, use to man as food and in general, and, finally, its literary and philosophical background.

Actually the citations of other authors are tiresome, and although the physical descriptions are fuller than those of previous authors, Gesner did not add much that was new. The material on fish was simply incorporated from Pierre Belon and Guillaume Rondelet. Gesner, therefore, has the virtue of being encyclopedic. In one case in which he was accused by later writers of fantasy, he has had a posthumous triumph. A careful description of the *Corvosylvatico,* waldrapp, or wood crow, created a picture of a fowl unknown in Europe. He tells us it is black, has greenish feet like a hen, a short tail, a gray stripe on the back of its head, a long reddish beak and long red legs. The description was copied by Aldrovandi and by the English ornithologist John Ray; Linnaeus even included it in his bird book. It was not until the eighteenth century that the bird was omitted from European ornithologies because no specimen living or dead could be identified. Years later, the *Ibis cornatus* was identified in Egypt, exactly answering the description. Apparently the bird had either died out in Europe or, because of climate changes, had changed its habitat.

Fossils were a great mystery and a controversial matter up to the nineteenth century. Xenophanes and other fifth-century B.C. Greeks had believed that they were the remains of animals which had lived in water once covering the sites where they were found. Later generations denied this, and some scholars even denied that they were the remains of living animals. When Gesner was confronted with a fossil hippopotamus tooth, he played safe. "Whether this or similar are hippopotamus teeth or rather those of some fearful animal, we make no decision."

Turning to Aldrovandi, who studied Gesner, we encounter a less critical mind and a less able thinker. The French naturalist Georges Buffon is quoted as saying that if the nonsense were filtered out of his work, only about one-tenth would stand up as useful. On the other hand, some of his classification is a little more analytical than that of Gesner. For instance, he divided his ornithology into birds of prey, then wild and tame fowl (now called gallinaceous birds), which he characterized as those that bathe in the dust, then pigeons and sparrows, which will bathe in both dust and water, then berry- and insect-eating songbirds, and finally waterfowl. The Italian also included a number of exotic birds unknown to Gesner.

Aldrovandi also kept in his treatise such fantasies as the basilisk and the many-headed hydra, and his classical citations are unbearably irrelevant. To take his Volume II, Book 2, *Of Domestic Fowls That Bathe in the Dust,* as an example, he began with the word "gallus" (cock) and rang the changes on it to show his erudition. Not only were the priests of Cybele (called Galli) cited, but mention was made of every Roman of significance named Gallus. He went on to tell us that the Persians wore a rooster comb on their helmets and that Aristophanes said they were led by a cock. His philological prolixity then led him into the antecedents of the words "pullus" (pullet) and "ovum" (egg) and finally into a plenitude of chicken stories, including everything anyone had ever said about domestic fowls. One tale exemplifies his own superstitious credulity and ties up with his eventful pilgrimage to Compostela. It seems that a man, wife and handsome son came to view the body of the famous Saint James in Compostela. On the way they stopped at an inn in a small town, Domenico de la Calzada. The daughter of the innkeeper, a shameless wench, fell in love with the son. She propositioned him, but it turned out that his mind was on spiritual matters. A woman scorned, she determined to have her revenge and hid a silver dish in the young man's luggage. She then denounced him, the plate was found, and he was crucified on a two-pronged fork. The heartbroken parents went and prayed to Saint James for help. On their return they stopped at the place of execution and were met by their son alive and well. They immediately went to the mayor of the town to announce that their son had come back to life. The skeptical official said that the son was as much alive as the rooster, roasted crisp and brown, he had just placed on the table. Whereupon the cooked rooster not only crowed but jumped off the platter with all his feathers on. The girl confessed and was fined by the mayor, seemingly a very mild punishment. Aldrovandi tells us he was sure the story was true because he stopped at the same inn, the rooster had been preserved, and pilgrims used to take home a feather pulled from its body. No doubt, he took one himself.

After three hundred and fifty pages of chicken nonsense, our author got down to short, one-paragraph descriptions of different breeds of birds to accompany each picture. These are factual and accurate. Of course, he did take time out to tell us that a horse's cough may be cured by softening eggs in vinegar and dropping them down its throat.

The interest in the marvelous crops out in the freaks which are carefully illustrated. There are double-headed, double-bodied chickens, chickens with extra feet here and there, and finally a bird with no comb, hair on its head, and a lion's tail.

Aldrovandi on fossils demonstrates the difference between his mind and that of the more objective Swiss. He tells us he saw them "not as the remains of normal living forms but as incomplete animals in which spontaneous generation has failed of full accomplishment."

The two Frenchmen Guillaume Rondelet (1507-56) and Pierre Belon (1517-64) were, on the whole, more progressive. They abandoned the tiresome load of quotations from ancient authors and thus once and for all shed a good deal of the medieval heritage. These authors also impress us as having worked very closely with specimens they themselves collected, and they both show a germinal interest in comparative anatomy. Both worked with marine animals, thus laying the basis of ichthyology, while Belon also wrote on birds, thus becoming a worthy successor to Friedrich II. Both were travelers in search of specimens and both met in Rome. From this it will be seen that the four greats of the early Renaissance were all in touch with each other and scientifically friendly.

Rondelet was born in Montpellier, where he was educated in medicine and enjoyed the patronage of Cardinal François de Tournon. As personal physician to a traveling aristocrat, he was able to journey to Italy where, as has been mentioned, he met Aldrovandi in 1550. He returned to Montpellier as a teacher, built an amphitheater for anatomy, and published his first volume *Libri de piscibus marinis* in 1554. It was enlarged and brought out as *Universe aquatilium historiae* the following year. Thanks to his visits to both Holland and Italy, he was able to include two hundred and forty-four species, but in spite of some anatomical comparison—mostly of jaws and gills and a distinction between the shark family and bony fish—he did not advance much in the area of classification. Among fish he included whales and dolphins, seals, crocodiles and amphibians, crustaceans, mollusks, worms and invertebrates. He, too, was a collector of as many names as possible in the hope of identifying the target with a shotgun. Such marvels as a fish resembling a bishop covered with scales and a bald lion with scales, labeled *monstruo leonino,* were included in his book. When he died at forty-nine, his work had paralleled that of Belon and between the two there was probably an exchange of influence.

Belon was born to a poor family in Le Mans. His genius for science seems to have attracted the interest of the local bishop, and it also obtained for him the patronage of Cardinal Tournon. Thanks to financial aid from the bishop, he was educated at the University of Paris (which, it will be remembered, was more modern than the Sorbonne). He was, it seems, a licentiate in medicine but probably did not practice. He became the personal physician of the cardinal, who made it possible for him to travel. When he came to write his book on fish, he said he wished to show "by pictures what I have seen in various ports and on beaches in both Asia and Europe, especially of Constantinople, Rome, Venice, Genoa, Aquitaine, Flanders and England and in lakes, ponds and rivers." In Constantinople he saw a hippopotamus and was able to contradict the description of the animal found in Herodotus, who credited it with the size of an ass and the tail of a horse. Eventually Belon studied in Germany and roamed as far as Greece and Egypt. He collected much natural science material and also made notes on archaeology and ethnology, publishing a couple of travel books. When he settled in Paris, he was granted a pension by Henry II. He knew literary people in Paris, including the poet Ronsard, who dedicated verses to him. At forty-seven, he was murdered by highwaymen near the Bois de Boulogne.

His first book, published in 1551, devoted 38 of its 55 pages to dolphins and included ten other fish and the nautilus. In 1553 it was enlarged and published in French as *La Nature et diversité des poissons* to include 110 species, 103 of them illustrated. For Belon, fish was about as all-embracing a term as it was for Rondelet. He included, as usual, whales, dolphins, crustaceans, mollusks and various invertebrate forms, as well as the seal, hippopotamus, the beaver and the otter and a couple of land lizards whose habitats were innocent of water. Most curious, however, is a description of what, from the picture, is clearly a hyena. Identified as a *loup marin* (Belon used Latin, Greek and French names), we are told: "Although the English have no wolves in their country, nature has provided them with a beast on the shore of their sea similar to our wolf. Were it not that it prefers to devour fish and seafood, we would say it was exactly similar to our predatory beast, considering the corpulence, fur, the head (which is always very large), the tail much like that of the terrestrial wolf, but, as we have said, it only eats fish and was quite unknown to the ancient writers. The fact that it leads, as is said, a double life does not make it seem any less remarkable to me. Therefore, I was anxious to include its picture." Since Gesner also connected the hyena with water, we wonder how this misconception got started. Perhaps it arose from hearsay really based on the otter.

Belon, like Rondelet, included not the bishop fish but one in the shape of a monk. He said it was caught in Norway, but it only lived three days, didn't speak a word, only sighed and moaned. Belon's classification of fishes proper followed Aristotle's. The invertebrates were grouped together as fishes "without blood." He separated those fish with cartilaginous and bony skeletons and also those that lay eggs from those that produce live young. Related types of fish were also grouped together. Here is an example of his short and sober descriptions, in this case of the flying fish which he called the *hirondelle*:

> The swallow has four big fins without spines with which it is able to fly in the air somewhat as swallows do whence it has gotten its name in Greek, Latin and French. Its head is about as big as that of a mullet, its eyes large and wide, mouth small and without teeth, underpart of the head flat. It has only one fin on the back joining the tail. All other fish have a line on each side above the fins but the swallow has it below, a particular characteristic and it does not divide the fish into two equal parts because the part toward the belly is less. It is seldom less than a mullet in size and is covered with the same scales. Its tail is large and forked and another easily recognizable characteristic is that the lower fork is smaller than the upper. No other fish has such long side fins for they extend even beyond the tail while with all other fish they reach only to the anus. With us it is rare and little known but because it is one of nature's miracles, it is preserved and hung in cabinets among rare singularities.

Owing to Belon's extended trips, he was able to include Oriental species which were not known before to the West. Even in the book on fish his bent for comparative anatomy is evident. He reproduced a drawing of a dolphin's skull, remarking:

> It should be noted that, as has been said before, the bones of the dolphin from the navel up greatly resemble a pig's (I should almost dare say a man's). . . . As for the head you may see by this picture which I had made at the port of Arumino after nature, if it in any way resembles that of a pig.

Belon's *L'Histoire de la nature des oyseaux avec leurs descriptions et naïfs portraits,* 1555, is far more imposing. In this, a large, richly illustrated folio, he made a point of telling the reader that no one else had recorded "as we have succeeded in doing, the realistic portraits of serpents, fish and birds, and no one has yet shown them so true to life as we."

Belon made some advance in classification. He put the birds of prey and the magpies together with the birds that fly by night, among these the bat. This animal was a controversial figure, some still insisting it was a bird. Belon knew that it nourished its young with two breasts, and he had dissected bats and found the young within a womb much like that of a mouse. He listed the water birds, ducks, swans and cormorants together, crows, woodpeckers and thrushes formed another group, and the nightingales, swallows and sparrows another. Herons, ibis and storks were grouped, as were the ground birds, ostriches, peacocks, partridges and chickens. The weakness of the system was still the fact that habitat was relied upon rather than structure. Belon discussed nests, food, song, reproduction. His descriptions were factual on the whole, but he still included such bits of information as the belief that swallows gulp down stones to cure stomach ailments. He also recorded the fact that doctors believe a water distilled from swallows cures eye diseases.

On the anatomical side, his interest in comparative structure was once more emphasized. "We shall name each particular bone and confront it with those of other animals," he wrote, and published on two facing pages his famous comparison of a man's skeleton with a bird's.

With the authors we have just been discussing, the art of illustration assumed an important role in natural science. The pictures adorning the *Physiologus* were typically early medieval in that they were one-dimensional, symbolic and often quaintly distorted. By the twelfth century however the flowers, fruit and leaves adorning the capitals of columns in churches grew naturalistic; some species can be recognized. By the thirteenth century craftsmen in stone lovingly carved pine, oak, maple, buttercup, hop, bryony, ivy and hawthorn on the cathedrals of York, Ely and Southwell, while in France the strawberry plant, the snapdragon and the broom were also depicted.

Often cited is the notebook of the French thirteenth-century architect Villard d'Honnecourt who interspersed his designs and his plans for engines of war with excellent sketches of a lobster, a dragonfly, a grasshopper, parrots, ostriches, a rabbit, a sheep, a bear and a lion.

We have already had occasion to mention the accurate pictures of birds on the margins of Friedrich II's book on falconry. Three vellum fragments of a manuscript attributed to one Cybo d'Hyères, who may have been a fourteenth-century Genoese monk, are another link in the development of animal illustration. The margin of one fragment, a verse history of Sicily, is filled with delicately colored sketches of insects, spiders, marine invertebrates, reptiles and mammals. The other manuscript, a treatise on the vices, is quaintly decorated with birds.

Italian art was, of course, discovering perspective, a landmark differentiating the Renaissance from the medieval view of the world. Likewise the great Florentines were insisting that the artist must serve an apprenticeship in anatomy. Leonardo da Vinci's famous notebooks with their many sketches of animals and anatomical drawings are a link between science and art, for he planned to collaborate on an illustrated anatomy text.

Since scholars in natural science were already concerned with the problem of identification of species and accurate records of the creatures they studied, they eagerly supplemented words with the new three-dimensional art. The technique of woodblock printing made reproduction possible; color was limited to a few tones added by hand. The illustrations to Renaissance books of zoology have for us today a certain quaint, decorative quality. Those made by various artists for Gesner and Aldrovandi play in and out between fantasy and realism. The often reproduced rhinoceros from Gesner (attributed to Albrecht Dürer) exhibits a truly Baroque treatment of the armored skin. The pictures which Belon had drawn under his critical supervision are fairly naturalistic. The collaborations of graphic artists made these great folios beautiful and exciting books.

Much of the vigor of the Renaissance art, reflecting the fresh vision of the period, was derived from a delight in organic movement and a fascination with the interplay of the muscles of the naked body. The new men in both science and art were intoxicated with the excitement of discovery. Advances in the subscience of anatomy, therefore, are another step forward in zoological study.

6

Inside the Animal Body

☙ SINCE the study of the internal structure of animals was not yet undertaken for itself, despite Belon's tentative remarks about comparative anatomy and since when they were dissected they were viewed simply as substitutes for human beings, the next steps in the investigation of the mammalian body lead us into human anatomy. This was a study which had languished until the sixteenth century when the proponents of new scientific points of view took a closer look at the miraculous systems and organs within living creatures.

If the revolt against the intellectual establishment is first dramatized by Galileo's overthrow of Ptolemaic cosmology and by his brilliantly experimental approach to physics, when it came to the study of the body, it was the authority of Claudius Galen that had to be challenged. Although he made great contributions, Galen never dissected human beings, and as a result much of his descriptive material was inaccurate. The whole problem of dissection of the human cadaver was surrounded with emotionalism. It seems clear that whether conscious or not, the old primitive feeling that the dead were dangerous had not entirely disappeared. The Christian objection rested upon the rationalization that man was created in God's image and to desecrate that image with the surgeon's tools was impious. Friedrich II, free-thinker that he was, had encouraged dissection; after his time there was pressure against it, even if it was not officially prohibited. Dissection, however, was occasionally done at the University of Bologna, as early as 1275. In Salerno, liberalized by Saracen influence, it was a part of the medical curriculum. Postmortems were carried out sometimes to learn if a patient had died of wounds and in cases where there was suspicion of poisoning.

A regular practice of dissecting publicly for teaching purposes was instituted in Bologna by Mondino de' Luzzi (1275-1326) who as professor of anatomy published a book, *Anatomia,* the first treatise even devoted exclusively to the subject. Mondino was a close follower of Galen and hence saw pretty much what Galen had described.

It must be remembered that there was a split between physicians and surgeons. The former, literary and theoretical, were the respected scholars and professors. Surgeons and barbers were of the same guild, and the illiterate barber was often called upon to let blood or amputate a limb. Thus these artisans were on the same level and often practiced both trades.

Public dissections, which took place in the open air in winter, when decay was less rapid (there were no preservatives), were something of a three-ring circus. They were attended by the students, other professors from the faculty of philosophy, and morbid curiosity-seekers. The professor of medicine, pointer in hand, lectured fluently, regurgitating Galen. Indeed Jacob Sylvius, a Paris authority, considered the classical author divinely inspired and infallible. Meanwhile, the clumsy artisan sawed and chiseled with highly primitive tools, creating a bloody mess.

Again, since there were no preservatives, the most perishable parts were examined first: the abdomen, the thorax, then the head and the extremities. Muscles, blood vessels and nerves were also supposed to be exposed, but not too many surgeons had enough skill for this, and besides, the students were likely to lose interest and drift away unless the philosophers got into an argument over disagreements between Aristotle and Galen, a frequent source of disputation. None of the them ever bothered to look more closely at the cadaver to prove a point by observation. The whole process of opening a body was supposed to take four days. Andreas Vesalius, the father of modern anatomy, was to comment on the practice of disputation: "Days are wasted in ridiculous questions so that in such confusion less is presented to the spectators than a butcher in his stall could teach a physician."

It is with Vesalius that new methods of investigation begin. Born about 1514 in Brussels of a family with a medical tradition, he received his early schooling in Brussels and Louvain. The family name which, variously spelled, seems to have been Witing or Wittag but was changed to Wesel, Latinized as Vesalius, because of three weasels on its coat of arms, which in turn was taken from the Rhine province of Wesel.

Andreas' grandfather was a doctor and also rector of the University of Louvain, but his father, who happened to be illegitimate, was an apothecary in the service of the Emperor Charles V. Owing to his post with the emperor, he was officially legitimized in 1531. With all this medical background, it is not surprising that Vesalius lost interest in his studies at Louvain and moved on to the medical school of Paris in 1533. Of his life in Louvain we learn mainly that he was critical of both Michael Scot and Albertus Magnus' work in embryology. Students in Louvain swore to be true to the Roman faith and were prohibited from bearing arms and dicing in public; they were also supposed to obey a curfew of nine o'clock in summer and eight in winter. Louvain was on the whole not a particularly progressive institution. Unfortunately, Paris was not much better. Under the Galenist Sylvius dissection of humans was rarely practiced. Andreas worked on dogs. The curriculum included pharmacology, pathology, physiology, surgery and a little botany. After four years the candidate for a degree took examinations and wrote a thesis. Andreas never finished because that same war between Charles V and Francis I, which made life difficult for Aldrovandi, intervened, and Vesalius, since he was a Fleming, found himself an enemy alien in Paris.

Back in Brussels, where he had the opportunity of being present at the autopsy of a girl, we learn that he had hitherto only watched two dissections of the human body. He returned to Louvain where the medical establishment was less enlightened than the burgomaster of the town. Vesalius tells us that he went at night to the place of execution and stole part of a corpse by standing on the shoulders of a friend, for the body was chained to the gallows. He buried the extremities, went back a second night, and loosened the thorax. Eventually he cleaned and articulated the whole skeleton. Although the medical authorities frowned on his activities, the burgomaster made it possible for him to dissect the bodies of other condemned criminals. He seems to have obtained his degree, and since the war still continued, he next matriculated in Padua. Historically this university was an offshoot of Bologna, but being under the jurisdiction of first the dukes and then the Republic of Venice, its policies were more liberal than those of the parent institution which was accountable to Rome.

Andreas Vesalius evidently had already acquired notable anatomical technique, for although he did not possess all the requirements for teaching at Padua, after a few months he was made a member of the faculty as professor of surgery. He spent seven years occupying this post and at the same time acquiring experience in local medical practice. His fame as an anatomist grew rapidly. He performed the essential work with the aid of his students, banishing the ignorant surgeon.

He perfected new instruments, studying the tools of various trades and adapting them to his profession. Instead of the usual four days, his demonstration lectures went on for three weeks. He began with a demonstration of the skeleton, then one corpse was prepared to show muscles, blood vessels and nerves, while a second body was opened to show the internal organs and the brain. He seems to have been something of a theatrical performer, or at least an inspired teacher, for he was able to hold audiences of as many as five hundred. His fame spread to other universities to which he was invited to give lectures. The Duke of Florence offered him six hundred crowns and as many bodies as he wanted from the scaffold if he would teach at Pisa. For some years he worked feverishly, watching men dying from fatal diseases, robbing graveyards to do autopsies, sometimes even hiding corpses in his bed. For three years he did not depart from Galen, but then he began to compare monkey and human anatomy and to annotate Galen. More and more he found that direct observation gave the lie to the classical scholar, and gradually his corrections became a thick volume. Meanwhile he began to make drawings of his own; one of the venous system was so much admired that he was urged to do more. So it was, at the age of twenty-five he embarked upon his great work, *De humani corporis fabrica* (Concerning the Construction of the Human Body), a huge folio of more than seven hundred pages, which described all the systems and structures of the body, as he had seen them, with copious illustrations. For these he used capable artists, one of them, Jan Stephen von Calcar, a pupil of Titian and Raphael. The famous flayed "musclemen" were the first accurate representations of their kind, and the blood vessels and nervous systems were treated with equal detail. The title page showed him lecturing on a dissection surrounded by a crowd of onlookers.

He also prepared a thirty-one-page shorter version of the work, cheap enough to be used by students, called the *Epitome*. He took a year off from the university and saw both books through the press. They were printed in Basel in 1542. Since these books incorporated his corrections of Galen, they raised a storm. The conservatives were furious; his old master Sylvius wrote long refutations, even suggesting that men's bodies might have changed since Galen wrote. Other polemical treatises were more violent. Vesalius was treated with all the abuse that radicals draw down upon themselves in our own time. Aside from being all wrong, he was sordid, godless and he cut up men alive. The last he did not do. It was true that he practiced vivisection upon animals, but since they had no souls, pain inflicted upon them did not count.

The year after the publication of his book he took a step which radically changed his life. It is not clear why, but he gave up his professorship and became court physician to the Emperor Charles V. He could certainly have weathered the storm at Padua, but perhaps the tradition of imperial service in his family carried weight. It is easy to imagine him becoming fed up with the pettiness of the academic scene and perhaps thinking he would have more freedom in court circles. He dedicated the *Fabrica* to Charles and the *Epitome* to his son, Philip, evidently to pave the way to his new post. At any rate for Vesalius, the scientist, the new life did not work out well. Shortly after taking up his new duties, he burned some of his unfinished manuscripts and gave up research. It is thought that the other conservative court physicians may have created difficulties for him. Thereafter, he accompanied his royal master on his various journeys and wars. Most of his work now consisted of treating soldiers wounded on the battlefield. In this he did not distinguish himself. Like most people Vesalius was inconsistent; both he and William Harvey, the other great man of anatomy, intellectually exhibited a mixture of clear, progressive thinking in some areas and in others routine conservatism. As a physiologist and a doctor Vesalius did not depart from Aristotle and Galen. Other physicians had discovered that when wounds were washed with wine (which amounted to using alcohol) and kept clean and protected, they healed well. Vesalius adhered to the theory that wounds were poisoned and they must be cured with turpentine or by cauterizing with hot oil. The suppuration which followed was thought necessary. Strangely enough, the reactionary treatment was to prevail and cause the death of wounded men for hundreds of years.

Charles, Vesalius' employer, was often in poor health and suffered from gout. Although he knew that he was allergic to seafood, he was lacking in self-control and addicted to eel pies. Twenty-course heavily spiced dinners put him on the sick list. Rather than submit to the diet prescribed by his personal physician, he had recourse to quacks. When he was at death's door, he stopped gorging, listened to proper advice and recovered, only to start the same cycle over again. Vesalius fortunately was not his personal physician, for he was not the man to achieve much success in treating the royal glutton, who suffered from asthma to boot.

During the period of imperial service, Vesalius got out an improved edition of his great work with refutations of his opponents. He could be combative and arrogant when he knew he was right. The only authentic portrait we have of him shows a dark intense face with a wildly bushy head of black hair. He wrote bitterly: "They ought to be grateful to me as the first who has dared to attack man's false opinions, to lay bare the extraordinary frauds of the Greeks and to provide our contemporaries with an unusual opportunity for searching out the truth. Such, however, is not the case, and because of Galen's authority you will find many who, having glanced at my efforts only superficially and without investigation of the cadaver, still maintain what Galen wrote was wholly correct."

In 1556 the ailing emperor cut down his staff preparatory to abdicating in favor of his son, Philip II of Spain. Vesalius transferred his service to Philip but also got a pension from Charles. Philip, one of the most piously reactionary monarchs of his time, was a worse choice than Charles. Vesalius spent eight years in Spain and seems to have suffered from the nationalistic hostilities of the local doctors. In 1564 he visited Venice, perhaps with the thought of accepting his old professorship, just left vacant by the death of his pupil, Gabriel Fallopio, who is known for his work on the human reproductive system. In any case, Vesalius embarked on a pilgrimage to Jerusalem, was taken ill, and died on the way home in 1564.

Vesalius' great book created the method and technique of modern anatomy. Not only did he point out mistakes in detail such as the human jaw being one bone, not two, but his instruments, method and plan of work are still used today. Skeletons are mounted as he mounted them, anatomical illustrations scarcely improve on his, and his descriptive method is still ideally clear and simple.

The last chapter of the *Fabrica,* which deals with the vivisection of animals, almost approaches comparative anatomy. It is a good example of his method of work. He tells us that students must first study dead animals so that they become acquainted with the number and position of each part. "Vivisection sometimes plainly shows the function itself, and sometimes supplies helpful arguments leading to its discovery." By raising the skin from the forearm and paw of a dog and in the armpit, the nerves could be seen. Then when these nerves were ligated, the loss of motion in the muscles of the leg could be demonstrated. He pointed out that if the artery leading into a dog's groin was exposed and ligated, the part of the artery below the ligature no longer pulsated. He drew no conclusions from this, but it was this sort of approach to anatomy that was to lead to the great discovery of William Harvey. Significantly enough, Harvey was also to be exposed to the Padua medical tradition.

Harvey's father was a merchant and alderman at Folkstone, England. The boy, William, born in 1578, was given a grammar school education and sent to Cambridge, where he learned Latin and Greek and little natural science. After graduating, he went to Padua, which had certain ties with Cambridge's medical school. Padua not only had a considerable colony of English undergraduates but, with its traditional tolerance, did not discriminate against Protestants, to which sect Harvey belonged. Interestingly enough, there was now a rival college at Padua, run by the Society of Jesus, which was narrowly Catholic and hostile to Padua's liberalism. Student life was still far from comfortable. Lodgings customarily had linen over windows because glass was too expensive, and the young scholars complained that candles and oil for lamps were so expensive that their studies were hampered. In addition, there was a certain amount of brawling between students of different institutions and nationalities and between students and townspeople. Harvey, who seems on the whole to have been a very temperate character, nevertheless learned to wear a little dagger at his side, a habit he carried into later life, and he was accustomed to play with the pommel as he talked. The great Galileo was teaching at Padua while Harvey was a student, and the medical authority under whom he studied was Geronimo Fabrizio, Fallopio's successor (1537-1619), a very productive natural scientist. He worked in embryology, comparing that of birds, mammals and sharks. He compared the motion of animals and made some tentative attempts at animal psychology. He also described the valves of the venous system, but made no theoretical study of circulation. Instead he decided the venous valves were merely intended to prevent the force of gravity from accumulating too much blood in the extremities.

Although there is no record of Harvey and Galileo's meeting, Fabrizio was Galileo's personal physician and certainly the astronomer-physicist's approach to science must have affected the climate of the University of Padua. Galileo and Harvey were akin in their insistence on testing theory by repeated experiment.

Having earned his degree, Harvey returned to England, was licensed by Cambridge University to practice, and became a candidate for membership in the College of Physicians and Surgeons of which he was made a fellow in 1604. Three years later he married Elizabeth Browne, the daughter of a doctor. The union was happy, although childless.

His reputation as a doctor grew steadily; he was made a censor, an office similar to that held by Aldrovandi, which empowered him to expose quacks and check the purity of apothecaries' medicine. In 1615, he was appointed professor of anatomy, a post which was held for life, to give a series of lectures endowed by Lord Lumley. They covered the treatment of wounds, dissection and surgery. The problem of dissection continued to trouble the medical profession. The college supplied the bodies of only four executed felons a year. Clandestine dissection could be punished and postmortems were not allowed. The latter prohibition he found frustrating, for, as principal physician of St. Bartholomew's Hospital, he knew that the examination of patients who died would be of great value in confirming diagnoses and extending in general his knowledge of specific diseases. It is probable that he made use of the authority of his position to perform autopsies quietly. All during this time, Harvey had continued to work in the direction of comparative anatomy, a study in which he had been well trained by his teacher. His statement of principle is important:

> Those persons do wrong who, while wishing, as all anatomists commonly do, to describe, demonstrate and study the parts of animals, content themselves with a look inside one animal only, namely man —and that one dead. In this way they merely attempt a universal syllogism on the basis of a particular proposition like those who think they can construct a science of politics after exploration of a single form of government or have a knowledge of agriculture through investigation of the character of a single field.

Particularly interested in the heart, Harvey examined such vertebrates as lizards, frogs and fish but also extended his investigations to snails, shrimps and insects. His notes for his Lumley lectures, 1615-16, show that he had already arrived at the idea of the general circulation of the blood. Interestingly enough, there were philosophical reasons why he was prepared to accept a circular course of the life-giving fluid which tie up with the intellectual climate of the period. A staunch admirer of Aristotle, he was impressed by the Greek scholar's dictum that the course of the planets was circular and that circular motion was the noblest and ideal form. A precise parallel was the fact that a wave of Neoplatonism in the Renaissance, largely literary and poetic, had an effect upon Galileo. Plato glorified mathematics, feeling that geometrical forms in a mystical way reflected the ideal unchanging forms which were for him the substance of the universe. Galileo transformed this concept by working to show that the nature of the world was mathematical and thus brought mathematics into physics. Progress in science develops in devious ways, and thus certain elements in the thought of the past were still a stimulus while at the same time experiment was rejecting large areas of obsolete ancient philosophy.

The mystery of the motion of the blood was waiting for a mind that could make use of known facts and organize them into a new pattern. Galen had proclaimed that venous blood was produced in the liver from food and thence flowed out to all parts of the body; very little of it was supposed to enter the heart. Some was supposed to pass from the right to the left ventricle to be converted into arterial blood. In diastole (expansion) blood was conveyed from the veins to the heart. Also in diastole the left ventricle was supposed to take in air from the lungs, and from the right ventricle sooty waste products were expelled as a result of combustion in the heart. There were supposed to be porous passages through the wall between the right and left ventricles. All these ideas had created problems for later anatomists. In the first place, many of them (Vesalius was one) could not find any passages between the two ventricles. Another anatomist had noticed that a ligatured vein swelled up on the side away from the heart. He drew no valuable conclusions from this. We have seen that Fabrizio noticed valves in the arteries but failed to understand that they prevented backflow toward the heart. A thirteenth-century Syrian physician Ibn al-Nafis al-Quarasha had maintained there was no passage between the ventricles and that venous blood had to pass from the right ventricle through the pulmonary artery, through the lungs where it mixed with air, and that it then went back to the left ventricle of the heart through the pulmonary vein. His work seems to have been unknown in the West, but as early as 1553, Miguel Servet y Reves, a Spanish doctor and theologian, rather casually, in the midst of heretical theological opinions, had indicated a similar theory of lung circulation. Servet, who baited Calvin, was burned with enthusiasm in the flesh by that Protestant dictator and in effigy by the Catholics that same year. Realdo Colombo, a pupil of Vesalius, picked up the idea of pulmonary circulation and also pointed out that systole coincided with arterial *expansion.* He also opened the pulmonary vein and found no sooty wastes or fumes but only blood.

Thus all the pieces of the puzzle were lying around; they only needed to be put together. Harvey's book, published in 1628, *De motu cordis* (Of the Motion of Heart), records the precise, logical proofs by which he was able to show what really happened within the living beast. It is a short book, fewer than a hundred pages, but clearly, economically written, using graphic homely images. He first demolished the notion that respiration and pulse did the same thing—that is, draw air into the blood vessels. Experiment showed there was only blood in blood vessels and only blood in the left ventricle which was supposed to be full of air. He also pointed out that each ventricle possessed three valves which prevented the backflow of the blood. Prevailing theories were erroneous. What really went on?

Harvey contemplated the hearts in toads, snakes, frogs, snails, transparent little prawns, small fish. In all of them the heart moved and was still, moved and was still. On opening dying dogs and pigs, he could see the slow deceleration of the beat until the organ had "fallen in upon itself." Patiently studying the pulsations, he established that when the heart contracted, it became long and narrow, the circular muscles contracting, the walls thickening. When felt by the hand, it was hard and tense. On contraction it whitened, on expansion it reddened, and whenever a ventricle was injured, blood came out in spurts. Even Galen had admitted that from a wounded main artery all the blood would drain out of an animal's body.

Then Vesalius, with his theory of longitudinal contraction in which the heart shortened and became a cup, drawing in blood and "vital spirits" as a bellows draws in wind, was wrong. The image was wrong. "The arteries increase in volume because they are filled up like leather bottles or a bladder and do not fill up because they increase in volume like a bellows." The expulsion of blood from the left ventricle into the arteries resembled blowing into a glove and increasing the volume of the fingers. By the use of his senses Harvey showed that the reverse of the traditional view was true: The heart does not suck in blood but pumps it out into the body. This in itself was a great step forward.

As he wrote of the heartbeat, Harvey became lyrical. He particularly delighted in discussing its existence in the small invertebrates he did not scorn to study. He found it in slugs, crabs and even in wasps, hornets and flies "with the aid of a lens to distinguish the very small objects, I have at the top of the portion called the 'tail' seen the heart pulsating, and have pointed it out for others to see."

Harvey continued his comparative investigation. He watched the development of the heart in a chicken embryo from a tiny red dot to the fully developed organ. He compared the structure of hearts of fish, which had no lungs and a single ventricle and auricle, with that of vertebrates. He noted that in many cases, after removal from the animal bodies, the organ continued to beat. He even performed an experiment which foreshadows heart surgery of today:

> In an experiment carried out upon a dove, after the heart had completely stopped moving and thereafter even the auricles had followed suit, I spent some time with my finger, moistened with saliva and warm, applied to the heart. When it had, by means of this fomentation, recovered—so to speak—its power to live, I saw the heart and its auricles move, and contract and relax, and—so to speak—be recalled from death to life.

Eventually, he drew a complete picture of the heart action in man. First, the ventricles filled with blood from the auricles, then the organ tensed and gave a beat, blood was forced from the right ventricle to the lungs through the pulmonary artery and from the left ventricle through the aorta through the arteries into the whole of the body.

The two movements of first the auricles and then the ventricles took place successively and so harmonically that it was "comparable to what happened in machines, in which, with one wheel moving another, all seems to be moving at once."

Harvey communicated his admiration and delight at this wonderful mechanism, which he further compared to the firing of a gun in which trigger, flint and steel all operate in succession and so smoothly that all happened in the wink of an eye.

Parts of the pieces of the puzzle were falling into place. The mechanism was clear, but what happened to the blood after it was driven from the heart? Harvey already had certain clues; in animals without lungs and in embryos, before the lungs developed, blood simply passed from one side of the heart to another through both veins and arteries. Now the flow of the blood from the heart was continuous, and by simple arithmetic he showed that by the organ's capacity and rate of beat, in one hour it pumped more than the whole weight of the body. What then became of the blood? It could not be simply discharged into the tissues as Galen and Aristotle had thought. By various experiments in which he ligatured the vessels, he was able to show that the blood returned to the heart through the veins, and in support of this he could now demonstrate the real function of the valves. Those of the arteries prevented backflow into the heart; those of the veins prevented backflow away from the heart. The conclusion was unavoidable: The movement of the blood was circular. To Harvey this was philosophically admirable. He quoted Aristotle who said that the air and rain emulated the circular motion of the heavenly bodies. Thus as a result of countless rigorous experiments, faultless logic and unflagging enthusiasm, his picture was complete. It was a beautiful example of method paralleling Galileo's breakthrough in physics and astronomy. It is true he was a little vague as to how the blood got from the arteries to the veins. Having only a low-powered microscope, he could not see the capillaries, and he also still subscribed to the notions of "cooling" and "heating" as functions of the vital fluid. Chemistry would eventually reveal the secret of respiration, however, and his admirable pioneer work once and for all put an end to vague theorizing without rigorous proof as to how the body functioned.

Harvey's book was published in 1628 at the Frankfurt book fair, where it received wide distribution. As was to be expected, the establishment attacked it. Harvey was quoted as saying that at first his practice fell off and that the vulgar thought him crack-brained. Younger scholars, however, whose minds were attuned to the new world of science, were convinced. In about thirty years his work was universally accepted in the academic field.

As the leading medical man of England, Harvey was physican in ordinary to Charles I. He also attended Francis Bacon for whom he had no great respect as a scientist. In his own house he kept a small menagerie of experimental animals—newts, reptiles, fish, birds, rabbits, and a privileged parrot which dined with the Harveys and liked to nestle in Mrs. Harvey's breast.

His portrait shows him with fine-drawn, rather delicate features, wearing the mustache and Vandyke beard of the period, and gives the impression of a sensitive intellectual. Like Vesalius, he could become impatient with his detractors, and once, when demonstrating his theory to one Dr. Caspar Hofman at Nuremberg, he became so disgusted with the German's refusal to acknowledge what was being shown him that Harvey threw down his scalpel and walked out of the dissecting theater.

One interesting anecdote from Harvey's career deals with witchcraft. In 1634 seven unfortunate women were rounded up and accused of practicing the black art. Apparently they were so cruelly treated that three of them died. The other four were examined for bodily abnormalities, called witch's teats, by midwives under Harvey's direction. All four were cleared and pardoned, which speaks well for the rationality of his approach.

The conflict between Parliament and the king was growing more intense and outbreaks of violence became common. Simply because Harvey was the king's physician, his house was broken into and pillaged by Parliamentary troopers. All his notes on animal anatomy were destroyed. When the civil war broke out, he followed the king, withdrawing from St. Bartholomew's Hospital, which was taken over by Parliament. The spectacle of Harvey's being persecuted as a reactionary by the Roundheads is one of the ironies of history. In his own field one of the most progressive minds in Europe, his royalism was of a simple unquestioning sort. He was, of course, personally acquainted with the king. His rather charming dedication of the *Moto* to Charles displays his innocently apolitical character:

> Most Serene King!
>
> The animal's heart is the basis of its life, its chief member, the sun of its microcosm; on the heart all of its activity depends, from the heart all its liveliness and strength arise. Equally is the king the basis of his kingdoms, the sun of its microcosm, the heart of the state; from him all power arises and all grace stems.

Harvey often used the reverse metaphor of the heart as king of the body; his politics were, in short, Aristotelian.

The autumn of Harvey's life was saddened by the national conflict. He attended the wounded and later resided at Oxford where he became warden, resigning with the fall, imprisonment and dethronement of the king. He practiced for a while in London and retired in 1649, spending the rest of his days with one or another of his five well-to-do brothers on their country estates. In later life he suffered from gout, soaking his bare legs in ice-cold water to relieve the symptoms. He died of a stroke in 1657.

Harvey's later years were passed in preparing his treatise on reproduction, *De generatione animalium* (Concerning the Generation of Animals), published in 1651. Although he offered much well-observed detail concerning the evolution of various types of embryos, he did not advance beyond Aristotle theoretically. Sperm contributed a mystical vital force in reproduction. All animals, including man, are evolved out of the egg, he maintained—a curious intuition, since he was not equipped to observe the eggs of mammals and mistook the pupae of insects for their eggs. Nevertheless, his formula is important —*omne vivum ex ovo.*

7

The Animal as Machine

☙§ THERE is no doubt that the most characteristic philosopher of the new science was René Descartes. Where Bacon had prophetic insights and was acutely critical of the past, Descartes was the creator of a system. Bacon made no important contributions to specific areas in scientific research; Descartes was a mathematician and a good one. His interests roved over the whole field of investigation, and his ambition was to synthesize the thought of the "moderns." To an extent he succeeded, and it is generally agreed that he developed certain lasting concepts.

His father was a councillor of the parliament of Brittany; his mother died when he was born, in 1596. The family was wealthy, and although René was considered destined for an early grave because of pulmonary weakness, he was nevertheless given an education in a well-known school, the Collège de La Flèche, and afterward spent two years at the University of Poitiers studying law, in which he obtained his licentiate. He was critical of the academic philosophy of his time as taught in the celebrated Jesuit Collège Royale de La Flèche. He said it was "mainly a means of speaking plausibly about all things, and of making ourselves admired by others less learned." Mathematics pleased him most because it had formed clarity and its principles were not simply matters of disputation. La Flèche was apparently rather middle of the road, for there is record of a sonnet being read at a College Commemoration Feast in 1611, celebrating new astronomical discoveries made by Galileo. At La Flèche he mastered Latin (Greek was not taught) and acquired some taste for poetry. At twenty-two, he spent some time in the army of Prince Maurice of Nassau and in 1618, while in Holland, met Isaac Beeckman, rector of the college at Dortmund, who was active in natural science. From this scholar he gained a new insight into physics and began to envisage a connection between this study and mathematics. He continued to travel, visiting Poland, Hungary, Austria and Bohemia. One cold day, which he spent alone by his stove in a small German town, he had a kind of combined emotional and intellectual experience, a mathematical insight which he referred to as a "wonderful discovery." It seems to have been a concept that all sciences would be unified, the first glimpses of his method. When he returned to France in 1622, the family expected him to marry and settle down. Instead he sold his estates in order to provide himself with enough income to live where he chose. Marriage, he maintained, would interfere with the cultivation of reason and the practice of scholarship. This was the official reasoning whatever the private motives, for elsewhere he went on record with misogynist remarks.

His next journey took him to Switzerland where he studied glaciers, was impressed by the men of learning, but always prudent and even timid, dared not settle in any of the towns because he felt them inadequately policed. If we remember Harvey's little dagger and keep the violent and brawling nature of Renaissance life in mind, however, he may be credited with good reason to be nervous. He returned to Paris and spent the years up to 1628 in developing his method, although publication of some of the work of this period did not take place until fifty years after his death. In 1628 he settled in Holland. He may have chosen Holland to be near his friend, Beeckman, or he may have appreciated the atmosphere of intellectual tolerance. It had been pointed out, however, that Dutch tolerance was not so much from conviction as from the fact that the country was so full of small independent sects that it was politically expedient to let them all alone.

At any rate, in the peaceful tree-lined city of Amsterdam he could propound intellectually revolutionary ideas, stroll in isolation among the crowds, and pursue his meditations. Eventually he settled in a small town, near Leiden, where he resided in a small château with a comfortable staff of servants, a garden and an orchard. All the major cities of Holland were no more than a day's travel by canal. There he worked on his scientific treatises and took considerable interest in anatomy, for he visited the slaughterhouses of Amsterdam to study the anatomy of animals. Meanwhile, he was arriving at certain ideas dependent on the Copernican view of astronomy. He was about to publish his *Traité du monde* (Treatise on the World) when, in 1616, he learned of the censorship of Galileo's work by the Inquisition. Since he already knew that Galileo had anticipated certain of his ideas, he was profoundly disturbed. Like too many thinkers of the Renaissance, he naïvely supposed that natural philosophy could be divorced from theology as long as one proclaimed oneself an orthodox Catholic. But the establishment had suffered from divisive heresies and was now weakened by the great schism of Protestantism. Like all self-perpetuating bureaucracies it reacted with force and violence. Galileo was fined, imprisoned, and then confined to house arrest for life. Others, besides Giordano Bruno and Servet, had been burned. Descartes was not of the temperament of martyrs. He withdrew the manuscript from publication, officially "in obedience to the Church," actually out of prudence, privately hoping that in time it would be safe to publish it. From statements made to his friends, we know that he did not read the Bible literally, did not believe in the literal inspirations of the Scriptures, and did not accept the Mosaic version of the Creation of the world. It has also been pointed out that many of his friends were heretics, atheists and Rosicrucians.

Descartes prudently turned to work on optics and in 1637 published his famous *Discours de la méthode* (Discourse on Method). His contributions to geometry in this period are of permanent significance. The fact that Descartes wrote in French is also significant. He wished to reach the intelligent man in the street and even to be read by women. "Those who avail themselves of their natural reason alone may be better judges of my opinions than those who give heed only to the writings of the ancients." He thus aligned himself with the "moderns" and threw down the gauntlet to the medieval, scholastic type of scholarship with its deadweight of traditional authorities.

Nevertheless, even in tolerant Holland, the conservatives attacked him. His philosophy was officially rejected by the University of Utrecht thanks to the pressure of its president. He was summoned to answer charges brought against him, fled to The Hague, and even appealed to the Prince of Orange for protection and got an order from the States General which saved him from being arrested. A few years later, Queen Christina of Sweden, who had a habit of summoning intellectuals, invited him to Stockholm to instruct her in philosophy. Disgusted by his Dutch experience, this seemed to him a welcome alternative to controversy, although he was dubious about the climate. Unfortunately, these doubts were well founded. The queen's mind was freshest at five o'clock in the morning, at which time the poor French philosopher had to appear to give instruction. One freezing

February morning he contracted pneumonia and died, in 1650, at fifty-five.

Descartes started with universal doubt and decided that he could begin by proving the existence of the thinking ego. From there he had considerable difficulty in escaping from solipsism, but like most philosophers, with the aid of verbal ingenuity he proved, to his own satisfaction, the existence of God and the outside world. This was pretty much the opposite method from medieval thinking which started with the world as given and was not very much concerned with the ego.

Descartes' cosmology endeavored to be geometrical, for mathematics, he believed, was the fundamental aspect of nature. Matter was simply extension. In creating a universe of infinite extension, God also gave it motion. Motion could neither increase nor decrease in total amount, but could only be transferred from one body to another. The universe continued to run as a machine, and each body persisted in a state of motion in a straight line, the geometrically simplest form of motion in which God set it going, unless acted upon.

Although Descartes felt he had proved the existence of the ego and of material bodies, including one's own, what was the connection between them? This was really the reef upon which the unity of Descartes' philosophy foundered. The common-sense intuitive belief that they are connected was not satisfactory: there must be a causal connection. Unfortunately, having divorced the material from the nonmaterial, he did not succeed in putting them together again.

He assumed that all bodies, inorganic and organic, were subject to his mathematical theory. Hence men and animals were basically machines; he even made use of the metaphor of a watch. This mechanistic philosophy was of course consonant with and inspired by the great seventeenth-century advances in anatomy. In particular, Harvey's description of the circulation of the blood, he felt, substantiated his point of view. Unfortunately, he still clung to the notion of animal spirits flowing in the blood which became an important link in his physiology. He considered them "the most animated and subtle portion of the blood." Indeed, as he described them, they became semi-mental, semi-material. The soul, a little man operating in the heart or head according to classical and Christian philosophy, was now dissolved into a fluxion in the blood.

Since animals had no souls, their physiology was no problem. Descartes even held that they had no feelings. In the case of man, he had already proved the existence of an ego that involved rational activities and the possibility of free will. Intuition insisted that there was a connection between these activities and the mechanical functioning of the body. Descartes, on the basis of his establishment of the ego, maintained that man had a soul but its only connection with the body was through the pineal gland, embedded in the brain. There it received certain impressions from the animal spirits flowing through the nerve tubes. How there could be a connection between this gland, which was material, and the nonmaterial soul, he never successfully explained.

On the other hand, his mechanical approach to the body provided a rough sketch of reflex action. Animal spirits were conveyed from the brain to the nerves and from them to the muscles, which were then set in motion. This could take place in the absence of thought. In the case of sense perception, when an object approached the eye, for instance, it agitated the cord in the optic nerve, which in turn through animal spirits affected the brain, which then forced the animal spirits back through the nerve tube to move the eye muscles. The Cartesian theory of emotion was, however, inadequate. Emotions constituted changes in the body which affected its action, but they were also connected with awareness, with the nonmaterial functioning of the ego, and this could not be explained by his system.

What was most heretical about the Cartesian philosophy was the absence of moral qualities in its God, who was reduced to a first cause which set the world running and thereafter only served to aid Descartes in solving his verbal problems. Despite this aspect, his thought had a strong appeal for his younger contemporaries. In fact, the rector of the University of Leiden was driven to attack him precisely because of its popularity among students.

The importance of Cartesian thinking for physiology and science in general was that it provided a convenient working approach which has continued to be useful. In life sciences the mechanistic view covers certain vital processes, particularly reflexes. To quote his biographer, S. V. Keeling:

> Though it would be too much to credit Descartes with having elaborated the idea of the reflex as understood by the contemporary physiologist . . . it may be allowed that his views of reaction and glandular activity were its forerunners. Indeed much in present-day behaviorist psychology, in method and result, is Cartesian.

Moreover, Descartes can be credited with the use of the model that will help explain effects. The model functions in an analogy that suggests an interpretation of phenomena. In a sense it is a metaphor, like that of the watch in relation to the animal body.

On the debit side, the mechanistic view of existence carried to extremes has frightened various modern writers of the brave-new-world type with nightmares of a manipulated, regimented society. All of which is testimony to the vitality of the great Renaissance thinker's contribution to Western culture.

8

Who Sees What No One Saw . . .

⸎ THE four investigators grouped together here all dealt with a miniature world. Insects, even smaller invertebrates, the one-celled animal, subjects not yet or at best only vaguely described by zoology were now engaging men's attention. In addition, three of them—Leeuwenhoek, Swammerdam and Redi—were interested in the problem of the origin of life while all contributed to our knowledge of human anatomy.

No one is more typical of the rising bourgeoisie or more clearly an example of the solidly materialist mentality of that class than the father of microzoology, Antony van Leeuwenhoek. A native of the charming old town of Delft, famous for its beer, china and Vermeer, Leeuwenhoek was born the same year as the celebrated painter (1632) and in all probability knew him, for when the artist died, he was appointed executor of the estate. Leeuwenhoek's father was a basket maker, his mother the daughter of a brewer; they were all solid tradespeople. Antony's father died while the boy was still young and his mother remarried. Antony was given a grammar school education, which would have included some arithmetic and a touch of natural science, and then sent to Amsterdam to learn the draper's trade or, as we should say, the dry-goods business. He became cashier in a shop in Amsterdam but by 1654 was back in Delft, where he remained all his life. He set up his own drapery shop near the Oude Kerk and sold silks, cottons, ribbons, buttons and braid successfully, for he seems to have made an adequate living. He married at twenty-two and had three children of which only one, Maria, grew up. She never married and outlived him. His wife died in 1666; he married again five years later.

As he prospered in business he accumulated small municipal offices. He was chamberlain to the sheriffs, a job which seems to have been a sort of sinecure. He was supposed to light the fire in their office and keep the room clean. He also became wine gauger, which required him to check on the quality of the wine and the honesty of the measures used in selling it in Delft. At some time fairly early in life, he became interested in the microscope.

It must be remembered that the development of instruments for measurement and observation was accelerating in the seventeenth century. The medieval study of optics and the invention of spectacles had prepared the way for the telescope and microscope. The earliest two-lens microscope was developed in 1590. The thermometer was invented around 1603. By 1614 Galileo had both his own telescope and microscope. A more accurate time measurer, the pendulum clock, was devised in 1625. Harvey and the Italian investigator Francesco Stelluti, who published on the anatomy of the bee in 1625, had used microscopes, as did Marcello Malpighi, who made many contributions to anatomy in the second half of the seventeenth century. Robert Hooke, a British inventor and scientist, had published *Micrographia* in 1665, in which he described what he christened "cells" in a sliver of cork. It was Leeuwenhoek who made a hobby of grinding lenses, which he brought to such a high art that he was able to magnify two hundred and seventy times.

By the snobbish standards of his time, Leeuwenhoek was uneducated. He knew no philosophy, no language other than Dutch, could not communicate with the learned world in Latin, and always worked on his own in his own way. He was possessed of great patience, an insatiable curiosity and an urge to experiment. Here we can clearly see how the material interests of the solid tradesman, who had to measure goods accurately and keep accounts, who had taken an examination which licensed him as a surveyor, and who also came from a craftsman's environment, relate to modern science. Leeuwenhoek approached everything directly, with great attention to detail and mostly without preconceived theory. In his case, absence of philosophy was an asset, and his craftsman's handiness helped him devise practical do-it-yourself techniques.

Until 1673 nothing was heard of him outside Delft. In that year, the physician Regnier de Graaf, who had made detailed studies of anatomy and isolated the follicle containing the mammalian egg, wrote a letter to the British Royal Society. Ironically enough, that same year Graaf fell ill and his scientific career was cut short, but by means of that generous letter, in which he told the Royal Society that Leeuwenhoek had made better microscopes than anyone else, he started his fellow townsman on his career. Graaf included a letter by Leeuwenhoek describing his observations of mold and of the mouth parts of the bee and the louse. The president of the Royal Society was patronizing, but some interest was stirred up and he answered the letter asking for more details. Leeuwenhoek wrote back that he could not draw, but he would find someone to do some drawings. He also seems to have made friends with the noted writer and diplomat Constantijn Huygens. Huygens was an aristocrat and a man of the world. He had evidently come to look through the draper's microscope, for he wrote to the Englishman Robert Hooke that Leeuwenhoek was "unlearned both in science and language but of his own nature exceedingly curious and industrious." Huygens' son, Christian, was first skeptical, thinking that Leeuwenhoek imagined what he saw, but later he was convinced and even undertook to repeat some of the microscope maker's experiments.

At any rate, Leeuwenhoek embarked on a series of letters to the Royal Society, which continued all his long life. Since he could not use Latin, he never wrote a scientific paper or a book, but letters flowed from him in profusion. Those to the Royal Society in his homely colloquial style might discuss his own health, point out that he drank coffee for breakfast and tea in the afternoon, and shaved twice a week, and solemnly recommend many cups of tea the morning after drinking too much wine. His observations, however, were written out in great detail but without organization; several different subjects might be included in one letter. The secretary of the Royal Society translated or made abridgments of them and they were published in the Society's journal for seven years. At the end of that period, someone realized that all those years Leeuwenhoek had never been invited to join! He was rapidly elected a fellow of the organization in 1680. When he received the news, he was overjoyed. Everyone who was anybody in Delft came to visit him and look through his wonderful lenses. When one of those group portraits, which Dutch organizations (in this case, the town doctors) went in for at the time was painted, they put Leeuwenhoek in the middle! The uneducated draper had arrived. Strangely enough, although his letters were now sometimes translated into Latin and published, Leeuwenhoek seems to have been almost alone in his diligent pursuit of microbiology. In 1692, Hooke said that the microscope had "become almost out of use and repute. So that Mr. Leeuwenhoek seems to be the principal person left that cultivates those inquiries."

Although he was glad to show visitors his bag of tricks (at least until he became old and tired), Leeuwenhoek never showed them his best and most high-powered instruments, which he kept jealously hidden. His instrument consisted of two thin oblong plates of brass or silver with a hole through the center. These enclosed the lens, and the whole assembly was fastened upright to a rod with a collar, which could be fastened to another upright. Upon the latter the subject was secured in one way or another; a test tube, for instance, was attached by a coil of wire.

His interests ranged from noting the blood vessels in the tail of an eel to examining the faceted eye of a fly. But an epoch-making date in his career was the year 1674 when he took a specimen of water from a somewhat marshy lake, which in summer had green clouds floating in it which countryfolk said were caused by dew and consequently named by them honeydew. The pond was inhabited by fish that were excellent eating, Leeuwenhoek pointed out. After filling a test tube with water, he was able to observe the long trailing streaks of spirogyra (noted for the first time), the green algae which constituted the pond scum. But on looking closer, he saw tiny animals that moved, some round, some oval. "On these last I saw two little legs near the head and two little fins at the hindmost end of the body." From this description, these were probably rotifers, tiny invertebrates of the seventh phylum in modern classification, less than one-sixteenth of an inch long, which possess two ciliated organs near the mouth, two finlike appendages at the tail. They abound in all fresh and stagnant water. No one had ever seen them before. The letter went on: "Some were white, some transparent, others green and glittering with little scales, others again green in the middle and before and behind white, others were ashen gray. And the motion of those animals in the water was so swift and so varied, upward, downward, and round about that it was wonderful to see and I judge that some of these little creatures were above a thousand times smaller than the smallest one I have ever seen upon a rind of cheese, in wheat flour, mold and the like." The green animal was probably *Euglena viridis* whose body contains organs which manufacture chlorophyl. It is called a flagellate because it draws itself through the water by means of the lashing of a long cilium. The little animal belongs to the phylum *Protozoa,* one-celled animals possessing the simplest structure of any organism that deserves the name of animal. Leeuwenhoek was also to describe another creature "which stuck out two little horns which continually moved after the fashion of horse's ears" and had a tail four times the length of its body. This tail got entangled with inert particles in the water and wound itself around them. This was clearly a vorticella, another protozoan. Leeuwenhoek had discovered an unsuspected microscopic world, and we now know that these tiny creatures, which cannot be seen with the unaided eye, are the most numerous organisms on the planet.

Leeuwenhoek had no vocabulary and no accepted standards of measurement for describing his discoveries. He was obliged to compare his *diertjes* to a grain of sand or the tiny crustacean *Daphnia,* which his contemporary, Jan Swammerdam, had christened the water flea. Leeuwenhoek collected fresh rainwater but found no animals in it. He poured pepper in the water which contained protozoans and discovered that they died.

The notion that there was a host of tiny invisible living beings in almost any type of water the scientific world found difficult to believe. Leeuwenhoek stuck to his guns. He invited clergymen to view his little animals through the microscope and forwarded their written testimonials to the Royal Society. The scientists were convinced, but many persons still made fun of the microbiologist. Robert Hooke, the English scientist, however, repeated some of his observations.

The procession of tourists who came to look through the microscope became something of a burden. In four days, twenty-six people arrived with introductions which he could not ignore. One royal visitor was Peter the Great of Russia, who came up the canal to Delft in a "canal yacht." It so happened that the monarch had learned good colloquial Dutch. Leeuwenhoek was summoned to bring his exhibits to the boat, for Peter wished to avoid crowds which might have gathered in the city. Since there were no language barriers between the emperor and the draper, they spent a congenial two hours together.

In 1677 Leeuwenhoek arrived at a new epoch-making discovery. A man named Johan Ham had seen what looked like tiny animals in human sperm. His specimen came from a man infected with gonorrhea and hence he concluded they were abnormal, symptoms of disease. Leeuwenhoek persisted in more observations and discovered spermatozoa in the semen of dogs and rabbits, as well as man, and once and for all established them as the male principle in sexual reproduction. This had an important bearing upon the whole problem of sexual theory. Aristotle, it will be remembered, was on the side of the male and considered the male principle to be truly creative, the female passive. Harvey, with a mixture of insight and error, announced that all life came from the egg; the male principle was almost abstract. Now Leeuwenhoek, impressed by what he had seen, announced that the male principle was basic, the female simply contributed a nutritive function. The argument was to go on for centuries.

A related problem, that of spontaneous generation, a process accepted by Aristotle, was also discussed by Leeuwenhoek. He took some mud from his gutter and kept it five months in his office (he called it his *comptoir* and it seems to have been his laboratory). He then put it in some boiled rainwater, and in three hours he saw many protozoans. From this experiment he concluded that these animals had the power of sustaining life in the bottom of dry ponds in summer. He was convinced, and rightly, that the animals in some form were present in the refuse from the gutter all along. Weevils were popularly supposed to arise by spontaneous generation from wheat because they appeared in new granaries in which wheat had not before been stored. Leeuwenhoek said they could have been carried in on men's clothing or shoes. He put weevils and wheat in a test tube, saw the weevils peacefully eat the wheat, couple and produce eggs. He concluded by writing that he did not see how people could cling to "the ancient belief that living creatures are created out of corruption."

In 1716 the University of Louvain sent him a medal and a eulogistic poem in Latin; it was the equivalent of an honorary degree. Leeuwenhoek replied that "it brought tears to my eyes." The portraits show us Leeuwenhoek's broad, blunt features, the eyes looking at us steadily and seriously. He sported a hairline mustache and wore the long ringlets of the period. Toward the end of his life, his legs troubled him and he had attacks of dysentery. Always the scientist, he examined his own feces and found what were probably bacteria.

The first mention of his discovery of bacteria is in 1675. He described small creatures shaped like eels which swam backward as well as forward, a characteristic of thread bacteria.

This indefatigable man was active up to the day of his death, at the age of ninety-one, from pneumonia. At the very last he gave his daughter instructions to send his final letter to the Royal Society and bequeathed to them a cabinet containing twenty-six silver microscopes. The Society cherished them for a hundred years and then managed to lose them.

Another area in which Leeuwenhoek made pioneer discoveries was that of parasitology. He found a flagellate, *Trichomonas,* in a frog's blood. While the Italian Malpighi and Swammerdam had discovered capillaries, it remained for Leeuwenhoek to describe and draw them in a tadpole, showing conclusively how the veins and arteries were united; thus, in 1668, he added one more proof to Harvey's demonstration of the circulation of the blood. Finally, he made an observation which, although he misinterpreted it, was actually the first mention of a dominant characteristic in the field of genetics. He had noticed that rabbit breeders often mated wild gray bucks with white, spotted, or black tame females. The immediate offspring all were gray. This he took to be a proof of his theory that only the male element counted in reproduction.

On the whole, Leeuwenhoek did not often venture into theory, but when he did, he was always careful to preface his ideas with the words "I imagine . . . ," strictly separating speculation from observation. He also maintained a truly scientific readiness to respond to new facts. ". . . 'Tis my habit to hold fast to my notions only until I am better informed, or till my observations make me go over to others: and I'll never be ashamed thus to chop and change." Yet like other pioneers, when he was convinced, he could be stubborn. He wrote: "I well know that there are whole universities that don't believe there are living creatures in the male seed: but such things don't worry me, I know I'm in the right."

Before leaving Leeuwenhoek, we must record one more observation in the field of physiology that pointed the way toward an understanding of digestion. He introduced gastric liquor from a calf into milk and discovered that it curdled the milk. By questioning butchers, he also discovered that calves, in whose stomachs the milk was not well curdled, were not properly nourished. Although chemistry was still in its infancy, and Leeuwenhoek was no chemist, he had really indicated the chemical nature of digestion.

It was just this type of diligent experiment in which Leeuwenhoek's Italian contemporary Francesco Redi excelled. He was born in 1626 in Arezzo; his father was a well-known doctor and his mother was related to a noble family. His early education took place in a Jesuit school in Florence; his medical degree was obtained in Pisa in 1646. He visited Rome and Naples and came back to Florence to practice, where he distinguished himself as a poet, as well as a doctor, being elected to the most important literary academies in the city. By 1660 he became physician to the Grand Duke Ferdinand of Tuscany, an office he held for the rest of his life. He also held a minor diplomatic post at court and was obliged to follow the court around when it traveled, an obligation which cut into his scientific work. Redi reminds us of the zoologists of the past, being a practicing doctor and, like Vesalius and Harvey, enjoying royal patronage. Like the earlier scholars, he was well versed in Hebrew, Arabic, Latin, French and Spanish and wrote in Latin. Intellectually, however, he belonged to the progressives, for he wrote: "I have taken the greatest care to convince myself of facts with my own eyes by means of accurate and continued experiments before submitting them to my mind as a matter for reflection." His first publication was on vipers and their venom, an investigation carried out on the order of the grand duke. Viper poison was popularly supposed to consist of the animal's bile. Redi discovered that a snake's bile was not harmful when drunk or injected. He then isolated the venom in the fangs and proved that this was not lethal to animals or man when taken by mouth. This study took him into viper anatomy. He wrote *Osservazioni intorno alle vipere* in 1664. Curiously enough, opponents arose who derided his work and attacked him with a mixture of history and myth. Redi repeated his experiments and added some in which he showed the efficiency of the poison related to the size of the animal bitten and the proximity of the bite to a vein or artery. Like Leeuwenhoek, Redi communicated with the British Royal Society and to this institution Thomas Platt, an English doctor, wrote confirming Redi's experiments.

In Redi's next book, *De generatione insectorum* (Concerning the Generation of Insects), he set out to attack the theory of spontaneous generation. Leeuwenhoek, not being a scholar, was not affected by the opinions of Aristotle whom he had not read. Redi, conversant with his contemporaries, as well as the classic writers, showed great independence of mind and did not hesitate to disprove their statements and branded them "fables." Since flies were supposed by everyone, from Aristotle on, to be bred spontaneously from decayed meat, he put meat and fish "in a large vessel closed with fine Naples veil which allowed the air to enter. For further protection against flies, I placed this vessel in a frame covered with the same net. I never saw any worms on the meat, though many were to be seen moving about on the net-covered frame." He observed that flies even dropped their maggots from the air above the net. He also proved by experiment that some flies laid eggs and some were viviparous. He exposed meat, saw the maggots deposited, watched them turn into pupae and eventually into adult flies. He then repeated his experiments by putting a snake, some fish, some Arno eels and a slice of milk-fed veal into four wide-mouthed flasks which he then closed and sealed. He filled other flasks with the same materials and left them open. The open containers bred maggots, the closed did not. He watched carefully how the flesh in the closed vessels decayed, the fish became liquescent, the eels turned to glue, the veal dried up. He was now thoroughly convinced that putrescent matter did not breed flies and had "no other office than of serving as a suitable nest . . . in which they also find nourishment. . . ." He found considerable satisfaction in refuting Father Athanasius Kircher, author of *Mundus subterraneus* (The Subterranean World) who said that dead flies, soaked in honey water and then heated, would breed microscopic worms which then grew wings. The same authority had announced that caterpillars spontaneously generated from ox dung. Having disposed of Kircher, Redi called Paracelsus, one of the founders of chemistry, a charlatan and went on to tackle other authorities.

He patiently observed the viviparous birth of scorpion young. "I perceived there is no truth in the reports of Aristotle and A. Caristio that the mothers are killed by the newborn young, nor, as Pliny states, that the young are killed by the mother with the exception of one more clever than the rest, who runs up on his mother's back out of reach of the sting and afterward avenges his brothers' death by killing his parent."

He next experimented with the scorpion bite. Pigeons, he found, generally died of it, but their flesh could be eaten with impunity. When a deer was bitten, it experienced no bad effects. He disproved Avicenna and Aldrovandi's notion that a dead scorpion would come to life if sprinkled with hellebore juice. It was, he remarked, "an old wife's tale." He even held flies underwater to disprove the notion that they came to life after drowning. Among his other observations was the fact that spiders spin from the ends of their abdomens, not their mouths.

Actually his work on the breeding habits of flies was important in laying the basis for the preservation of foods in an age which knew nothing of sanitation. "Know then that as it is true that meats, fish and milk products kept in a protected place do not breed worms, it is likewise true that fruit, vegetables, raw or cooked, in the same way, do not grow wormy. . . ."

Redi's book contains large, imposing woodcuts of mites, lice and ants. In it he discusses tapeworms and worms in the cavities of the frontal bone of deer and sheep. This led to his next study, *De animaculis vivis quae in corporibus animalium vivorum reperiuntur,* 1684 (Concerning Animalcules Which Live in the Living Bodies of Other Live Beasts), which, on the whole, can be considered the first book devoted mainly to parasitology. Amusingly enough, he began with the dissection of a two-headed snake, which he found stretched out on the banks of the Arno. He examined the intestines of this snake and found worms in them and also in a lizard with three tails, in the lungs of wolves, in dogs' kidneys. The work, which is illustrated, is mostly anatomical, with drawings and descriptions of the part of the host in which he found worms. His interest in this type of creature led him to earthworms, which he carefully dissected. The volume also contains anatomical studies of snails, shellfish, and water fleas. Redi died in 1667.

One other Italian scholar, Marcello Malpighi, 1628-94, whose work was diversified, nevertheless made a contribution to the exploration of the miniature world with a monograph on the silkworm. He was born in a small town near Bologna, and since his family was in comfortable circumstances, they were able to send him to the University of Bologna in 1646. His career in this institution is a contrast to those of the scholars who located in Padua. Bologna was under the jurisdiction of the Pope, and although the local senate had an administrative voice, the papal legate was the final authority. The foreign students were organized into nations with a councillor at their head. When one of these group of students criticized a member of the faculty and their councillor spoke for them, the legate overruled him and told the students to obey. Literature and ideas were censored, and no one was able to forget that the Inquisition was ready to investigate serious heresy.

In 1649 Malpighi lost both parents in an epidemic of fever. He took charge of his brothers and sisters, thereafter becoming head of the family. While at medical school he joined the *coro anatomico;* this was the "modern" group at Bologna who rebelled against Galen and the "ancients" and devoted themselves to experiment. Malpighi seems to have suffered annoyance all his life because he was a progressive. While he was an undergraduate, the students were stirred up against him and his life was endangered. The situation was intensified by the fact that one of the conservatives, Tomasso Sbaralia, was a member of a family which had personal grievances resulting from a conflict over some land. An example of the atmosphere at Bologna was the fact that a candidate for a doctorate in medicine was required to swear, on penalty of revocation of his degree, that he would not for more than three days attend a patient who had not confessed and professed himself a devout Catholic. He also had to swear he would not agree to grant a doctorate to anyone who did not take the same oath. Despite all this, Malpighi got his degree in 1656 and also was granted a chair in logic. He left for Pisa the next year for a professorship in theoretical medicine, perhaps hoping for more liberal working conditions. On his way to Pisa, he stopped off at Florence and Padua, where he met Nils Steensen, the anatomist, and probably Francesco Redi, with whom he had a warm scientific correspondence. At Pisa he made friends with the anatomist Giovanni Borelli (1608-79), in whom he found a kindred spirit. Borelli's scientific platform was thoroughly Cartesian and Galilean. His great work, *De motu animalium* (Concerning the Movement of Animals), was not published until 1679. He endeavored to apply strictly mechanical analysis to the movements of animals, using the principles of the screw, the lever, the pulley, scale and wedge, to demonstrate the actions of the muscles and the mechanics of locomotion. He and Malpighi worked together at dissection, and the friendship cemented at Pisa lasted for many years; indeed much of what we know of Malpighi's early life is drawn from a series of letters to his brother scientist. Under Borelli's influence he studied the effect of a dog's gastric juice on bones in the animal's stomach and discovered that the bones were much reduced in size. At times the older scientist was a trifle ungenerous. Malpighi wrote: "Dissecting living animals at home and observing their parts, I worked very hard to satisfy his very keen curiosity. While investigating the inclination and propagation of the fibers of a macerated beef heart, I hit upon their spiral structure." Borelli took credit for this, but Malpighi does not seem to have held it against him.

Family problems and the stabbing of Dr. Tomasso by Malpighi's nephew brought the scientist home in 1659. He eventually got the chair of practical medicine at Bologna and published his discovery that the lungs consisted of small inflatable membranous globules. Although he could not solve the relation of the lungs to the blood, he did decide that the air drawn to them produced fluidity and volatilization of the blood and this was a step beyond the Aristotelian notion of "cooling."

Malpighi was still unhappy at Pisa, and aided by Borelli's influence, he accepted a professorship at Messina, where his friend was teaching. There were still influential Galenists at Messina, however, so he shuttled back to Bologna to be appointed there in 1666. The following year, at the age of thirty-nine, he married the sister of one of his old teachers, Francesca Massari, who was fifty-seven. Apparently she had looked after his sisters and probably helped marry off all but one, who was put in a nunnery. She also brought him a little money. Although this marriage was certainly not romantic, Malpighi showed a real affection for his wife and indeed he seemed to have been as upset as she about her possessions when his house accidentally burned, in 1684, and she lost all her jewels and linens. He, too, had losses, his microscopes and notes. Redi, just publishing his book on parasitology wrote consolingly about the notes: "Your illustrations are those of master painters who work in broad strokes with a truly masterful freedom of the hand." He added that he hadn't a moment he could call his own and had merely put together some small worthless observations about insects.

The new Pope, Innocent XII, decided that he needed Malpighi as a personal physician. The old scientist was suffering from kidney stones and not anxious to travel, but the new honor also had the virtue of confounding his enemies who never ceased to snipe at him. He set off for Rome with his aged wife, a maid and a manservant in 1693. Both he and his wife died in Rome the following year.

Malpighi's anatomical contributions (both animal and vegetable) were a series of careful but isolated studies; his detailed study of the structure and life history of the silkworm was so detailed that it was considered a model of its kind. He discovered the tracheae used in breathing and the animal's excretal organs, which were characteristic of the phylum *Arthropoda,* observed the butterfly's evolution from the pupa, and, in general, paved the way for further work in invertebrate anatomy and for the first real entomologist, Jan Swammerdam.

Swammerdam's grandfather, whose name was James Theodorus, took the name Swammerdam from a small town near Leiden. Jan's father was a successful apothecary who took an interest in "curiosities," for he collected a private museum of Indian relics, fossils, insects and animals. Born in 1637, Jan was destined for the church and set to studying Latin and Greek. The boy, however, soon showed a disinclination for the ministry and a preference for the profession of medicine. His father allowed him to study medicine and, at the same time, set him to work in the museum. This had a profound effect on Jan's life, for he became fascinated with insects and began collecting on his own. He roamed the countryside, scouring the hedgerows, the fields, the forests, ponds and ditches, and began perfecting methods of preserving his specimens by blowing out the internal organs and drying the shell. In 1651, he entered the University of Leiden and successfully took preliminary examinations in medicine and surgery two years later. At the university, he made friends with the Danish anatomist Nils Steensen and the physician Regnier de Graaf. He seems to have acquired anatomical technique early in his career, for it was he who discovered the valves in the human lymphatic system. He went to Paris with Steensen, where he made a friend who was loyal to him all his life, Melchisédech Thévenot, a former ambassador to Genoa and now the king's librarian. Through him Swammerdam met members of the French nobility, entertained them by showing them his minute dissections of insects, and made one friend of importance, Conrad van Beuningen, Dutch minister to the court of France. Beuningen was to serve as burgomaster of Amsterdam and, in that capacity, made it possible for Swammerdam to dissect the bodies of executed criminals.

In 1688, having finished his studies in Paris, where even the conservative Sylvius was impressed with his talents, he returned to Leiden and received his degree in medicine. Unfortunately, in the same year he contracted malaria, a disease which never left him and also may have accounted for some of the psychological crises he suffered. At any rate, he felt himself so hampered by his illness that he concerned himself much less with human anatomy and concentrated more on insects. Swammerdam met Grand Duke Cosimo III of Tuscany, the son of Ferdinand, Redi's patron. Redi continued to serve Cosimo who, unlike his father, was a religious bigot. The duke, on seeing Swammerdam's collection and observing the dissection of a butterfly pupa to disclose the developing adult, offered the scientist twelve thousand florins for his collection on condition that Swammerdam join his court. The Dutchman, used to the liberal atmosphere of Holland and not liking court life, refused. All this time he had been living in his father's house, devoting himself to research. His book on insects was published in 1669. Thévenot wanted him to come to Paris, but now the conflict with his father, which must have been going on for some time, came to a head. His father cut off his funds, insisting that he practice medicine. This seems to have made his illness worse, for he was sent to the country to recuperate. He, of course, merely collected more insects. He then spent considerable time cataloging his father's collection in hopes of pacifying the hostile old man. The tug-of-war dragged on between them until Swammerdam's psychological condition deteriorated. He became a prey to acute religious depression, accusing himself of seeking fame and neglecting his soul. Actually he had always been a passionate man, prone to intellectual disputes and conflicts about priority of discoveries. At this time (1672) he had the misfortune to meet Antoinette Bourignon, an ecstatic and mystical reformer, who had a background of torturing children to drive out the devil. Persecuted by Protestants and Catholics alike, she contracted a spiritual marriage with a clergyman who owned an island. This was temporary haven for her band of followers. She then lived in Amsterdam where Swammerdam came under her spell. His new state of mind is reflected in his work, for his treatise on the ephemera is full of melancholy piety. Indeed, after 1673 he did no more scientific work. In 1676 he went to Denmark to see if Bourignon could find a haven in that country, but his mission was unsuccessful. On the one hand, his mentor, Bourignon, was urging him to retire from the world, and on the other, his old friend Steensen was urging him, once more, to turn Catholic and take shelter with the duke. Steensen, himself a convert to Catholicism, was in the Tuscan ducal court. He, too, suffered from his religious conflicts, for he eventually gave up science, took holy orders, became a fanatic and such an ascetic that he undermined his health and died at the age of forty-eight.

Poor tortured Swammerdam found his situation complicated by a final revolt on the part of his father. The old man would no longer endure his son's dedication to research and experiment. Jan was in his forties and had never been self-supporting. The conflict was this time brought to a head by the marriage of Jan's sister. She had been the housekeeper, so now the father decided to give up his house and cut his son off with two hundred florins a year. Since it was not enough to live on, Jan tried unsuccessfully to sell his collection. Within a year, his father died, leaving him some money, but the invalid scientist was further harassed by quarrels with his sister over the inheritance. He died in 1680 at the age of forty-three. His last important study, on the honeybee, cleared up, once and for all, misconceptions which had stemmed from Aristotle. Thanks to Swammerdam's anatomical understanding, he discovered that the mythical "king" was a female, that the drones were males and "ordinary bees" neuter females, which he named workers.

Swammerdam bequeathed all his papers and unpublished works, including that on the honeybee, to his friend, Thévenot, who had to go through litigation to get them. They were not published and, after Thévenot's death, went through many hands until they were finally bought and published, in 1737, with the earlier work on insects by the famous and wealthy Dr. Hermann Boerhaave, himself a biologist. He contributed a biography of Swammerdam and gave the whole work the rather charming title *De Bijbel der natuure* (The Bible of Nature). The book, which is a huge folio of about three hundred pages, half of which contain elegant illustrations, is particularly significant, for aside from the excellent anatomical descriptions, it embodied Swammerdam's theoretical attitudes and was a first attempt at classification of insects. He attempted to do this on the basis of their development. There were, he said, four groups:

1. Those that come out of the egg with all their limbs, grow to their proper size, change to a nymph and shed their skin once. These were spiders and mites, but he also included scorpions.
2. Those hatched with six legs which, when their wings are gradually perfected, change into nymphs. These are ephemerids such as mayflies and dragonflies.
3. A worm or caterpillar that comes out of the egg with no legs or six or more. Its limbs grow under its skin imperceptibly until it casts its skin and resembles a nymph. This group included ants and bees.
4. A worm that emerges from an egg either without legs or with six or more legs and invisibly grows under the skin. It does not shed its skin but acquires the form of a nymph under it. This applied to certain flies.

Although this was not a wholly satisfactory classification because it was not based on form, it was at least a beginning. Indeed in his attitude toward growth he differed sharply from scholars who had written in the past. Aristotle had believed that pupae were eggs, which Swammerdam disproved by dissection to show the fully formed insect under the pupal skin. He insisted that nymphs (not fully formed insects which are able to move about) and pupae (which do not move and are sometimes enclosed in cocoons) were the same thing and used the word "nymph" for both. He disagreed sharply with Harvey who talked about epigenesis, the addition of limbs or wings while the body was in a state of "putrefaction." Nothing was added, said Swammerdam, everything was essentially present, waiting to develop. In a sense he was intuitively talking about the genetic predisposition of the species. This was why he compared the growth of the chicken in the egg and the development of the tadpole into the frog with insect histories. He was determined to show that insects, although small and hitherto despised, paralleled in interest the characteristics of other creatures. "How is it possible but we must stand amazed when we reflect that those animals whose little bodies are smaller than the finest point of our dissecting knife have muscles, veins, arteries and every other part common to the larger animals, creatures so very diminutive that our hands are not delicate enough to manage or our eyes sufficiently acute to see them. . . ." Swammerdam had a more convenient microscope than Leeuwenhoek, for it had six different lenses and a tripod which left his hands free. For instruments, he made use of needles and knives so finely pointed they had to be sharpened under the microscope. He also employed tiny glass tubes drawn out into minute points. The delicacy and accuracy of his technique, which he often demonstrated, were truly amazing. For preservatives he used wine and turpentine. It was typical of this dedicated entomologist that he let a louse bite his hand and then vainly endeavored to watch the action of the mouth parts, for the louse, he insisted, sucked and did not bite. He wrote: "I had then almost wished I had three hands."

For Francesco Redi he had great respect; his appendix on the anatomy of the cuttlefish is dedicated to Redi as the "most indefatigable searcher into the miracles of nature." It will also be remembered that Leeuwenhoek referred to his water flea. Here is a bit of his description of the movements of that tiny crustacean: "A second motion is like that of the sparrow for as these, by the expanding and contracting of their wings, pass with an uneven motion through the air, and sometimes descend, and immediately after are carried aloft again; so this little animal, by striking the water now and then with its branching arms, obtains a like unequal motion, and sometimes it dives as it were to the bottom, and again rises to the surface. . . . Since therefore the motion of this little creature is not at the time very regular, it happens that it is continually seen to jump in the water, its head always tending toward the surface, and its tail stretched downward."

Swammerdam exhibited in his life and character the transitional conflicts of the seventeenth century. On the one hand, he could become absorbed in precise detail, like his contemporaries, the Dutch painters. On the other, the emotional attachments to a religious synthesis which was already fragmented resulted in that same melancholy that infused much of the literature of the period. His summation of the life of the mayfly is tinged with cosmic sadness: "It grows, it is born into the world, it is a worm, it sheds its skin twice, it becomes an adult, it lays eggs, it grows old and dies at last—all in the brief period of five hours."

We see that in this period zoology was becoming gradually and partially an independent study. Of the men just discussed Swammerdam comes close to being a specialist; he can rightly be called an entomologist.

9

The Book of Nature

☙ THE two British clergymen considered in this chapter differ in their temperaments and contributions to zoology, but both belong to the tradition of the country squire, the gentleman naturalist. John Ray was more of a philosopher, more of a scholar; Gilbert White, who is not often included in histories of biology, was the gifted amateur whose style has won him a place in histories of literature.

Ray, with whom we begin, is significant, too, for his concern with classification which leads to the work of the founder of modern nomenclature, Linnaeus. Ray was born in 1627 in the small town of Black Notley near Braintree, on the highway which led from London into Suffolk. His father was a prosperous blacksmith and his mother something of a country herbalist who was accustomed to minister to the illnesses of her neighbors. There was, therefore, in Ray's family background the tradition both of the artisan and of folk botany. Since he was a good Latin student in the local school, Ray's talents were outstanding enough to gain the support of the vicar who helped him obtain a scholarship to Cambridge. In 1664, while the civil war was still in progress, the atmosphere in that institution was Puritan. The college was not strong in science and little mathematics was taught, although Descartes did have a following. There were no laboratories in Cambridge, and hence the dissection and collection of both botanical and zoological specimens were carried out by groups of students on their own. So far, Bacon and Harvey in England were the only outstanding champions of science, although the Royal Society, founded in 1613, as we have already indicated, was a most important clearinghouse for scholars from many nations. Ray and some of his undergraduate friends dissected birds, probably working in their lodgings. He held college offices, gave lectures in Greek and mathematics, and was eventually ordained. The most important event of his academic career was his meeting with Francis Willughby, who became his student. Willughby, who was eight years younger, came of a wealthy aristocratic country family. He became wholly dedicated to natural science and strengthened Ray's inclination toward this field of study.

Ray's university career was cut short when Charles II came to power and, in 1660, required all English clergymen to subscribe to an Act of Uniformity, which to many, including Ray, seemed to deprive them of liberty of conscience. This resulted in a purge of Cambridge, Ray being among those who would not sign. For the rest of his life he was a clergyman without parish and a teacher without pupils, except for his faithful disciple, Willughby. The friendship between the two men lasted all the pupil's life and was of great significance to Ray, for the young squire became his patron. Between them they had projected a great work, no less than a history of the natural world, an attempt to follow in the footsteps of Aristotle and Pliny. They had traveled together in northern England, Scotland and to the Isle of Man, always collecting specimens. In 1663, when Ray no longer had a future at Cambridge, the two set off for Europe, visiting the Low Countries, Germany, Italy and France. At Naples, Ray left Willughby and went on with another of his disciples, Philip Skippon. At Montpellier, they met Steensen, who was at that time teaching in the university, and watched him dissect an ox's head. In 1666 the King of France ordered all Englishmen out of the country, which resulted in the travelers' return to England in April of that year.

Ray settled down in his friend's house as tutor for the children, chaplain and collaborator on the natural history project. Willughby was concentrating on animals, Ray on plants. Ray had published a catalog of plants in the Cambridge area in 1660 and had contributed to the Royal Society's publications. In 1667 he was elected a member of that institution. An interesting sideline is his collection of English proverbs, published in 1670, a first attempt at gathering this particular kind of folklore. By this time he had a reputation as a scholar, his life seemed organized, his future secure. But once more all his plans were upset, this time by the death of his patron, in 1672. The unfortunate young squire succumbed to pleurisy at the age of thirty-seven. This was a great personal blow to Ray. Willughby had appointed him one of his executors, left him sixty pounds a year for life, and placed the education of his children and the completion of his scientific work in his friend's hands. All went well as long as the squire's mother, Lady Cassandra, was alive, but there was no doubt that the future was precarious. Willughby's wife had little use for Ray and even less for science, although she agreed to pay for the publication of the unfinished ornithology. Nevertheless, Ray decided to marry Margaret Oakeley, a governess in the household, in 1673, he being forty-six and she only twenty.

A list of problems written on the margin of a letter is almost the only suggestion of his personal reactions. First, his old mother was still alive and occupying his family cottage. Then, "You brought up in a different way and not likely to love my prayers" suggests that the young lady was more frivolous than he. "The children will never delight in my company for that I shall be old before they come to years of discretion" is understandable, for he was marrying rather late. Nevertheless, he was to have four daughters. He also wrote, ". . . inconvenient for me to die here for want of attendance," a rather enigmatic and melancholy statement for a man in the prime of life and apparently in perfect health. Finally, "I shall have nobody here to converse with" suggests that he was feeling some hostility in the household, and perhaps wanted to make sure he did not lose the girl. The blow fell in 1675, when Lady Cassandra died. Ray was promptly dismissed as tutor of the children and he and his wife had to leave. (Was Mrs. Willughby jealous of the close relationship between her husband and Ray?)

At first, a friend lent him a house. When old Mrs. Ray died in 1679, the cottage in Black Notley became vacant. The Rays moved in to remain there until 1705, the date of Ray's death.

Although he was offered the secretaryship of the Royal Society, Ray would not accept and preferred to work on his books. The cottage of lath and plaster was small, but it had a garden and an orchard. Ray had some books on botany and not much else in the way of a library. There was little room for natural science collections. He worked cheerfully, however, although he complained that there was no one with intellectual interests in the vicinity except for a local doctor, Samuel Dale, who took some interest in botany and insects.

Ray published extensively on botany, a catalog of British plants, and finally his most important work, *Historia plantarum generalis,* 2,860 folio pages summing up the botany of the world. He classified according to seeds and leaves and distinguished between monocotyledonous and dicotyledonous forms (single and double seeds) before Malpighi. He also made an advance in nomenclature; he characterized genera by one name, the species by a few words of description.

The ornithology, on whose title page he put Willughby's name, was of course a joint work. On their trip to Europe, the two men had explored markets where they looked at birds for sale, visited museums, studied exhibits and bought books and plates wherever they could find them. The older works on the subject were scrutinized, and whenever possible their mistakes corrected. Although habitat was relied upon to some extent in classification, far more attention was paid to the shape of the beak and feet. Diet was also used as a distinguishing factor. The crooked-beak and talon group of birds, for instance, was divided into fruit eaters (parrots and macaws) and the carnivorous subdivided into the day fliers (hawks, shrikes, birds of paradise) and the night fliers (owls).

On the whole, it was the best of its kind to date. It was used by Gilbert White, Linnaeus copied it, and Buffon borrowed anatomical material it contained. Despite the solid achievement of the book, it still contains some quaint passages. The one on the Brazilian vulture (mostly drawn from Spanish sources) is a good example. Ray gave the local names *urubú* in Brazil, *Menscheneter,* the Dutch version, *tzopilotl* in Mexico, and *aura* elsewhere in the world. We learn that:

> If anyone pursues them they empty themselves presently that they may be more light to fly away; with the like haste casting up whatever they have swallowed. The ashes of their burnt feathers take away hairs so that they come not again; which faculty is also attributed to the Dung of Pismires and the Bloud of Bats. Their skin, half-burnt, heals wounds if it be applied and the flesh withal eaten; which is wont also to help those sick of the French Pox. The Heart dried in the sun smells like Musk. The Dung dried and taken in any convenient vehicle to the weight of a Drachm is profitable to melancholy persons. The Barbarous people say, that where they lay their eggs, they compass their nests with certain Pebble stones, which promote transpiration: but the more probable opinion is, that they exclude their Young underground, and take them out when they feed them and again cover them in the earth.

After the reader is referred to Harvey and Malpighi on the development of the hen's egg, the following sensible statement occurs: "That the Lion is afraid of a Cock, can not endure the sight of him, yea is terrified by his very crowing, hath been delivered and received by Ancients and Moderns with unanimous consent and approbation, and divers reasons sought and assigned for this antipathy: When as the thing itself is by experience found to be false." The book is unhappy in its illustrations, which were a varied lot; some were taken from the older works, some were lent by Sir Thomas Browne, some freshly drawn, some purchased here and there. As a result, although the page is folio, too many figures are crowded together. The grand Baroque charm of Gesner and Aldrovandi's cuts is lost in reduction and the unevenness of execution spoils the general effect of the page.

Another quotation shows the authors discussing a bird with which they had firsthand acquaintance, the small heron known as the bittern, noted for its booming voice:

> They say it gives an odd number of bombs [*sic*] at a time, *viz.*, three or five; Which in my own observation I have found to be false. It begins to bellow about the beginning of February, and ceases when breeding is over. The common are of the opinion that it thrusts its Bill into a Reed, by the help whereof it makes that low humming note. Others say that it thrusts its Bill into the water or mud, or earth, and by that means imitates the lowing of an Ox.

Ray commented that he thought it sounded more like the braying of an ass.

Ray finished the treatise on fish, but Mrs. Willoughby refused to pay for its publication. She had married again and was farther away than ever from her former husband's interests. Always helpful, the Royal Society finally agreed to pay for the work in conjunction with the Oxford University Press. The amiable Samuel Pepys, who was president of the Society, personally covered the cost of sixty plates. Nevertheless, the work did not sell well, perhaps partly because it was very expensive and partly because there was political unrest. In the end the project left the Royal Society bankrupt.

The book was almost all Ray's work. Few new species were added, but Ray's pupil, Sir Philip Skippon, contributed a detailed treatise on the herring industry. As always in Ray's work, the descriptions were outstandingly accurate.

Ray's *Synopsis methodica animalium quadrupedum* (Synopsis of Four-Footed Beasts), published in 1693, is his most important contribution to zoology. In it he attacked Descartes' theory that animals were machines without sensation and disproved it. He opposed spontaneous generation but did not contribute much to sexual theory. Fabulous animals inherited from classical authors he discussed and rejected. His classification started with what was sound in Aristotle. He then further divided the vertebrates into those with two-ventricle hearts: mammals and birds; and single-ventricle hearts: frogs, fish and reptiles. His further classification of the "hairy quadrupeds" into the hoofed and clawed groups was an advance. Still better was the division of the hoofed into one-hoofed, horses; pair-hoofed, cattle and pigs; and the multihoofed, rhino and hippopotamus. This type of grouping is close to the modern treatment of the related classes *Perissodactyla* and *Artiodactyla,* both being partly based on the number of toes.

Toward the end of his life, Ray was working on insects. Old and ill, since he suffered from diarrhea (which he cured with biscuits boiled in milk) and chronic leg ulcers, he could no longer venture farther than his garden. He used his daughters as his legs and sent them forth into the fields and woods with nets. Rather charming is his baptizing certain insects by the girls' names, "Jane's Chickweed Caterpillar," for instance. Ray died in 1705 before the work was finished. His friend, Dr. Samuel Dale, who had contributed to it, was too modest to undertake the task of editing and finishing it. As a result, it was finally brought out by the Royal Society three years later in a rather fragmentary state.

Although Ray did not set up a system of classification, his importance for taxonomy is his sense of structural similarities and his distinguishing between the concept of species and genus (a group of related species). He was aware of the problems involved, and if he did not always achieve a "natural" grouping, some of his insights paved the way for this development.

Fossils and the matter of geologic time were destined to be a sensitive area in the history of natural science. As long as the literal authority of the Bible remained unchallenged, advance in geology was blocked. Such philosophers as Descartes remained in sufficiently abstract areas so that they did not come into head-on collision with the Bible. In the case of Ray, possessor of a first-rate seventeenth-century mind, but also a clergyman, we have an interesting record of his struggles to reconcile the Mosaic dogmas with the facts of natural science. Ray collected fossils from quarries and had read the work of Nils Steensen. Steensen, it will be remembered, had made contributions as an anatomist, but he also concerned himself with fossils. He had studied the geological strata in Tuscany and proved that they had been laid down by water, a fact he felt was confirmed by the number of animals and plants found in them. This was a most important step in the investigation of the earth's crust and the creation of sedimentary rock formations. Having thus laid the basis for paleontology, he shied away from further investigation, for his fanatic conversion to Catholicism effectually blocked such unorthodox thinking. Ray, however, as a visiting preacher, delivered some sermons at Mary le Bow, in London, which gave rise to two books, *The Wisdom of God* in 1691 and the following year *Three Physico-Theological Discourses.* These were efforts to reconcile natural science and religious authority. Ray searched for a good materialistic explanation of the separation of the land from the water in Genesis; by using "secondary causes" he was able to move one step away from the Michelangelesque concept of a bearded patriarch gesturing majestically. Subterranean fires, he said, could have thrown up the mountains and, quoting Psalm 7, suggested that "voice of thunder" could mean earthquakes. If God created only individual animals, how did the earth become populated? Ray concluded there must have been two of each kind and in the ovaries of each female were contained in miniature the innumerable animals that were to come. ". . . If a grain of sand were broken into 8,000,000 equal parts, one of those would not exceed the bigness of those creatures as Mr. Leeuwenhoek affirms." He was not agreeing with Leeuwenhoek on the function of the sperm, however, for Ray was a staunch believer in the importance of the egg.

Steensen's fossils and the ones he himself had seen were a problem, however. He rejected the arguments that they were not the remains of organic forms, and he could not swallow the theory that they were deposited upon mountains by Noah's Flood. He had seen shells of sea urchins in rock stratas, flattened and compressed out of shape, proving that they had been acted upon by pressure. If the water had covered high mountains, even the Alps, and such natural forms were deposited in sand and mud and covered by the latter until they were compressed into rock, "Now this could hardly be the effect of a short Deluge. . . ." He went on: "If the mountains were not from the beginning, the world is a great deal older than is imagined, there being an incredible space of time required to work such changes." Furthermore he was "puzzled and confounded" by fossil species which could not be recognized as contemporary. Putting these facts together, Ray had to admit, "There follows such a train of consequences as seem to shock the Scriptural history of the novity of the world; at least they overthrow the opinion generally received, and not without good reason, among Divines and Philosophers that since the first creation there have been no species of animals or vegetables lost, no new ones produced." Ray was already grappling with the problem which, a century and a half later, Darwin was to bring to a head. Ray's theological geology proved popular and his two books on this subject, in ironic contrast to the financial failure of his scientific works, became, for their period, best sellers.

There is a gloomy seventeenth-century sense of drama in Ray's final discourse on the destruction of the world. This could happen by the burning out of the sun or the eruption of the central fires within the earth (there he accepted Descartes' theory that the earth was once a sun). It would be the work of God and would result in either total destruction or refinement and purification. Having faith in God, he believed the planet would be refined and purified. But what would happen to the "material heaven" situated somewhere in the empyrean and the "material hell" which was supposedly located in the center of the earth? For this he had no solution. After the holocaust, "There may be a new race of rational animals brought forth to act their parts upon the stage which may give the Creator as much Glory as Man ever could or did." The words have an ominous sound for twentieth-century ears.

In the fifteen years between the death of Ray in 1705 and the birth of Gilbert White we have already made the transition from the melancholy duel between the flesh and spirit that inspired dark poetry in Donne, Ford and Webster, akin to Ray's vision of the dissolution of the world, to the easy, optimistic view of life held by the celebrated Addison and Steele. Indeed White is a blood brother to Sir Roger de Coverley.

White never had to worry about money, nor was he dependent on a patron as Ray was, because his mother was an heiress. His father, although a justice of the peace, does not seem to have counted for much in the household. White was born in his grandfather's vicarage in Selborne in 1720. After moving to another village, the family settled in a country house called Wakes, also in Selborne. When Gilbert's mother died in 1739, his grandmother, Rebecca, took over the domestic responsibilities. Gilbert was tutored, probably by the local clergyman, and well grounded in Latin, Greek and religion in preparation for Oriel College, Oxford. As a young man in college White hunted, but in his maturity he laid down his gun to become a bird watcher and a naturalist. His academic years seem to have been frivolous; he flirted with girls, rode around the countryside, attended parties in the houses of friends. He also contracted a mild case of smallpox. In 1749, when he was ordained a clergyman, he had enough to live on, but showed no sign of wanting to assume the responsibilities of a solid citizen. He continued to ride around the country and socialize, but he always came back to Wakes, where he spent time working in the formal garden. In 1751 he started a diary, and the following year he was appointed proctor and dean of Oriel

College, a post which seems to have been fulfilled in absentia. The same year his brother, John, was expelled from the college for disorderly conduct. Eventually John married and became chaplain of the garrison of Gibraltar, which kept him out of mischief.

In 1755 White's grandmother died. The following year he accepted the curacy of Selborne and an absentee living in Northampton, in a nearby town, which brought him thirty pounds a year. After his father died, his sister took over the household for a few years until she married. Gilbert was finally left in Wakes with a maid and a gardener. He had been hopeful of getting a good living, but his application was not granted. With his various small sinecures and his fellowship at the college he was perfectly comfortable. He settled down to a cheerful bachelorhood during which he savored the simple pleasures of rural existence and lived an unshadowed life.

In his forties, his house was always full of young people. There was his romantic younger brother, Henry, and his school friend, John Mulso, and various young ladies who came to his house parties, which ended with dancing to the harpsichord until three in the morning. We have a record of this cheerful period as a country squire in the diaries of Kitty Battie, one of three attractive sisters who used to visit White and threaten his bachelorhood. The young people took tea in the Hermitage, described as a hut with a thatched roof, built on top of a nearby hill which was reached by a zigzag walk. Gilbert wrote of this scene in winter:

> Is this the scene that late with rapture rang,
> Where Delphie danced and gentle Anna sang;
> With fairie step where Harriet tripped so late.
> And on her stump, reclined, the musing Kitty sate?

Writing heroic couplets was one of the gentlemanly accomplishments of the time, and in accord with eighteenth-century pastoralizing, the young people sometimes got themselves up as shepherds and shepherdesses. Kitty wrote that she remembered one of these house parties as the happiest day of her life, "agreeable company, fine day, good spirits, all combined to make it of all days the most agreeable." Evidently, Gilbert was a talented host.

He began his nature study as another dilettante activity on a level with his poetry, but he gradually became seriously involved. The parish of Selborne lay in the extreme eastern corner of the county of Hampshire, fifty miles southwest of London. There was much woodland and farmland with a couple of small streams. The roads were cut deep into the native freestone and in wet weather became streams. The village itself contained about six hundred inhabitants, some living in stone or brick cottages. Many of the people were poor, finding employment growing hops and cutting wood. The women spun wool for making a cloth used in summer wear. The King's Forest of Wolmer seems to have been largely deforested, but in it were marshes supporting many game birds; at one time, it contained red deer, but the last of these had been captured and taken to Windsor.

In 1761 White met Thomas Pennant, who was writing a book on British zoology. Pennant was not a fieldman, but must have shown a keen interest in White's local notes. A correspondence began in which the country clergyman chatted about the little world of Selborne, recording all the small doings of his animal neighbors. During the next twenty-six years his journal became more and more detailed and richer in zoological data.

White occasionally went up to London where he met his friends, was shown collections, and kept up with the intellectual life of the city. He was never happy in the metropolis, generally catching cold or developing rashes. On one of these trips, he met Joseph Banks, Captain James Cook's scientific mentor, just before the travelers set sail for Tahiti, and was much impressed by the perils of their expedition.

Another important correspondent was Daines Barrington, who was also writing on British fauna. White's letters became his most important occupation, along with his journal. He also corresponded with his brother, John, who was writing a natural history of Gibraltar. John, in turn, corresponded with the great Linnaeus, something Gilbert was too modest to do. Barrington read White's letters on the swift and the martin to the Royal Society. Ironically enough, the country naturalist was too modest to apply for admission to this institution. Field study was not yet considered "scientific."

His friend, John Mulso, also a clergyman and a worldly one, now the holder of a half dozen benefices, had repeatedly urged White to publish his material. The naturalist, who was a bit lazy, delayed, good-naturedly giving his time to reading the proof of Pennant's book. Finally, after more prodding, he got the letters together with transitional material from his journals and brought out his masterpiece, *The Natural History and Antiquities of Seborne,* in 1788.

The eighteenth century had developed the reflective essay to a high point of perfection and thus White had a literary form ready to hand in which he was able to express his own sunny personality while painting his picture of the little universe of Selborne. In an age of rather fussy style, White wrote with limpid simplicity. Birds were his great preoccupation; he identified the lesser whitethroat and distinguished between other warblers that had not been well described. He also identified the largest English bat and the smallest mouse. The great value of his work lies in his painstaking and carefully dated notes concerning animal behavior. Hitherto eyewitness observation had been casual; in White's own time, the preoccupation was with classification. The Hampshire parson, however, was a model ethologist. He was much preoccupied with migration, for in this prebirdbanding era there was no agreement as to where birds went in the winter. While he did not accept the fable that swallows wintered underwater, he mulled over the possibility that they might hibernate in a dormant condition and once even excavated a clay bank in hope of finding some swallows wintering there. Evidence supplied by his brother went far to suggest what really happened:

> This is certain, that many soft-billed birds that come to Gibraltar appear there only in spring and autumn, seeming to advance in pairs toward the northward for the sake of breeding during the summer months, and retiring in parties and broods toward the south at the decline of the year: so that the rock of Gibraltar is the great rendezvous and place of observation from where they take their departure each way toward Europe or Africa. . . . It is presumptive proof of their emigrations.

White made use of musical neighbors with pitch pipes who assured him that some owls hooted in D-flat, some in A, and that cuckoos varied from D to D-sharp. One letter described the characteristic flight of all the birds of Selborne and another characterized their songs. It was just this systematic observation making for clearcut identification of species that in his time was a process not yet standardized. Indeed, he pointed out that the compilers of natural histories often gathered inaccuracies because their sources were nonprofessional and careless. He thought that a series of local monographs, compiled by specialists, could contribute to more satisfactory general works.

White was not without an inclination to experiment. When he read that Monsieur Herissant explained a cuckoo's irresponsible habits by the fact that its crop lay behind its sternum and that this would cause discomfort on the nest and prevent it from incubating, he got hold of a fern owl (the goatsucker *Caprimulgus europaeus*), which he knew from experience incubated successfully, and dissected it. Its crop also lay behind its sternum. In consequence, Herissant's theory "seems to fall to the ground."

A hundred years before Darwin, White turned his attention to the humble earthworm. "The most insignificant insects and reptiles are of much more consequence and have much more influence in the economy of nature than the incurious are aware of. . . ." After this ecological statement, he pointed out that although a small link in the chain, the earthworms would be a great loss if destroyed. These creatures "seem to be the great promoters of vegetation, which would proceed but lamely without them, by boring, perforating and loosening the soil. . . ." He stated that farmers who ignorantly blamed earthworms for eating their corn, a crime committed by slugs and beetle larvae, should learn to appreciate the innocent night crawler. Worms, he said, were "much addicted to venery" and added some notes on their amatory activities in his diary. When they lay out at night on the turf, although they extended their bodies a great distance, they left the ends of their tails fixed in their burrows so that they could immediately pull back within. "Even in copulation, their hinder parts never quit their holes: so that no two, except they lie within reach of each other's bodies, can have any commerce of any kind; but as every individual is an hermaphrodite, there is no difficulty in meeting with a mate. . . ."

Although we have so far encountered a couple of hints concerning the importance of territory, it is Gilbert White who first made a clear statement of the basic fact in relation to birds:

> I have always supposed that that sudden reverse of affection . . . which immediately succeeds in the feathered kind to the most passionate fondness, is the occasion of an equal dispersion of birds over the face of the earth. Without this provision, one favorite district would be crowded with inhabitants while others would be destitute and forsaken. But the parent birds seem to maintain a jealous superiority, and to oblige the young to seek for new abodes; and the rivalry of the males in many kinds prevents their crowding in on one another.

Another of his outstanding essays is on the habits of the field cricket. At first, he attempted to dig this creature out, but nearly always destroyed the burrow. He did, however, discover the eggs and learned to distinguish between the sexes and identify the female's ovipositor. He then discovered that he could coax them from their holes by putting down a stalk of grass. The crickets appeared at the mouths of their cells, which they bored open about March 10. They were in a pupal state, with the rudiments of wings lying under the skin. They ate whatever herbs grew in front of their burrows and dropped their dung on a little platform close by. Males fought with their serrated jaws. They were most active in July, and by August their holes began to be obliterated and they disappeared.

An old lady, a neighbor of the naturalist, kept a land tortoise in her garden, feeding it for thirty years. White noted that the creature, which was called Timothy, knew its mistress. He kept an eye on its habits. Timothy hibernated in the winter, digging itself a burrow in the earth. It also became very active before a storm, serving as a weather prophet. In 1780 White became the owner of the animal and dug it out of its hibernation in March. It was annoyed but soon reburied itself in his garden. In April, when the thermometer stood at 50 degrees, it poked its head out of the earth. White tried to take the animal's pulse and put it in a tub of water to see if it could swim. He also played a trumpet to it, but it seemed not to react to music. Actually the animal was not indigenous; it was an Iberian species which was known to live from one hundred to a hundred and twenty-five years.

As White grew older, his house was made gay by the visits of his many nieces and nephews—in all he had sixty-two. He continued to live peacefully among the birds and beasts of Selborne, while Louis XVI was executed and the newly constituted Republic of France declared war against Holland and England. In that same year, 1793, his friend, Mulso, came for a visit. White wrote, "June 15, 1793: men wash their sheep. Mr. Mulso left us." He had numbered the days of the following week, June 16-22, but was never to go on with his journal. On June 10, he buried a parishioner, Mary Busbey, age sixteen, caught cold, and died two weeks later.

White's book lives on as an example of both pioneer field observation and as literature. Perhaps a good part of its perennial charm arises from the envy which the modern reader feels for a man so unquestioningly at home in the world of nature. White, the hunter, laid down the gun to become a naturalist; White, the country gentleman, showed what an inexhaustible and absorbing drama took place every day in the fields, woods and streams, for the delight of those who took the trouble to look around them.

10

Method Is the Soul of Science

☙ IN Sweden, a backward boy who disappointed his parents in their efforts to make him a clergyman grew up to become the Adam of the natural world and to bequeath a system still used by the entire biological community. Carl Linnaeus, whose family took its name from the linden tree, was born in 1707 into a singularly narrow-minded Lutheran family. His father was a poverty-stricken clergyman in a small town near Lund. He was, however, an amateur botanist. The boy was destined for the ministry as the only worthwhile profession. Unfortunately, he played hooky throughout his years in the grammar school in order to collect plants and read botany. When his clerical masters were interrogated on his theological progress, they had very little to report. The rector of the school, however, had allowed him to use his botanical library and was not unsympathetic. His father, on the other hand, was furious. His mother (the daughter of a clergyman), who seems to have been simply stupid, felt she must withdraw her love if her son could not make it as a parson. Fortunately, the state doctor, Rothman, a friend of Carl and the family and also his physics teacher, pointed out that botany could lead to medicine.

Carl was lucky in encountering a series of kindly patrons throughout his career. Rothman took him into his house, taught him physiology and botany, and generally prepared him for the university at Lund. His school recommendation was lukewarm, and he set out with almost no money, but he had the good fortune to board with Kilian Stobaeus, a member of the faculty. On being caught late at night sneaking into his landlord's library, he was questioned by Stobaeus, who discovered that in the areas that interested him he showed remarkable application. He was given the run of the professor's library and considerable encouragement. It is therefore somewhat curious that he next turned up at the University of Uppsala, again with no money and apparently without having said a word to his amiable mentor.

The medical faculty in Uppsala proved disappointing and negligent, but Carl absorbed more botany and also attended some lectures on birds. He must have tutored in order to live and he obtained the lowest class of royal scholarship in medicine. The Swedish academic world seems impoverished after the Renaissance Italian universities we have encountered. The only record we have of Linnaeus' attending a dissection is an episode in which he rushed to Stockholm because a woman was to be hanged and subsequently handed over to the Stockholm medical faculty.

Carl, however, studied and starved; the fact that he suffered from scurvy is testimony to his miserable diet. It was at the botanical garden of the university that he made the acquaintance of the dean, Dr. Olaf Celsius, Sr., who happened to question him on his knowledge of plants and was so impressed that he took him into his house and gave him the run of his library, a pattern which had now occurred for the third time. It was in 1729 that he met Peter Artedi, also the son of a clergyman and also a refugee from theology, studying natural science on his own, who was two years younger than Linnaeus. The naturalists were complementary and became close friends. Artedi was drawn to zoology, especially fish; Linnaeus was already a promising botanist. They exchanged ideas, studied and starved together, Linnaeus' lively optimism a foil to Artedi's calm critical disposition.

Carl's energetic self-instruction had brought him to such a point in botanical knowledge that he was made assistant to the botany instructor. He was already drawing up lists of plants and wrote a short treatise on classification in terms of their sexual organs, a method which had been tried before. His ability brought him sufficient recognition so that he received a grant from the government to study botany on the islands of the Swedish coast. We next find him living in the house of Olaf Rudbeck, professor of botany at the university. This happy state of affairs broke up apparently because of the behavior of Rudbeck's wife who had an eye for young men.

Another grant from a scientific society made possible Linnaeus' famous trip to Lapland in 1732. This country was in many ways a primitive area into which the young scientist plunged, alone, with very inadequate funds. In five months he traveled more than 4,600 miles, exploring most of Lapland and part of northern Sweden and Norway. He carried a small bag containing a couple of extra shirts and vests, an inkstand, microscope, telescope, comb, his diary, paper for drying plants, unpublished manuscripts, a light gun, and wore a sword at his side.

His family felt he was going to his death, but he pluckily endured heat, cold, fog and mosquitoes, climbed precipices, forded torrents, and traveled on rivers in light boats which had to be carried around rapids on the boatmen's heads. It was roughing it, but he was young and resilient and obviously had the time of his life. Seasickness sometimes troubled him, but he had a remedy: The sufferer should be tied to the mast and allowed to smell warm bread; after that he should drink seawater and wine. True to his pious upbringing, he always managed to find a church to attend every Sunday, even in the wilderness. When he ran out of money, he starved as usual, and indeed, the Lapp diet often consisted of no more than water and raw fish. With luck he was able to buy a reindeer cheese. He took many anthropological notes on the seminomadic Lapps, who lived in tents of reindeer skin or huts and were dependent on a reindeer economy. Reindeer milk, thick as cream, was passed around in a skin bag as a special treat. Linnaeus rather lost his appetite when he saw that the communal spoon was each time licked clean before being passed on.

Linnaeus discovered a hundred new species of plants and accumulated many zoological notes. He rather emphasized his hardships because he had messed up his expense accounts and, in any case, insisted he had spent 122 dalars more than he had been given. To his annoyance, the Uppsala Academy, after much bureaucratic haggling, came up with only 40.

The young scientist also played up the romantic side of his expedition, appearing in Lapp costume, a skin outfit with pointed-toed boots and a cone-shaped hat, not forgetting a sorcerer's drum. An often reproduced portrait shows him thus attired. Back at the university, Linnaeus gave private lectures on assaying and botany and also, quaintly, set forth seventy-five rules of hygiene, some sensible, for he believed in bathing, on the whole not an eighteenth-century indoor sport.

Uppsala was so backward that it was not prepared to grant a medical degree. While in Falun, Sweden's second city, Linnaeus fell in love with a doctor's daughter, Sara Lisa Moraeus. His future father-in-law not only was in favor of the match but even helped Linnaeus financially, making it possible for the young man to visit a small Dutch university, at Harderwijk, where he could obtain an inexpensive medical degree. Engaged, the scientist set off, wrote a dissertation on "intermittent fever" which he attributed to living on clay soil, and in a couple of weeks got his degree, including a diploma, gold ring and silk hat.

During his travels, he tactlessly offended the burgomaster of Hamburg, who was the proud possessor of a stuffed "hydra," the many-headed beast of antiquity. Linnaeus pointed out in a newspaper article that the seven heads were those of weasels and the body was made of snakeskins sewn together. The fake, which had come from a church in Prague, was about to be sold for a large sum. It was an example of the lingering love of the miraculous in zoology that a fanciful portrait of the exhibit had appeared in various natural histories.

Linnaeus ended by going to Amsterdam and Leiden where he met the patriarchal Dr. Hermann Boerhaave, who was just publishing Swammerdam's belated masterpiece. The Dutch physician, finding him conceited, never took him personally. He did, however, recognize his talent and helped him get patrons to cover the cost of the first edition of the *Systema naturae* (1735). This consisted of only fourteen folio sheets but, as he developed as a scientist and revised and amplified it, was to make his name. He noted a charming habit of Boerhaave's who always raised his hat to an elder tree because of his respect for its medicinal qualities. In 1735 Artedi, who had had a grant which permitted him to study zoology in London, arrived destitute, hoping to get his medical degree in Leiden. Linnaeus had friends in Amsterdam by now and therefore was able to get a publisher for Artedi's ichthyology and also find him a position cataloging a wealthy apothecary's collection of fish. Unfortunately, Artedi, after a sumptuous dinner with his publisher, Albert Seba, fell into a canal on the way home and drowned. No details are known of this mysterious misadventure.

Linnaeus was, of course, much affected. Loyal to his friend's memory, he edited the ichthyology and saw to its publication in 1738. Linnaeus could have stayed in Holland; indeed Boerhaave offered him the position of a government doctor in the colony of Surinam. The French naturalist Charles Du Fay made him a corresponding member of the French Academy of Sciences, and there was talk of a job in France. During these years, Linnaeus also visited England but acknowledged that he had no time or talent for languages and never bothered to learn a word of French, Dutch or English. What brought him home was news from Falun. His patient fiancée, while enduring an engagement of eight years, may well have gotten restive. Linnaeus says that some vile academic character was trying to cut him out. Of course, Sara Lisa might well have decided that something was needed to bring the situation to a head. Linnaeus rushed home, foiled his rival, married the girl—and went off to Stockholm, leaving her in Falun. In the capital he practiced medicine, becoming a specialist in gonorrhea, a disease afflicting a large percentage of the Swedish gallants. He could have had a professorship in Göttingen, but he kept his eye on Uppsala where, at long last, he got what he wanted, the professorship of botany. This was in 1741, and since he now had a son, Sara Lisa must have breathed a sigh of relief, at last being able to settle down. He was off to Gotland the same year, however, on another expedition about which he wrote his first book in Swedish. Indeed, his Latin was never elegant; his own notes were written up in a mixture of Latin and Swedish. His new book covered plants and animals and also minerals. Once more, he racked up a hundred new species of plants.

Always the debunker of frauds, Linnaeus noted that he visited a church where the bones of a giant were preserved. "They were really whale's bones." A peasant remedy for the cure of warts was one of his folklore gleanings. "Take a grasshopper and set its mouth on a wart, when it bites asunder and spews a black burning fluid over it which destroys the wart."

Linnaeus soon became the most popular and famous professor at Uppsala. At his university lectures he drew 230 students; another 165 came to private lectures. He spoke in short, simple sentences and displayed a keen sense of humor. Indeed, although he was a bit vain and liable to magnify the jealousies of his rivals, Linnaeus appears to have been a rather hearty personality. He is described as taking part in an athletic Swedish polka, danced by young friends and students, when he was a man of sixty.

He was also accustomed to lead as many as three hundred on nature walks, which went on all day. He lectured on everything the students encountered while one of them took notes on any new specimens observed.

In 1758 the tenth edition of the *Systema* was published. This version of the work, which had grown into several volumes, best incorporates his system and is the standard edition referred to by later naturalists.

Interesting is the rather simplistic piety which unrolls in the work from the very title page:

O Jehova
How ample are Thy works!
How wisely Thou hast fashioned them!
How full the earth is of Thy possession!

There was in Lutheran Linnaeus none of the sharp skepticism of the average eighteenth-century intellectual. He was convinced that the order his system imposed upon the natural world was essentially what Jehovah had put there for him to discover.

The introduction goes on with a further invocation of God, a rapid summary of astronomy, a dividing of the earth into four elements and further into minerals, vegetables and animals and, finally, the statement that man is created to worship God. Elsewhere he wrote: "The knowledge of nature leads us into moral theology itself, and how completely it depicts for us the Creator's magnificent works."

Turning back to the *Systema,* we discover Linnaeus' basic scientific creed. He explained:

> Wisdom's first step consists in knowing things in themselves. Knowledge consists in a true idea of objects, and, by the properties with which the Creator has endowed them, to distinguish the similar from the dissimilar so that this knowledge can be communicated to others by affixing a name to each thing to distinguish each from the others; for if the name be lost the knowledge of the thing is also lost. For these shall be the basis of literacy without which none can read for, in ignorance of the particular subject, no accurate description can be transmitted, or accurate demonstration made but only errors committed. Method is the soul of science. . . .

All his life Linnaeus had been making lists and arranging groups. The outcome of his thinking was that the world conveniently had five subdivisions. Rather charmingly he used the parallel of province, territory, parish, pagi, home and legion, cohort, manipuli, contubernia, soldier to support his fivefold: class, order, genus, species, variety.

Ever since natural science began, scholars had been wrestling with problems of identification. Mathematics had numbers and symbols that all nationalities could recognize, but animals and plants acquired all sorts of local names which were sometimes used indiscriminately for various different creatures. Scientists desperately needed to know what they were talking about in order to communicate with each other. The sixteenth-century investigators had listed all the names they could think of in various languages for each species, but this was a clumsy system. John Ray and several botanists had already used descriptive titles, often consisting of two Latin words. For instance, *rana* is the ordinary Latin word for "frog." Linnaeus used this for the frog group and for a specific species added *temporaria.* Since there have been many changes since his day, it is now customary to add an abbreviated form of the name of the man who first designated the species. A common type of frog is, therefore, *Rana temporaria, Linn.* Once this system was generally accepted, naturalists found it much easier to communicate. Taken together with the four other larger classifications, a basic scheme was set up for organizing the entire world of nature. Order was, of course, a group of related species, and class a still larger grouping. The binomial system still holds and also the larger grouping with the addition of phylum, a group of classes and family, a group of genera. Of course, the actual animals included in the groups have changed with the development of the science and, indeed, are still changing, for there are many possibilities of disagreement. Who is to say whether a group of genera placed in one family are equivalent in their likeness and unlikeness to those placed in another family? It can be seen from this that the system, far from being ordained by Jehovah, is mechanical and often depends on subjective judgments. Nevertheless, it has been useful, and so far, no one has thought of anything better.

Linnaeus himself, although he tried to use a few simple natural characteristics such as no upper teeth, chewing of cud and split hoof for cattle, which he labeled order *Pecora,* failed in other cases to find the best characteristics, and when it came to the invertebrates, he did not even try, for he created order *Insecta* and then rammed all sorts of heterogeneous creatures such as mollusks and starfish into *Vermes,* worms. This makes him less progressive in this area than Aristotle. But then, Linnaeus didn't like invertebrates and cold-blooded animals. His motto for the amphibians was: "Terrible are Thy works, O Lord!" His six classes were:

Quadrupedia	Mammals
Aves	Birds
Amphibia	Frogs, lizards and snakes
Pisces	Fish
Insecta	Insects
Vermes	Everything else

It is to his credit that he put the whales and dolphins into their own order, *Cetacea,* under mammals. The importance of Linnaeus' method was the boiling down of identification to a few details or sentences which made it possible to construct a manual for identification rather than a discursive picture book like so many natural histories both before and after his time. *Elephas maximus* is described as follows:

> Habitat Ceylon, eats foliage, seeds, fruit. Eyes small, elongate upper canines, long hanging ears, skin very wrinkled, very thick, two breasts on chest, toes on edges of feet. Flexible knees, short neck.

This is followed by a paragraph on the beast's habits which repeats some of the Plinian fables.

Linnaeus bravely grouped man with the great apes and coined the species name *Homo sapiens.* His description of the varieties of man was quaint:

Wild man—	four-footed, mute, hairy
American man—	erect, choleric, obstinate, gay, free. Lives by custom.
European man—	gentle, clever, inventive. Governed by rites.
Asiatic man—	melancholy, rigid, severe, discriminating, greedy. Governed by opinions.
African man—	crafty, indolent and negligent. Governed by caprice.

A finer set of stereotypes can hardly be imagined! Nevertheless, our nomenclator included the orangutan in the same genus with man, calling him *Homo troglodytes.* The "wild man" listed above was probably the result of some vague reports of the gorilla.

Linnaeus settled down as "the prince of botanists," honored by everyone and petted by Queen Lovisa Ulrika. She collected butterflies and was fond of blindman's buff. Linnaeus, who was bored by this game, peeked and caught the queen as soon as he could in order to put an end to it.

His wife, who bore him two boys and four girls, was a domestic type, who loved card games, but had to forgo them when her husband was visited by his serious scientific friends. The house was enlivened by a number of pets, including a monkey, a parrot, a raccoon, dogs and even crickets, which were frowned upon by the mistress of the house. The parrot sat on the professor's shoulder at dinner and if the meal was late would call out "Twelve o'clock, Mr. Carl!" mimicking one of the servants.

Linnaeus seems to have undergone a psychological crisis in his sixties, for he became unusually touchy and quarrelsome. He made his son, Carl, his alternate in the botany instructorship, although the boy had not taken qualifying examinations. Oddly enough, his one commercial venture came late in life. He devised a method of creating artificial pearls by placing a silver wire tipped with a tiny ball of plaster in a live mussel shell. The government took over the rights to this method and paid him £450.

The great classifier never wavered in his belief in the immutability of species. "There are just as many species as there were in the beginning," he wrote. "There is no such thing as a new species." It is obvious, of course, that in this assertion his pietistic leanings affected his point of view.

Linnaeus died in 1778, leaving a collection of specimens and a library so extensive that he had had to build a special museum on his country estate to house them. His family put them up for sale, but the only offer came from England, where a wealthy young manufacturer, James Edward Smith, was prompted by Sir Joseph Banks to bid on the material. Smith bought everything for £1,000. At the last minute, Swedish chauvinism was aroused and pressure was put on the king to keep this national treasure from leaving the country. Before official machinery began to move, however, the vessel carrying the collection passed through customs and was off. A subsequent fable pictured it pursued by a warship which failed to catch up.

Smith founded the Linnaean Society and became its president in 1788. Linnaeus' contribution was after all international. Indeed, one of the seminal functions performed by him was the training of both Swedes and foreigners who became collectors and explorers. Many of them died in such distant places as South America, the Near East and Spain. Others listed the fauna and flora of North America and Japan. Perhaps the best known of Linnaeus' traveling pupils was Daniel Solander, who really merited the position as his assistant. It so happened, however, that just as Linnaeus was making his decision, Solander was invited to join James Cook's expedition in the *Endeavour* by Sir Joseph Banks. Since this offer carried with it an annuity of £400, Solander accepted. He is the link between Linnaeus and the first important expedition in natural science. Cook's voyage is significant for zoogegraphy and also leads us to consider the role of one of the great patrons of natural science, Joseph Banks.

11

Gentleman Amateur

ONE summer afternoon in 1758, a young Etonian went swimming with a group of friends. He came of a good family; he was handsome, of a lively disposition, and also a remarkably poor student of Latin and Greek. His friends decided to leave, but he prolonged his swim. As he strolled back to school, he was suddenly struck with the beauty of the afternoon. The flowers by the roadside took on a new splendor. He was aware of the teeming forms of life which populated the landscape. It was a turning point in the life of Joseph Banks (born 1743) who was to become in later years the arbiter of science in England.

Banks ties in with all that was going on in natural science in the latter part of the eighteenth century. He is an important example of the wealthy patron and gentleman amateur and throughout his life typified the virtues and the failings of his class. He is particularly remembered for his connection with James Cook and the beginnings of scientific exploration.

Banks paid the old wives who acted as folk herbalists to teach him the names of plants. At home he found a dog-eared book on botany and devoured it. He converted his schoolmates to his new enthusiasm, leading them on expeditions to collect plants, beetles and fossils. He attended Christ Church College of Oxford and discovered that the botany professor had not given a lecture in thirty-five years. Prodded by Banks, the old gentleman looked around and turned up a teacher from Cambridge whose salary young Banks paid. When his father died the following year, his mother moved to London. The future Earl of Sandwich was his neighbor, a young man with whom Banks fished and botanized and who was to be helpful later in his career. Banks came into his money (£600 a year) in 1764, bought a house in London, and met all the people worth knowing. Among them were

Thomas Pennant and Daines Barrington, Gilbert White's correspondents, and White himself. He also met Linnaeus' pupil, Daniel Solander, a plump, cheerful little man, who was cataloging collections according to his master's system and had become a general favorite among the best people. Although young Banks had written nothing, he was elected to the Royal Society. After all, he was rich, he was nice, and he was in a position to do things for science.

In 1766 a vessel was being sent to Newfoundland to investigate the fishing industry. Since Banks knew one of the officers, he was taken aboard, for he wanted to collect plants and meet "Indians."

He was seasick, saw icebergs, netted jellyfish and enjoyed the wild, imposing scenery. He also learned something about the Newfoundland Indians' method of scalping and even secured a scalp. He collected specimens, was bitten by mosquitoes, got a fever, gave up the flute for the guitar.

Back in London, he considered paying a visit to his "master," Linnaeus, but changed his mind and went botanizing in Wales instead.

In 1768 the Royal Society was anxious to send several astronomers to various points to observe the transit of Venus (across the sun's disk) in order to obtain data that would help calculate the distance between the earth and the sun. Since Venus would not be in transit again for another hundred years, the opportunity was not to be missed. The Society asked the crown for funds to send a ship to the Pacific. The request was granted, less on scientific grounds than because the government saw a chance to explore and possibly add new colonies to the British Empire. From then on, scientific exploration became an adjunct of the far-flung activities of colonialism.

The Royal Society had a certain Alexander Dalrymple in mind as a commander, a man with some scientific background but little seamanship. The Admiralty picked Captain James Cook, a solid, self-made man, one of the best navigators and map makers in England. Cook was also unique, since in an age in which seamen died like flies from scurvy, he had discovered the simple fact that fresh vegetables and, lacking these, sauerkraut would keep his men healthy. When they balked at sauerkraut, he served it only to the officers' mess and presently had his men demanding their share.

It was Banks who made of Cook's voyage in the *Endeavour* a natural science expedition. Thanks to the fact that his friend, Sandwich, was about to become First Lord of the Admiralty, he was allowed to join the expedition with his friend, Daniel Solander, and a party of eight, which included two artists and four servants, two of the latter blacks. Banks put up £10,000 toward the expenses of the expedition.

It meant close quarters for everyone, but Cook was an admirably balanced character, and Banks, at twenty-five, was resilient, highly enthusiastic, and not afraid of roughing it (as long as he had four servants). In a flurry of last-minute advice and excitement (Thomas Pennant advocated oilskin umbrellas for the tropics and impetuous Banks had gotten himself engaged to Miss Harriet Blosset) they were off. Miss Blosset grieved during Banks' absence and spoke of death.

The explorers, however, enjoyed themselves from the beginning. They fished and caught birds, cataloging them according to the Linnaean system. Cook and "the gentlemen," as he always referred to Banks and Solander, became good friends. Banks saw much to admire in the Yorkshireman, who had a special instinct for navigation, aside from having taught himself all there was of this science, and who handled his men with such firmness and moderation. Cook, on his side, learned to respect scientific investigation and was influenced by Banks' writing style, even occasionally borrowing from Banks' journal for his own account of the journey.

Banks, who didn't punctuate much, had nevertheless a lively talent for description. At night, the sea off the coast of Brazil, he wrote, was full of brilliant lights:

> They proved to be a species of medusa which when brought on board appeared like metal violently heated, emitting a white light—on the surface of the animal a spot was fixed the exact color of it which was almost transparent not unlike thin starch in which a small quantity of blue is dissolved. In taking these animals 3 or 4 species of Crabbs were taken, but very small, one of which gave full as much light as a glowworm in England . . . indeed the sea at night seemed to abound with light in an uncommon manner, as if every inhabitant furnished its share, which might have been the case tho none kept that after being brought out of the water, except these two.

Rio de Janeiro turned out to be frustration. The Portuguese Viceroy refused to believe the *Endeavour* was a government ship. Count Lauragais was sure the vessel was full of smugglers or spies and referred to the scientists as *"foutus philosophes,"* refusing to let anyone land. The *Endeavour*, with the aid of British diplomats, managed to refuel and reprovision while Cook and Banks drafted indignant letters. Solander and Banks sneaked off the vessel a couple of times at night and collected a few specimens. Needless to say, Banks reported that the colony was in a condition of "abject slavery."

Tierra del Fuego, a little-explored area, provided Banks' first glimpse of natives. The Indians (Onas) approached, bearing sticks in their hands which they threw away as a sign of peaceful intentions. Beads and ribbons were distributed. Some natives came aboard, and one, "who seemed to be a priest or conjuror," shouted as loud as he could every time he saw something new. When offered food, "they eat bread and beef" but would not touch liquor. Their houses consisted of poles lashed together tepee style and covered with grass. They had no furnishings, nothing but bows with flint-tipped arrows and woven satchels which the women used when they waded in the sea to dislodge shellfish with a stick. They went naked except for cloaks of guanaco skin. Their bodies were "of a reddish color nearly resembling rusty iron mixed with oil." The women wore a G-string, but the men "have no regard to that kind of decency." They wore shell bracelets and painted their faces in black-and-white stripes. They were without boats and, as Banks surmised, seminomadic. Actually, this was a pioneer collection of data. The unfortunate Onas were studied but little and were later exterminated by the kind offices of the missionaries around the turn of the nineteenth century.

On a trip inland, the first tragedy took place. Although the temperature was low, the weather seemed fine when they set out. A sudden blizzard set in and the cold became intense. They were obliged to struggle through thick undergrowth. Part of the group went ahead toward some woods to make a fire. The two blacks, Daniel Solander, and a seaman insisted they could not go on. Banks remonstrated that it was death to lie down. One black announced he "would lay down and die." Solander was finally got up and made to travel on. When they came to the fire, they revived and sent back for the others. The two blacks were past saving. Later it was discovered, since they had stolen and consumed a bottle of rum, that drunkenness had contributed to their death.

In 1769 the expedition anchored in Matavai Bay, which they called Port Royal. Natives came out in boats; the Europeans traded for provisions and finally went ashore. Crowds of Tahitians came to greet them, crouching low and carrying green boughs in their hands as a sign of peace. Since the French explorer Louis Antoine de Bougainville, had spent some weeks among them, they were not entirely unused to Europeans. Later some dignitaries came aboard the *Endeavour,* and finally Banks and Cook were taken to the old chief Tuteha. The chief's ugly wife waited on Banks, but he was only interested in a pretty girl whom he singled out in the crowd. Meanwhile, Solander and another member of the party had their pockets picked of a snuffbox and a pair of field glasses. After considerable commotion and threats, the articles were returned.

From this point on, the Europeans had a hard time keeping their tempers, for the Tahitians seemed to have no sense of property rights and took everything they could lay hands on. One man who snatched a musket from a soldier was shot. Thanks to Cook's discipline and self-control, such incidents were smoothed over, but poor Banks, after sleeping in a native hut one night, awoke to find all his clothes stolen.

Banks, however, had an eye for the young women and seems to have had an affair with a girl by the name of Otiatia. His journal abandons all attempts to record botanical or zoological discoveries and turns to ethnology. His descriptions of Polynesian life are vivid and interesting. With remarkable good nature he took part in a mourning ceremony:

> I was next prepared by stripping off my European clothes and putting on me a small strip of cloth around my waist, the only garment I was allowed to have, but I had no pretensions to be ashamed of my nakedness for neither of the women were a bit more covered than myself. They then began to smut me and themselves with charcoal and water, the Indian boy was completely black, the women and I as low as our shoulders.

They then met ten Indians, "ran at them and dispersed them." The procession crossed the river and passed among the houses which were all deserted. They then cried out, "There are no people!" and went back to the river to scrub each other. Banks and the others in the group were playing the part of mad fools, supposedly out of their minds from grief.

Cook had drawn up rules for barter, all of which was to be done through Banks, for the commander wanted to keep up the value of nails, which was an important commodity. Banks did an excellent job and learned a good bit of Polynesian, the only language he was ever to master. In his journal, he described surfing, the preparation of breadfruit, the Polynesian form of clambake, and the temples, or marae, sacred groves with a stone platform at one end, which was decorated with tall boards on which men were carved as if standing on each other's shoulders, memorials to departed chiefs.

It was in Tahiti that one of the artists, Buchan, had an epileptic fit and died. This was unfortunate for the recording of Polynesian people because he was supposed to draw "figures and clothing." The other artist, Parkinson, was an excellent botanical draftsman and ventured as far as birds, but did not handle the figure.

Banks made no mention of animal life in Tahiti but had more to say of New Zealand and Australia. The New Zealanders proved warlike, and although Tupia, a Tahitian the Europeans had with them, was able to communicate, a clash occurred during which four Maori were killed. Both Banks and Cook recorded this as a black day in their journals.

Banks believed that the dogs and rats seen in New Zealand were imports. "Of seals we have seen a few and one Sea Lion." He rightly decided there were no other quadrupeds. He mentioned ducks, hawks, owls, quails, albatrosses, shearwaters, pintados and penguins. He also noted butterflies, beetles, flesh flies, sand flies and mosquitoes. "For this scarcity of animals on land, the sea makes abundant recompense. Every creek and corner produces abundance." This included mackerel, hakes, breams, blue cod, lobsters or sea crayfish, and elephant fish.

After landing in what they named Botany Bay because of the abundance of plants, they saw a quadruped the size of a rabbit which was probably a bandicoot. Banks' description of a kangaroo is less adequate than Cook's: ". . . the full grown of it is as large as a sheep, yet it always goes on its hind legs. . . ." Cook tells us it was faster than the ship's greyhound:

> In form it is most like the jerboa but as big as a sheep. The head, neck and shoulders are very small in the proportion to the other parts. The tail is nearly as long as the body, thick near the rump, and tapering toward the end. Its progress is by successive hops, of a great length, in an erect posture. The forelegs are kept bent close to the breast. The head and ears bear a slight resemblance to those of a hare. The animal is called by the natives *kangaroo*.

By navigating the Endeavour Strait between Australia and New Guinea, Cook showed that these two great land areas were separate.

The voyage home was tragic, for the expedition was obliged to put in at Batavia to repair the ship and reprovision. Although Cook had not lost one man from scurvy, there the port was riddled with disease. Malaria and dysentery attacked the members of the party; of the scientists only Banks, Solander and the two white servants survived. Their prize Polynesian exhibit, Tupia, also died.

When the travelers returned to England, they were, of course, lionized. Meanwhile, Harriet Blosset waited in the country. Banks, it seems, distracted by his Polynesian amour, had lost interest. At length, she wrote a letter asking for an explanation. Was he planning to behave like a gentleman? Banks wrote back, protesting undying love but explaining he was not the marrying kind. Miss Blosset came to London, and there was an interview lasting all day during which the injured girl fainted. Banks promised to marry her immediately. Alas, no sooner did she return to the country than he wrote another letter, protesting his affection and again explaining his disposition was too volatile for him to settle down! The wits about town suggested that he should at least compensate Miss Blosset for the waistcoats she had embroidered for him during his absence. Likewise, the Grub Street hacks wrote satirical poems about the offspring Banks was supposed to have left behind in Tahiti. The Blossets let it be known that they would not take money to compensate for their injured feelings. They decided Solander had influenced Banks.

All in all, the expedition had been a landmark in natural science. One of the most positive achievements was a collection of 212 insects, which were given to a pupil of Linnaeus to catalog and duly included in the 1775 edition of the *Systema.* Linnaeus, however, was biting his nails with impatience for a full report of the project. Unfortunately, in this area lack of professionalism was a drawback. Cook's and Banks' journals were turned over to a stuffy academician who published a miserable selection of material overlaid with his own moralizing. The dried plants, at least, went into the Natural History Museum of South Kensington. A full report on the activities of the scientists was not made. Solander was supposed to write up the botanical collection on which he seems to have spent ten years of work. He died in 1782, and the hand-written folio volumes with the illustrations still repose in the British museum. Banks became president of the Royal Society in 1778 and seems to have neglected the whole matter. His journal, perhaps out of a kind of gentlemanly feeling that publishing anything so intimate was vulgar, was not printed. A cut version was brought out in 1896, and not until 1962 was it finally properly edited. If Banks had published, on the basis of his Tahitian and New Zealand material, he could well have earned the position of the world's first field anthropologist. Cook's excellent diary did not see the light of day until 1893.

Nevertheless, a precedent had been set. Government funds had partially supported a scientific expedition. The pattern of exploration and collection of specimens paved the way for Cook's second voyage, the botanizing of Matthew Flinders in Australia, eventually Darwin's celebrated voyage in the *Beagle*, and finally the Challenger Expedition of 1872-76 which studied oceanography.

Banks, when he became old and gouty, was considered a rather despotic president of the Royal Society. Nevertheless, he was personally helpful to scientists all over the world. Even during wars, he endeavored to gain the release of those who were interned in enemy countries. He made his house in Soho Square a gathering place for intellectuals and men of distinction. At his Thursday morning breakfasts, where coffee and rolls were served and scientific journals lay around on tables, such men as the anatomist John Hunter, the chemist Joseph Priestley, the botanist Matthew Flinders, distinguished foreigners and travelers like Mungo Park gathered. On his second voyage, Cook brought back a Tahitian, Omai, on whom Banks kept a paternal eye. When Omai got lost, he found that all he needed to do was to mention the name Banks to a passerby and he would be taken to Soho Square.

Banks' role as middleman in the securing of the Linnaean collection has already been mentioned. He was made a baronet in 1781, and after his death in 1820 he bequeathed his house and library to the botanist Robert Brown, who appropriately rented quarters in it to the Linnanean Society.

The Banks to be remembered is the figure in the portrait by Sir Joshua Reynolds, painted shortly after the return of the *Endeavour*. He is handsome, alert, his hair unpowdered; the globe of the world is behind him. It is easy to imagine that he is about to leap out of his chair and rush off on another expedition to exotic parts of the earth.

12

Giraffe in the Hallway

☙ AMONG those who joined the convivial group at Sir Joseph Banks' breakfasts was a rough-hewn Scotsman whose earthly personality triumphantly proclaimed his rural background. He once remarked that people had tried to teach him Greek and Latin, but he got rid of such pests very quickly and illustrated his remarks by the pantomime of cracking a louse on his fingernail.

This was John Hunter, the most famous London surgeon of his time, who, together with Petrus Camper of Holland, once and for all established the technique of comparative anatomy.

Hunter was considered a dull child by the Scottish farm family into which he was born in 1728. Since he was not making much headway in the Lanerk school, he was taken out when he had acquired only the rudiments of an education and put to work in the fields. As a boy, he became interested in nature study, but he was trained for no profession, and finally, in hopes of making him an artisan, he was sent to his brother-in-law, a cabinetmaker, to learn the trade. When his brother-in-law went bankrupt, John returned home with still no foreseeable future. Fortunately, an older brother, William, had become a successful doctor in London and had been commissioned to give examinations to prospective army surgeons. William was kind enough to send for his brother, who became his assistant in 1748. John proved to have a natural flair for anatomy. Entirely self-taught, except for what coaching he got from William, he soon took over the direction of the anatomy course. He even gave lectures (illegally, for he had no degree), but in 1753 he was elected master of anatomy at Surgeon Hall. In 1756 he was admitted to Oxford University but lasted only two months, for it was there that Greek and Latin proved too much for him. His enemy, Jesse Foot, maintained that he could not write six grammatical sentences and his spelling was admittedly atrocious. His lecture notes were scribbled on scraps of paper, which he was often unable to read. When this happened, he told his students to forget what he had been saying and immediately embarked on a new subject. Yet all along, as a practicing surgeon and certainly by reading, although he pretended to despise books, he perfected his medical education sufficiently so that he was appointed house surgeon to St. George's Hospital in 1756.

During the Seven Years' War he was active as a surgeon attached to the English fleet. Back in London, he resumed his practice and his connection with St. George's Hospital and was elected to the Royal Society. Unfortunately, in experiments with syphilis, he deliberately inoculated himself with the disease, which was treated with mercury but left him with a debilitated heart. By 1770 his fame had spread to such an extent that he was appointed Surgeon Extraordinary to the king, and the following year he married Anne Home, a poet who is remembered for the graceful song "My mother bids me bind my hair."

Hunter's London was the city of Thomas Gainsborough and Sir Joshua Reynolds (both painted his portrait), but it was also for doctors a city with a brutal underworld, familiar to us from William Hogarth's pictures. Gin was the chief remedy for all ills, and credulity was such that there was an official investigation into the case of a woman who was reported to have given birth to fifteen rabbits.

The red-haired Scotsman was a man of strong passions, who fought with colleagues and gave them the rough side of his tongue. In consequence, he made many enemies. When Jesse Foot, a rival surgeon, maliciously referred to his failure to acquire Greek and Latin, Hunter retorted that he knew no dead languages but he knew more about dead bodies than Foot did in any language.

In later life, he quarreled with his own brother to whom he owed so much, claiming that William did not give him credit for his contributions to a treatise, *The Anatomy of the Gravid Uterus.*

Hunter took up comparative anatomy fairly early in his career and obtained the concession to dissect all the animals that died in the Tower Zoo (the ancestor of the London Zoo). He wrote a treatise on the hearing of fish, having proved that they possessed this faculty by lying down to observe fish in a pond while a friend fired a gun. He also wrote a paper discussing the fact that dogs, wolves and jackals could interbreed. He wrote on earthworms and studied the development of the dragonfly. His country estate at Earl's Court was a regular zoo, for at the gate there were three cages which contained dangerous wild animals, such as leopards. Other creatures in residence were jackals, zebras, and bulls. In a special pond, he bred leeches, eels, mussels, frogs and various kinds of fish. The numerous quadrupeds plus a number of birds kept the place in an uproar. A picturesque touch was a pair of buffalo which he was able to harness and drive through the streets of London.

During his lifetime, Hunter dissected five hundred different species of animals and the culmination was the investigation of a whale. At his own expense, he had sent a surgeon on a whaling vessel to Greenland. When the seventeen-foot specimen arrived at Earl's Court, he was in his element. He wrote: "The aorta of a sperm whale measured a foot in diameter . . . when we figure to ourselves that probably ten or fifteen gallons of blood are thrown out at a stroke . . . the whole idea fills the mind with wonder."

The whale was not the only strange specimen brought to Earl's Court. Hunter had had his eye on O'Brien, an eight-foot Irish giant. The poor man, finding his end near and wishing to preserve his body intact, gave orders that he was to be placed in a lead coffin and sunk at sea. Hunter bribed the undertaker, smuggled the body down to his country house at night, and rapidly boiled it for the bones. These, somewhat brown, because of the rapid preparation, are still in existence in the College of Surgeons.

The problem of obtaining enough bodies for medical dissection continued into the eighteenth century, when anatomists had recourse to professional "body-snatchers," who profaned graveyards. Sometimes there was violent public protest. Indeed, a mob in the seventies once threatened to burn down Hunter's anatomy theater because he was accused of body-snatching.

Although Hunter was no theorist, he did take a position on a controversy which has run through all biology since the time of Descartes. It was the latter who first introduced the machine model as an approach to living bodies. Following the same line of thinking was Giovanni Alfonso Borelli, with an engineering treatment of the human body, and Hermann Boerhaave, who approached physiology from the point of view of chemistry. Said Boerhaave, who was a Newtonian and ascribed all motion to gravity: ". . . The motion, which we find in the world, is always dwindling and on the decay: so that there arises the necessity of recruiting it by active principles: such are the cause of gravity by which planets and comets keep their course and bodies acquire motion in falling: such the cause of fermentation by which the heart and the blood of animals are kept in perpetual motion. . . ."

On the whole, mechanists like Boerhaave generally kept religion in a separate compartment and paid lip service to the effect of the "soul" on the body. The sincere vitalists spoke of "life force" or sought for some special principle over and above mechanical processes. Hunter was one of the latter. He felt the life force had something to do with heat, since organic bodies become cold at death. He also wrote: "The actions and productions of actions both in vegetable and animal bodies have hitherto been considered so much under the prepossessions of chemical and mechanical philosophy that physiologists have entirely lost sight of life." He thus held that it was something common to all organic bodies, independent of structure. In a sense, he was groping toward the concept of protoplasm, although the word had not yet been invented.

The culmination of Hunter's life was the formation of his famous museum in the house on Leicester Square. Thanks to his fame, which brought him international contacts (and sometimes controversies, notably with Albrecht von Haller and Lazzaro Spallanzani, whom we shall be discussing), and his friendship with Banks, he was able to obtain specimens for his natural history collection from all over the world. He dedicated himself to building this monument, spending his money lavishly to obtain exotic animals. Moreover, he made preparations of every part of the human body "both wet and dry" and displayed them with the corresponding parts of whatever other animals possessed them. Thus he truly put his comparative approach into practice. He arranged his exhibits on an ascending scale from the polyp up to man. The Leicester Square house became so full of specimens that he had to cut off the legs of a stuffed giraffe and install it in a hallway. He also included as many fossils as he could get for his collection. The curator of this collection was one William Clift, a simple, dedicated soul, who arrived at the museum at the age of seventeen and never left. When Hunter died from a heart attack after becoming Surgeon General in 1793, his enemy, Foot, rushed into print with a poison-pen biography, which sought to prove that Hunter was the greatest fraud in history.

Hunter had willed that his collection be offered to the British government. The Prime Minister, the younger Pitt, had the true politician's reaction. How could the government buy a museum when it did not have enough money to buy gunpowder?

Most of the estate was tied up in the collection. Mrs. Hunter had little to live on as she waited for a buyer. She was able to allow Clift only seven shillings a week, but this devoted man for six years kept the collection in better order than it was when his master died until, with the aid of Sir Joseph Banks' influence, the government reconsidered and finally bought the material for £15,000. Clift had copied many of the manuscripts and it was well he did so, for Hunter's jealous and unscrupulous brother-in-law, also a surgeon, had been made executor. The wretch plagiarized some of the papers and published the gist under his own name, burning the rest to hide his depredations. Thus four volumes of notes on the dissection of animals were destroyed.

The collection was given to the College of Surgeons, where it became a model for the well-organized museum. Clift was appointed curator and stayed on until his death in 1849. Over the years, the museum doubled in size, only to suffer badly in 1941 during the bombing of London.

The wealthy and well-educated Petrus (or Pieter) Camper met Hunter on his first trip to England in 1748. The two anatomists were contrasting types. Camper's father came from a well-to-do Dutch merchant family and had been minister to Batavia. Pieter had the advantage of Hermann Boerhaave's friendship at an early age, and to his home came painters and other men of distinction. The young Camper first studied design and architecture, becoming a capable artist and always occupying his spare time painting in oil. Although Boerhaave was now old and ill, he nevertheless must have been an influence, for Pieter turned to medicine and took his degree in 1746 at Leiden. His London trip amounted to a kind of graduate study, for he visited famous surgeons and studied art for a time in that city. In Paris he also made contact with people in his field, including, of course, Buffon. He returned to Holland after spending some time in Germany to become professor of surgery at the University of Franeker in Friesland.

Camper was a polished, attractive, bewigged eighteenth-century gentleman, who spoke English, French and German fluently and knew his Latin and Greek. As his reputation grew, he accepted a professorship at Amsterdam and then one in Groningen, but he had become attached to Friesland, having married Jeanne Bourboom, the widow of the burgomaster of Harlingen. His first volume on anatomy was published in 1761, at which time he retired from the university to his Friesland estate. For a time, he was a deputy to the State Assembly and returned to academic life with a professorship at Groningen in 1763. He was also made an honorary member of the Academy of Design in Amsterdam.

During an epidemic of hoof-and-mouth disease in Holland, he set up a medical group to inoculate cattle. Unfortunately, this type of progressive medicine was beyond the comprehension of Dutch peasants. One of his medical officers was threatened by them and had to flee for his life. Subsequently, Camper published a pamphlet on the disease, hoping to educate the public.

A contribution to physical anthropology resulted from a speech he made to the Academy of Design, in which he discussed the measurement of the facial angle as a way of identifying different ethnic strains. This was a first attempt to use the skull shape as an anthropological diagnosis and was later to result in interminable head-measuring.

In 1786 Camper traveled in England, meeting Sir Joseph Banks and Hunter and, of course, visiting Hunter's museum. He had had a mild controversy with Hunter in which he pointed out that he had first discovered air in the hollow portions of birds' bones, even in the head and mandibles, which Hunter had not noticed. While in England, he read a paper on fossil fish to the Royal Society. Camper emerges as a handsome, highly moral pillar of the community. He died in 1789.

In his comparative anatomy, he made good use of his talent for drawing. In a talk before the Academy of Design, which is published in his collected works, he said: "A new Proteus, I will show you how one could by modifying certain traits metamorphose a cow into a horse, into a dog, into a stork and from the stork to a carp or any other species of fish." This is illustrated by a number of sketches in which he superimposed changes on various animal types. Although he did not carry the point out in great detail, he was nevertheless breaking ground in an area which was to be worked over by Étienne Geoffroy Saint-Hilaire in France and the nature philosophers in Germany. It was an emphasis on fundamental organic structural types which led away from special creation of species and in the direction of evolutionary concepts.

Camper did considerable work on the anatomy of the elephant. While Buffon was speculating whether a young elephant nursed with its trunk, Camper was stating that from the nature of the lips and tongue he knew the young pachyderm was able to suck from a teat.

Most interesting, in the light of his period, is his discussion of the orangutan. It should be remembered that even up into the nineteenth century a good deal of uncertainty existed as to the relation between man and the higher primates, and even as the tales of the abominable snowman persist today, there was a notion that some sort of wild man existed and this was confused with not very accurate reports concerning gorillas and orangutans. The situation was aggravated by the unsettling opinions of Julien Offroy de La Mettrie (1709-51), a French philosophical scientist who studied medicine under Boerhaave. Mettrie, a convinced atheist and Cartesian mechanist, expressed such heretical views in print that even Holland would not tolerate him, and he finally took refuge in the court of the sophisticated Friedrich the Great of Prussia. A complete *enfant terrible,* not only did he challenge the immortality of the soul but maintained that the orangutan was a kind of human being whom it should be possible to civilize. This enraged the pious just as much as Darwin's theory of the descent of man was to do decades later.

Camper, in his discussion of the orangutan (or pygmy as it was still being called), argued that from the shape of its backbone it was not formed to walk upright. He also pointed out that the skull revealed the intermaxillary bone which was not identifiable in the adult human skull. Camper spent some time on a hand a foot long, which was said to be that of a "savage man" from Batavia, supposedly proving that there was some type of giant hominoid in that area. Camper announced that it was a distorted animal's paw, probably a bear's, the nails were added bits of horn and the wrist bone was an artificial addition. Since the Academy of Science of Rotterdam owned the specimen, the spokesmen of this institution insisted that all European savants declared it authentic. Camper replied by asking permission to moisten it in order to distinguish the form more clearly. This he was not allowed to do. He wrote frostily: "Is the decision of a small number of people, little informed concerning natural history, sufficient to proclaim an authoritarian decision concerning such an interesting question?"

Camper made a progressive contribution to anthropology when he wrote concerning the color of the black human type: "Why should we look for a difference in race in mere color, which depends only on a slight alteration of the epidermis?" He was intent on disproving another contemporary myth, that blacks were a hybrid offspring of a white human and an orangutan or else of a white human and a "satyr."

His psychobiological remarks on the human race are quaintly

amusing. Philosophers, he believed, were, on the whole, long-lived in spite of the fact that their sedentary pursuits prevented sufficient voiding of phlogistic air and absorption of vital air. (Of phlogiston we shall hear more later.) He took a dim view of clerics, however: "Living a solitary life in which misplaced zeal leads them to renounce the sweet bond of conjugal union, and since they do not satisfy the impulses of nature, they fall into a profound melancholy which often degenerates into frenzy."

13

The Elegant Popularizer

☙ IN 1662, Louis XIV built a zoo on the Saint-Cyr road. It was planned by an architect in the shape of an open fan with a little summer house at the apex of the handle. Around this was an eight-sided court full of fountains. Farther out, on the ribs of the fan, were pie-shaped bird and animal courts. In the animal courts were fountains and little cabins in which the beasts could repose and there were tanks of water for the water birds. The first elephant was a gift from the King of Portugal. It was fed eighty pounds of bread a day, twelve pints of wine and two caldrons of potage. It was also given a shock of wheat from which it ate the grain, using the straw to brush itself. We wonder if it was always in a state of amiable inebriation from the wine.

The Versailles zoo was a pioneer institution which, in the second half of the seventeenth century, was open to both scholars and the public. It contained fifty-five mammals, seventeen ground birds and doves, twenty parakeets, fifty-one birds of passage, twenty-nine shore birds, thirty-nine runners and five reptiles. When new animals were received, they were painted, drawn or recorded in miniature on vellum by the official artists, Nicolas Rober and Jean Joubert.

Animal anatomical studies were stimulated by the founding of zoos. As far as the rarer animals went, Fabio Colonna in Italy dissected a hippopotamus for the first time, while Harvey in England had the privilege of investigating an ostrich. In 1669 the French Academy began to work on animals which died in the Versailles zoo, beginning with a beaver. When the famous elephant died in 1681, three scholars took over the mighty corpse: One dissected, one wrote up notes, one drew. When the king appeared one day and looked for the anatomist, he could not find him. Presently he emerged from within the elephant.

Versailles was the precursor of the famous Jardin du Roi, which was also to blossom into a zoo. One of the greatest figures of zoology is connected with its early history, Georges Louis Leclerc, who also assumed the name Buffon by which he is known.

Buffon is a strong contrast to the sober doctors, professors, and clergymen who have so far busied themselves with zoology. The rich Burgundian bourgeois, who loved money and sensual pleasure, who cut a dash in society, was a true eighteenth-century gallant. All his life, he did as he pleased and regarded moral verities as mere conventions. He lived grandly, could be at times grasping and greedy, at times tolerant and generous.

His tall, imposing figure with the rugged features and humorously arrogant expression has been recorded in various busts and portraits. Above all, this many-sided character could be a loyal friend, slippery in money matters, and rather a devil with the women.

His ancestors were mostly doctors and judges. His mother's uncle grew rich from holding the office of tax collector for the King of Sicily. When only seven, Georges Buffon inherited a couple of million francs from his mother's side of the family. Born in Montbard in 1707, he attended the Jesuit college in Dijon. On the whole, he liked sports and mathematics but was not much of a student. He did, however, become a licentiate in law and thereafter went to Angiers to study medicine. Medicine led to the study of botany, but before Buffon finished his studies, he got involved in a love affair which led to a duel with an officer whom Buffon unfortunately killed. To avoid being prosecuted, he fled to Nantes. There he happened to fall in with a kindred soul, the wealthy Duke of Kingston. Kingston was doing the continent with a tutor, a German by the name of Hickman, who took an interest in natural history. Buffon studied dissipation with the duke and natural science with the tutor. He also picked up some English. The trio traveled around for six months. Buffon said that the wine at Nantes was good, but the inhabitants were all shopkeepers. In Toulouse, he never saw an ugly woman. In Rome, the opera was magnificent, the theater agreeable.

It was while he was traveling that his mother died in 1732, an event which brought him home. Buffon then lived mostly in Paris, for he began to be on bad terms with his father who, at fifty, was courting a girl of twenty-two. Antoinette Nadault, in Georges' opinion, had nothing but youth in her favor. His father went ahead and robbed the cradle anyway. Georges sued for what was due to him from his mother's estate and remained on bad terms with his father until 1771. Finally they made up; Buffon was designated his heir and the prodigal son returned home.

Buffon had been dabbling as a dilettante in science; with the aid of influential friends and a paper on the laws of chance (no doubt, inspired by gambling in company with the duke), he got himself elected to the Academy of Sciences. Having a shrewd eye for profit, he made himself an expert on wood, performing various experiments to discover its properties, because he wished to sell lumber to the Navy. He published a treatise on wood with a collaborator, who accused him subsequently of plagiarism. Buffon got out of the charge somehow and, in 1737, went to England to help his friend, the duke, blackmail a certain Monsieur de la Touche into silence. The duke had stolen the unfortunate man's wife. Buffon, as is to be expected, became a member of the Royal Society. During his year in London, he studied mathematics, physics and botany. He also translated Isaac Newton's *Fluxions.*

Back home again, Buffon became a member of the botanical section of the Royal Academy of Sciences. About this time, he was also taking an interest in theories of reproduction.

In 1739 the director of the Jardin du Roi, which consisted of a botanical garden and a natural history museum, fell ill. The directorship of this institution was an important scientific post. One of the king's ministers had promised it to Henri Louis Duhamel de Monceau, Buffon's disgusted collaborator. On the other hand, the logical successor to the incumbent, Dr. Charles Du Fay, was Pierre Louis Maupertuis. The latter was a physicist who, after a trip to England, became interested in Newton's revolutionary new concepts. It was he who tutored Voltaire and Marquise du Chatelet in the new physics and made of them stalwart Newtonians. Maupertuis was interested in astronomy and in the current controversy as to the shape of the earth. The new thinkers believed that the globe was flattened at the poles. In order to substantiate this theory, meridial arcs had to be measured at various points. A series of measurements had already been made in Peru, near the equator. Now it was necessary to work in the polar region. The king, being convinced that the data would be of use in navigation, appointed Maupertuis to lead an expedition to Lapland to make the necessary measurements. Maupertuis showed himself an admirable pioneer. He ran rapids in light boats, slept on the ground wrapped in reindeer skins, lived on fish and wild fruit, suffered torture from flies, and at times was completely blocked in his operation by persistent fogs. Nevertheless, to while away the time, he wrote bad verse to a Finnish girl with whom he was having an affair. Finally, he got his measurements, was nearly shipwrecked on the way home, but arrived in triumph with the data which effectually established the shape of the earth and validated Newton's theories.

Maupertuis' polar expedition took place in 1736-37. Like many distinguished men, he had jealous rivals. Buffon used pull, Maupertuis was passed over, and taking advantage of the fact that Duhamel was in England, Buffon saw to it that the dying Dr. Du Fay was pressured into signing a letter, designating Buffon his successor. Poor Maupertuis, who never received recognition in France, went off to Berlin to found Friedrich the Great's Academy and to become an unhappy pawn in the quarrel between Voltaire and his imperial patron. In the biological area he is to be credited with the belief that new species could be gradually formed.

The post of director of the Jardin du Roi made Buffon a real scientist. He took his duties seriously. One of his first tasks was to draw up a catalog of the collections, which comprised animal and plant specimens, "costumes of savages," fruits, minerals and arms. The material was hung on walls or kept in drawers. The institution had been founded in 1635 as a pharmacological herb garden; Buffon set out to make it a headquarters of natural science. When he became director, some skeletons were lying in a heap on the floor and no particular system of display was used. Buffon wrote all over the world to travelers, explorers and doctors for new material. He also rearranged the displays. His method was more esthetic by modern standards than instructive. Insects were next to birds, eggs alongside the bivalves and dried fish, while reptiles peered out from beneath the legs of a stuffed zebra.

The catalog of the collection was the germ of the work that has made Buffon a great name in zoology. It was a *Histoire naturelle,* the first part published in 1749, once more an updating of Aristotle, planned as a work in fifteen volumes. Buffon was overoptimistic; when he died, he had published thirty-five volumes and thirty-six more were in preparation.

Buffon started off with man treated as an animal. Since he freely discussed puberty, reproduction, sterility, eunuchs and such matters, his book succeeded in arousing a slightly scandalous interest. Well illustrated with colored engravings and written in Buffon's polished periods, it became a best seller; fashionable ladies used it as bedside reading. Indeed Buffon, through the strength of his personality and the force of his position, plus his undoubted literary gifts, made animals popular and brought them into French society. Buffon's collaborator in this undertaking was Louis Daubenton (1716-1800), the leading French anatomist, who contributed anatomical descriptions of the animals until 1770. Buffon, however, was responsible for the general style and presentation. He began with his theories of life and the history of the world.

In this he was fairly revolutionary. It must be remembered that Buffon was a child of the eighteenth-century Enlightenment. Owing to his sojourn in England, he had become a Newtonian and the work of this cloistered, absentminded scholar who lectured at Cambridge, in rundown shoes, worn-out stockings, with uncombed hair, sometimes happily to an empty room, was crucial in the formation of the nascent modern temper. While Newton made no contribution to biology, the tangential effect of his thought marks a new epoch in Western scientific culture.

Born in 1642, the year of Galileo's death, Newton had come to Cambridge from a farm after a sickly, unpromising, rejected childhood. In the university he made such progress as a mathematician under the mathematician-astronomer Isaac Barrow that the master resigned in order to allow the pupil to take his post as soon as Newton obtained his degree. The single-minded, slightly cantankerous and ambitious Newton, in order to avoid the plague which raged from 1665 to 1666, retired to the country for eighteen months. In this short period, the basis of all his important discoveries was worked out. He perfected a method for differential calculus, the *Fluxions,* which was translated by Buffon; he discovered the binomial theorem; he perfected his atomic theory of color, developed integral calculus and began to think about gravity affecting the moon and the earth.

There is a parallel with Harvey, who gathered up much of what was already known and put the pieces together. The notion of gravity was not new. Johannes Kepler, arguing from the ebbing and flowing of the tides, had pointed out that the attraction between the moon and the earth was related to the masses of the two bodies. Unfortunately, he buried his best insights in mysticism. Besides, Cartesian cosmology, which was in the ascendant, determinedly refused to admit attraction at a distance. Such was the force of Descartes' reputation that his mechanistic speculation (he never attempted experimental proof) stood in the way of progress. About the same year that Newton was beginning to think about motion and attraction, the Italian Giovanni Alfonso Borelli talked of a natural tendency for satellites to approach the central body, carefully avoiding the prohibited term "attraction." Newton published his *Naturalis Principia Mathematica* (Mathematical Principles of Nature) in 1687, but it took fifty years for Newton's concepts to spread to the continent and to break the restricting formula of Cartesianism. Newton had revived atomic theory in modified form. What we now describe as waves of light, he envisioned more or less as a kind of atomic bombardment. Indeed, he considered the ultimate particles of matter to be held together by gravitational attraction.

Attraction as a basis for cosmology and for the structure of matter and the basis of life processes appealed particularly to Buffon. He was profoundly interested in love, both personally and in the animal kingdom. And what was love but sexual attraction? He wrote:

> Love! Innate desire of nature's soul! Inexhaustible principle of existence, sovereign power that is all-prevailing; and against which nothing prevails; through which everything is accomplished, all breathes, all is renewed! Divine flame, germ of perpetuity that the Eternal One has spread wherever the breath of life exists!

When Buffon described his view of the history of the world, he showed courage and originality. He pointed out that the earth was only partially investigated, that its composition was beginning to be known, and that for the first time, the various epochs of the material deposited in the past could be compared.

The oldest and most basic stuff was glass, vitreous material. This can be translated as igneous rock, whose shape and distribution he correctly attributed to fire, the earth's internal heat. Upon this, calcareous material was laid down by water. Finally, a third type of rock, such as marble, was formed from the debris of other stony material. These three divisions of the earth's rocks have remained basic in geology: igneous, sedimentary and metamorphic.

Buffon was also aware that elephant and hippopotamus skeletons were to be found in northern Europe and in North America, where no live specimens existed in his time. He also knew that fossil species of various life forms had been found which differed from the contemporary. He therefore went further than Steenson and showed a willingness to accept change, new species arising and possibly some dying out. He also ignored the Mosaic six-day account of the Creation and the more generous Bishop Ussher's chronology of six thousand years.

In a sense, by his flexible attitude Buffon was paving the way for the concept of evolution. (Maupertuis, we pointed out, went further and believed that a new species could be formed by gradual change.)

It has been said that the great conceptualizations of nature came late in biology. Preoccupied with description and classification, the zoologists achieved nothing comparable to the monumental achievements of Galileo and Newton. But here and there, as in Ray's puzzlement over fossils or Buffon's bold geological history, steps were being taken which would eventually lead to more profound contemplation of the nature of life.

Actually, the theoretical matter in Buffon's book brought down censure from the theological faculty of the Sorbonne because of the following statements:

> That seas produced mountains.
> That planets have been part of the sun.
> That the sun would eventually burn out.
> That there were several kinds of truth.
> That outside of mathematics and physics
> there is only seeming and probability.

Buffon published a polished disclaimer in a following volume, saying that he submitted to the authority of the Bible and the Church. In private, he said the whole thing was a silly comedy. His attitude toward the Church was mocking. Like many other leading figures of the Enlightenment, he was probably at bottom an atheist. The rationalism of the eighteenth century put its faith in the discovery of laws; in areas of religion it was not emotionally involved. Indeed, the astronomer Pierre Laplace did not mention God in his cosmology. The sharp skepticism of Voltaire and his attacks on the contemporary establishment were typical of the temper of the times, and although the Church, in various places, still managed to burn heretics, and he, himself, had to move from place to place to avoid harassment, the ecclesiastical monopoly was forever broken.

Buffon's essays on the animals included in his books were lively and colorful. The notes on the hedgehog, for instance, of which he had firsthand knowledge are very effective:

> The fox knows a great many things, the hedgehog but one, is a classical proverb. He knows how to defend himself without fighting and to wound without attacking. Having little strength and no swiftness in flight, nature has given him a spiny armor with the

> ability to roll himself into a ball in order to present his pointed defenses on all sides and repulse his enemies. The more they torment him, the tighter he rolls himself and the more his spines bristle.

The threatened hedgehog also releases his urine to cause disgust in his attackers. Because of the spines, Buffon said, the little animals couple face-to-face or standing up on their hind legs. In June, he often was given a mother with three or four young. The spines were just beginning to show. He put the family into a cask, but the mother ate the young. He accused the hedgehog of being malicious, for one got into his kitchen, took food out of a pot, and defecated upon it. When some were liberated in his garden, they made no trouble and subsisted on fallen fruit. "They ate cockchafers, beetles, crickets, worms and some roots; they delight in meat, eating it raw or cooked." In a natural state, they could be found in holes in rocks and in tree trunks. "I do not believe that they climb trees, as naturalists say, nor carry fruits and grapes on their spines." This last was an old fable from Pliny. The animals were supposed to roll on grapes in order to pick them up with their spines.

On Buffon's list of virtuous animals was the mole, whose sexual prowess aroused him to great enthusiasm. He criticized the bat for being ungainly, but praised the lion for its appearance and various noble character traits. The elephant was also extremely sympathetic:

> The elephant, by means of his trunk, which serves as both hand and arm and with which he can pick up large as well as small objects, carry them to his mouth, put them on his back, hold them firmly or throw them to a distance, has therefore the facility of the monkey and, at the same time, the docility of the dog, like the latter he is able to remember and capable of strong attachments; he gets used to man easily, submits to both force and kind treatment and serves with zeal, fidelity and intelligence.

Since Buffon had aristocratic pretensions and finally got to be a count (he threw down the family mansion and rebuilt it on a grander scale, using the ruins of the castle of the Dukes of Burgundy, which did not belong to him), he was happy to see the elephant in the role of faithful retainer. In this connection, he thought that even religion, for which he had no use personally, functioned well if it kept the lower classes in their place.

The elephant inspired some of Buffon's typically majestic, if somewhat ponderous, prose:

> We must remember that his steps shake the earth, that he pulls up trees with his hand [trunk] and with a blow of his body he can breach a wall; that since his strength is terrible he is invincible through the sheer resistance of his bulk and the thick leather with which he is covered; that in war he can carry an armed tower on his back containing several men, that he alone can move machines and carry loads that six horses can not move; that to this prodigious power he adds courage and strict obedience, and that he exercises moderation even in the liveliest passion, that he is more constant than impetuous in love, in anger he never fails to recognize his friends, that he only attacks those who injure him, that he remembers favors as long as injuries, that he has no taste for meat, eating only vegetables, that he is not an enemy of other animals, that, in short, he is loved by all since all respect him and none have reason to fear him.

Buffon repeated a story told by the "ancients" that elephants were so clever that they picked up grass trodden on by a hunter and handed it around so that all should be aware of the enemy's odor. Actually, some recent investigations of the elephant's intelligence, which indicate it can remember up to twenty-eight different symbols, suggest that intellectually it compares favorably to the overpublicized dolphins, although the latter seem to try to communicate with man.

Buffon speculated on the love life of the pachyderms. The coupling of elephants had been witnessed, for in this activity they ceased to be gregarious, the involved couple withdrawing into privacy. "They above all fear to be seen by their fellows, knowing better than we perhaps the pure pleasure of satisfaction in silence, occupied only with the beloved object." The classical authors speculated upon the technique of elephant coition, the *Physiologus* even declaring that the female entered the water so that the male could bouyantly mount upon her back. Buffon had another idea. He felt that the only possible solution was for the female to lie upon her back. This was proved by the need for privacy, since the posture was rather indecent and the natural modesty of the female dictated her withdrawal. The naturalist also speculated on whether the young nursed by trunk or by mouth and affirmed that the elephant's cry proceeded from the mouth and not the trunk. He ended with the information that the pachyderm loved liquor and "has such a horror of the pig that the animal's cry alone disturbs him and puts him to flight."

Buffon's information on seals was accurate, and he was even able to provide a good description of the three South American members of the camel tribe, the llama, alpaca and vicuña. His excursion into physical anthropology did not add much that was new, but his approach, treating man as a member of the zoological series, was fresh. He summarized what was known of anatomy and embryology and added a discussion of mental development which amounted to a primitive kind of psychology—for instance, he presented a lyrical description of the child's growing awareness of the world. Although he credited man with a soul, he also made such rationalistic statements as when (comparing man to the animals) he wrote: "We are superior to the animal only by a few characteristics granted us by the tongue and the hand."

In relation to the other primates he perpetuated a curious bit of sensationalism which had been handed down by travelers. These creatures were close to man because of "the vehement appetite of monkeys for women, the same conformation in the genital parts of both sexes, the periodic flow in the female, the crossings between the Negresses and monkeys, whose offspring join either one species or the other. . . ."

Among the animals whom Buffon felt to be a liability were the tiger and the cat. The tiger was unreliable and cruel; besides, its body was too long and its eyes were haggard. As for the cat:

> Although these animals are not without charm, above all when they are young, they possess innate malice, a false character, a perverse nature which age augments and education can do no more than mask.

Moreover, they were thieves, flatterers and above all could not look a man in the eye. Buffon was particularly scandalized by the female's lack of natural modesty:

> The female cat is more ardent than the male. She invites him, she seeks him out, she calls him, she announces the fury of her desires or rather the excess of her need, with loud cries and when the male evades or repulses her, she pursues him, bites him and forces him, so to speak, to give satisfaction even though the approach is always accompanied by intense pain.

In 1752, Buffon married Marie Françoise de Saint Belin-Malin, a sweet, pretty girl, only twenty and of a noble family. She made him a peaceful, obedient, amiable wife. Indeed, she seems to have been in a class with the more admirable types of animals. She had much to bear, for Buffon's trips to Paris were not without the solace of many light loves. One day his gaze fell upon Marie Blesseau, a pretty peasant girl who was working in the garden of his estate. Gravitational attraction took over. He installed her in the house and gave her a "lighter task." There were perhaps extenuating circumstances to account for Buffon's behavior. His wife had a fall from a horse two years after their marriage and remained an invalid until her death in 1769. She did, however, give him a son. She felt wounded at the Marie Blesseau affair and the parish curé was indignant, but Buffon, who had snapped his fingers at the Church, was capable of thumbing his nose at convention. He was rich and famous and made his own rules.

The volumes on birds were begun in 1765. After he broke with Daubenton (who did nothing of importance on his own), Buffon took as a collaborator Guéneau de Montbéliard, and when the latter became ill in 1776, he was replaced with the Abbé Gabriel Bexon. In every case, Buffon rewrote and polished his collaborators' contributions.

Buffon, as we have pointed out, accepted the modified atomic or corpuscular theory put forward by Newton. He applied it to living beings and it influenced his theory of reproduction. He knew that a piece of hydra could give rise to a new hydra. He believed it would take millions of his atomic particles to form such a bud. Since he did not accept the preformation theory, which we shall be discussing later, he worked out his own notion of pangenesis. His atoms were accumulated from all organs of the body and were then concentrated in the genital organs. The particles were first taken into the body with food. Both male and female sexual products consisted of vast accumulations of these minute atoms, all of which were mobile and able to combine to form new beings. His theory allowed him to bypass a deity, but it also involved a kind of universal creative element in the world, which was close to spontaneous generation. At any rate, he was not dogmatic about his theories and was careful to warn against confusing hypotheses with facts.

Buffon resisted Linnaeus' method of classification because he felt that the concept of species was arbitrary. His method of arrangement was to put the domestic animals first, and the more important wild creatures next. In other words, he felt that he was following the natural psychological tendencies of man as he explored his environment. Although there was a certain philosophical truth in his criticism, the

Linnaean system proved so useful that in his later volumes he began to be influenced by the great Swedish taxonomist.

We have an entertaining picture of Buffon's family life after his wife had died and Marie had become a self-effacing and loyal housekeeper. He slept in a vast columned bed with a floral baldachin, near which was a desk on which he kept his manuscripts and a writing table. He awoke at dawn, put on a silk dressing gown, called in his secretary, and dictated passages of his history. Dictation may well be one of the reasons why his style is somewhat oratorical. At eight, Marie Blesseau came to him and together they worked on accounts and finances. After that, the wigmaker arrived to arrange his coiffure and to retail the local gossip. His hair done, he had a *petit déjeuner* of bread and wine. At twelve, he did more literary work, often in his summer house if the weather was fine. From two to four was the time for an elaborate dinner after which a nap was unavoidable. During the afternoon he dictated his correspondence. After a light supper, if anyone interesting turned up, he appeared briefly in the salon, and if there were any pretty women, he shocked them with risqué stories.

Buffon was not liked by his neighbors, who considered him grasping in money matters, but as the spirit of the age changed and sentiment became fashionable, the cynical, pessimistic sensualist, in his old age, embarked on a "sublime" friendship with Madame Necker, and when he was dying of kidney stone, she visited him every day. Interestingly enough, he had a high opinion of Jean Lamarck as a botanist and sent his son to study with the younger natural scientist.

One of the last of the great synthesizers who sought to set down all that was known of nature, Buffon, who died in 1788, had an important influence on French scientific tradition, affecting such scholars as Lamarck, Georges Cuvier and eventually Erasmus Darwin, Charles Darwin's grandfather.

In a sense a complement to Buffon, for the great stylist got no farther than the vertebrates, is René Antoine Ferchault de Réaumur, his contemporary and the distinguished disciple of Swammerdam. His *Mémoires pour servir à l'histoire naturelle des insectes* (Memoirs to Function as a Natural History of Insects) which ran to six published volumes was to include work on invertebrates such as crustaceans, worms and polyps. Probably the greatest eighteenth-century entomologist, he and Buffon were rivals in their time. The son of a judge, he was born in 1683 in La Rochelle. Educated in Jesuit schools, he afterward studied law in Bourges. He was admitted to the Royal Academy of Sciences when only twenty-four. Interestingly enough, he made contributions to applied science, for he was a consultant to such industries as ropemaking and the manufacture of false pearls, which led him into a study of mollusks. He wrote a commercially important treatise on the manufacture of iron and steel at a time when the industry was in its infancy in France. He devised a certain type of hard white glass, introduced the artificial incubation of eggs, and invented a thermometer whose fixed points are still used today.

Réaumur seems to have had adequate independent means, for he never accepted an official post and spent part of his time on his country estate at Saintonge and part in another country house near Paris. He made a large collection of animal specimens from which his curator and collaborator, Mathurin Jacques Brisson, drew data for a book on birds and a six-volume ornithology. Brisson was to Réaumur what Daubenton was to Buffon. Réaumur began publishing before Buffon, from 1734 to 1742, whereas the *Natural History* was begun in 1749. Likewise, in the first part of the century, Réaumur was one of the arbiters of the Academy of Sciences, a position from which he was gradually ousted by his rival. Both scientists had their supporters. It seems likely that Réaumur was behind the attack on Buffon of the theological faculty of the Sorbonne. Réaumur wrote an article on the prevention of evaporation from museum jars which, allegedly, Daubenton obtained from the secretary of the academy before it was published and published it under his own name. This seems a rather petty theft; but when Buffon got the post of director of the Jardin du Roi and announced that he was doing an inventory, Réaumur, in a letter, spoke slightingly about Buffon's background and training. He probably would have liked the post himself. The two men's temperaments were at odds. Buffon was a bit of a popularizer, glib, fond of large generalizations, sometimes inaccurate. Réaumur was scrupulously careful, a sober stylist and a patient investigator who placed great emphasis upon firsthand observation. Knowing his own worth, for him it was a bitter pill to see himself eclipsed. In 1750, Réaumur wrote in another letter: "M. de Buffon's pompous manner will not persuade you to accept his queer notions. Had you read the first volume, you would have been no more satisfied with it than with the second. The three together can only impede the progress of natural history and physics in general, should the propositions they contain be adopted." Réaumur's religious opinions were apparently orthodox and thus his final attack on Buffon was through the publication of a friend, an Abbé Père de Lignac, who, at his instigation, wrote *Letters to an American* in 1751, in which he cast suspicion upon Buffon's orthodoxy. To even the score, however, it must be admitted that barring jealousy. Réaumur is described by his friends as generous, social, "his deportment was ever most refined and judicious."

He never married and during the latter part of his life did not finish the final volumes of his great book. This may have been due partly to his eclipse by Buffon and partly to his preoccupation with his industrial research.

Réaumur's work was pre-Linnaean, and indeed he cared little for classification and enumeration of the entire insect world. A practical scientific industrialist, he was fascinated by the "industries," that is, the constructive activities of insects:

> I confess that I am not in the least inclined toward a precise enumeration of every kind of insect, even if it could be undertaken. It seems to me sufficient to consider those kinds which prove that they deserve to be distinguished, either on account of their peculiar industries or because of their unusual structure or because of other striking singularities. It seems to me that the many hundreds and hundreds of species of gnats and very small moths which exhibit nothing remarkable than a few slight variations in the form of wings or legs, or varieties of coloration, or of different patterns of the same colors, may be left confounded with one another.

In an age when enumeration was paramount, Réaumur was a rebel. It is, however, the attitude of the true ethologist, of which Réaumur was one of the earliest. He belonged in the tradition of Gilbert White and was a precursor of J. H. Fabre.

A glance at his essay on ants, unpublished at his death in 1757, and probably meant to be a part of Volume VII, shows him at his best. Stylistically, he was a contrast to Buffon, for he wrote soberly. With a colloquial directness, he invited the reader to accompany him in his studies; indeed, his avoidance of rhetoric is strikingly modern:

> Although we may not always have occasion to praise the ants, we are generally well disposed toward them. For them we do not feel the aversion which we often harbor toward so many insects. One of the virtues most useful to society is the love of labor, we like industrious people and we are led to like small animals who labor with that intensity which we would wish all men to exhibit.

He was mostly concerned with the mound-building ant, *Formica pratensis,* an ant which makes nests, constructed of small pieces of wood or other objects piled above ground with tunnels below. He carefully examined the materials of the nest and found that in one case, where there were many barley kernels on the ground, they were used instead of wood. This led him to disprove some of the fables that had been handed down about the ants. One of these was that the ants stored up grain for the winter, a tale which went back to Aesop with the ant and the grasshopper. By keeping ants in a bell jar and offering them various types of nutriment, he showed that they preferred sugar, honey, animal matter, and though some raised fungi they not only did not store grain, but never ate it.

By means of another bit of observation, in which he probably preceded Linnaeus, he dispelled the mystery of formic reproduction. Large ants with wings had been seen and, as usual (because of patriarchal bias), had been called kings. He noted very similar large ants without wings. He also noticed small ants with wings. He then was alert enough to catch the winged ants in their nuptial flight on which he found the smaller male copulating with the females. Winged females, which he kept in captivity, were seen to bite off their wings.

He had, furthermore, made sure of the two sexes by pressing the abdomens; the large ants extruded eggs, the smaller a copulative organ. This type of field observation plus simple experiment is similar in quality to that of the twentieth-century entomologist Fabre.

Because of his interest in socially productive insects, his 1740 volume on bees is particularly rich in accurate observation. Réaumur devised glass observation hives, some no more than two plates of glass so that the bees could construct only two galleries: others were the familiar square hive with a glass front covered by a shutter. He found that ants often set up a colony between the glass and the shutter, thus affording him an observation post for two insects. With this apparatus, he was able to progress beyond Swammerdam.

He verified the fact that a swarm tolerates only one queen. He also experimented with introducing foreign queens into the hive and decided that in some circumstances they would be accepted. He destroyed the queen and saw how the workers fed an ordinary larva on royal jelly and produced a new queen. He studied the licking action of the tongue, which Swammerdam had not seen, and also the action of the sting. He drew off the poison from the sting and with it inoculated himself and also a doubting member of the Academy, who complained loudly when the operation was a success. The patient observer counted eighty-four thousand trips in search of honey from one hive in fourteen hours.

He was also curious concerning the number of bees in a hive and devised a neat method for counting them. By plunging the hive in a bath of cold water, the insects became so lethargic as to allow him, aided by his modest but helpful secretary, Mademoiselle Moutier, to count 26,486 workers, 700 males and one queen.

Réaumur never made the mistake of anthropomorphizing his bees. Indeed, in all his observations and methods he remained a fairly strict Cartesian, even to suspecting that the bees' solicitous licking of the queen was triggered by substances she exuded, which they found pleasant. He thus kept his theological and scientific points separate. Concerning the problem of whether animals possess intelligence, he expressed himself rather quaintly:

> Shall we reduce them to the simple status of machines? This is the great question of the animal soul, agitated so often since M. Descartes, and in regard to which everything has been said since it was first propounded. All we can conclude from the discussions that have arisen is that the two opposite views maintain only what is very probable but that it is impossible to prove which of the two is true. If somebody wishes to believe that God could make machines capable of growing, multiplying and doing everything that insects and other animals do, who would dare to deny that the Omnipotent could accomplish so much? But if someone should believe that God could endow insects with an intelligence equal or even superior to our own, without our knowing that He has thus endowed them; and if this somebody should maintain that an oyster, vile as it is in our sight, attached to a rock and condemned to a mode of life, which seems to us very gloomy, may nevertheless enjoy a very delightful existence, being constantly engaged in lofty speculations, it would be impossible to deny that the Supreme Power could go so far or even further.

Réaumur's experimental work led into the investigation of the phenomenon of reproduction, and he took part in the great debate concerning the nature of life and the origin of living beings, which we shall be discussing shortly. Apropos of this debate, it came to Réaumur's ears that a rabbit in the possession of a clergyman friend exhibited a perverse interest in a hen. Struck by the possibility that something could be learned of inheritance from the outcome of such a union, he sent for the rabbit and the chicken, and under the scientific eye of the experimenter:

> She crouches as does every hen, after having fled from the cock, consents to his caresses; she permits the rabbit to take whatever position he wishes . . . ; indeed the hen for him becomes a female rabbit; he remains actively upon her four or five times, as long as a cock would. But was the connection as complete as that of a cock with a hen, a rabbit with a rabbit? This I do not know, but whatever happened was enough to satisfy the rabbit.

The public was now beginning to see the philosophical problems in biological speculation, and all of Paris became interested in the experiment when the hen laid six eggs. Would there be furred chickens and feathered rabbits? The outcome was awaited with baited breath. Alas, all the eggs were infertile!

A highly significant experiment performed by Réaumur dealt with the reproduction of aphids. Leeuwenhoek had already discovered that they reproduced viviparously. Now the question was raised: Were these females ever inseminated by a male? Réaumur undertook to rear a plant louse from birth in isolation from the males. Each time he tried, the insect died before arriving at the age for reproduction. At this time, a young disciple and admirer of Réaumur, Charles Bonnet, was writing him from Geneva. Bonnet asked if there was any experiment in which he could aid his master. Réaumur suggested further work on plant louse reproduction. The next step takes us into eighteenth-century biology and the great debate.

14

The Great Debate

ON May 20, 1740, Charles Bonnet began work on an aphid which inhabits the spindle tree. It took only seven days to mature, and then he was able to observe parthenogenetic reproduction. Réaumur's species had taken longer to mature and died before it was ready to reproduce. Bonnet's results were obtained by several other experimenters and communicated to the Academy of Sciences, where they produced a sensation.

Experiments such as these had a bearing on the great debate that took place during the eighteenth and part of the nineteenth century and in which most of the important natural scientists took part. It was concerned with the process of reproduction and the development of new life. It signified that zoology was finally occupying itself with great basic questions, as astronomy and physics had already done much earlier.

Buffon, Réaumur, Bonnet, Kaspar Friedrich Wolff, Lazzaro Spallanzani and Albrecht Haller were some of the savants who took part in the controversy. The last three were concerned with physiology and embryology in particular, while Bonnet worked in various fields but is best known as a disciple of Réaumur.

Actually, the debate involved two specific problems, the first being the role of the two sexes in reproduction and the second being the dispute concerning preformation and epigenesis.

It was Leeuwenhoek with his discovery and identification of spermatozoa who touched off the question as to which sex played the major role in reproduction. Oddly enough, the overimaginative physician Paracelsus had made a contribution centuries earlier. He wrote:

> Let the sperm of man be putrified in a gourd glass sealed up with the highest degree of putrifaction in horse dung, for the space of

40 days or so long, until it begins to be alive, move and stir which may be easily seen. After this it will be something like a man yet transparent and without a body.

Although this was pure fantasy, even with the advent of the microscope there was room for curious interpretations. Leeuwenhoek had soberly described a tiny entity with a mobile tail which thrashed about and drove the object forward. One of his contemporaries decided he saw little men and published pictures of them, complete with tails. Leeuwenhoek had maintained that the spermatozoa penetrated the egg, and those who adhered to his theory felt that the female was merely a host for the development of new life. The male supremacists, therefore, were known as animalculists. Here is a description of reproduction by one N. Audry de Boisregard, who flourished in the latter half of the seventeenth century:

A spermatic worm seeks out the ovary, slips into an egg, closes the door behind him with his tail and proceeds to develop. If several attempt to enter the egg at the same time, they become enraged and strike at each other, breaking and dislocating their limbs and thus give rise to monstrosities. Even at this stage, the spermatozoa are endowed with the nature of the animal to which they will give rise for those of the ram already live in flocks.

The argument over spermatozoa continued; were they independent animals? Buffon, who disdained the microscope and used it carelessly, said they were aggregates of living organic atoms formed outside the body from the liquid semen. Other investigators tried unsuccessfully to find mouth parts and reproductive organs. Some insisted they were parasites.

On the other side of the fence were the ovists, with William Harvey a leading figure who felt that the germ of life resided in the egg while the male element played a secondary role. The whole question was complicated by those who believed in spontaneous generation and those who denied it.

Finally, these opposing positions were related to that of preformationists and epigenesists. Preformation, most roundly championed by Swammerdam, had a certain amount of religious sanction. It will be remembered that Ray speculated on the infinitely great number of preformed seeds of life or minute animals which were being enclosed in the sex organs of the first females created of each species. An ovist, he was also discussing *emboîtement,* or encasement. It was a mystical idea in which preformed dolls, infinitely minute and infinitely numerous, were present in the female animals as created according to the account in Genesis. These minute dolls were passed on to succeeding generations. All the attributes of the adult were present; the seeds had merely to increase in size. To complicate the picture, a few animalculists applied the encasement theory to the male sexual principle. In proof of this theory, the preformationists pointed out that a parent who had lost an organ or a limb could still give rise to perfect offspring. This, too, they claimed, disproved the pangenesis of Buffon in which atoms from all parts of bodies of both parents somehow merged to form a new being.

Malpighi claimed he had found traces of head and brain and spinal cord of a chick in an unincubated egg. However, acute critics pointed out that the egg had lain in the hot Italian sun and, therefore, could have started to incubate.

Let us return to Charles Bonnet, who was born in 1720 in Genf, Switzerland, of a French family who had left France after the St. Bartholomew's Day persecution of the Protestants. A semi-invalid, he suffered from deafness and hence did not do well in school. After tutoring at home, he read a treatise on natural history and began to observe animals. He studied ant lions and caterpillars. He found that many of the latter laid down a silken pathway and when this was disturbed, they tended to lose their way. He had been reading Réaumur with excitement, wrote him about his caterpillars, and was charmed to receive a reply from the master. From then on, they corresponded regularly and he became the older man's disciple. Although his father forced him to take a law degree (which he obtained in 1744), thus interrupting his nature studies, he never seems to have practiced, but spent most of his time in his country villa at Genthod, where he died in 1793.

At the age of twenty his discovery of parthenogenetic reproduction of aphids was for him a proof of ovism, preformation and encasement. It is notable that Bonnet was a pious Protestant. He corresponded with both Spallanzani and Haller and was elected to both the French Academy of Sciences and Friedrich the Great's Berlin Academy of Sciences and Fine Art.

In 1741, Abraham Trembley, a Swiss naturalist, discovered that if a polyp was cut into small pieces, each developed into a complete polyp. Stimulated by this, Bonnet cut up earthworms and found that all the pieces were able to regenerate complete worms. He then cut off the legs of salamanders and found that new legs grew, each time smaller, as the experiment was repeated. To a preformationist this proved that the germs of life were not to be found in the ovaries alone but in some animals in all parts of the body. Since, in his view, animals possessed an indivisible soul, soul rudiments also must be scattered through the body of the salamander. This idea owed something to the philosopher Gottfried Leibnitz who thought the living beings were composed of monads, or atoms of life, and also to Buffon's atomic theory. Buffon, however, had used it to bolster spontaneous generation, which Bonnet severely denounced.

Bonnet married a Mademoiselle de la Rive in 1756, who took care of him when his eyes gave out from too much work with the microscope, and indeed, after 1745 it was very difficult for him to read or write. His work became increasingly theoretical, and in 1769 he wrote *Contemplation de la nature,* a mixture of theology, reproductive theory and observation. Yet the following description curiously gives the lie to his own dogma. He was describing the pupa of a butterfly:

> Open it delicately with the point of a needle. You are surprised to find nothing but a mass of pap which cannot be differentiated. Only shortly before, the insect has been a wormlike form. How has it reduced itself to pap? How does it become an insect? Hold back your questions and open a still further developed cocoon. What do you find? A little mass of flesh, oblong and whitish, in which you do not even find any vestige of members or organs with the glass. In a word, you have before your eyes an oval mass. Don't think that this is an envelope which encloses a nymph: the oval is in itself a very disguised nymph. Press the oval a little and at once the legs begin to be visible, springing from a little depression in one of the extremities of the oval. Augment your pressure by degrees and you will force all the parts of the nymph into sight. They already exist, this you cannot doubt. They are buried and folded in the interior of the oval somewhat as the fingers of a glove are folded into the palm.

The mass of pap in the newly formed pupa scarcely supports the idea of an image of the animal in miniature.

When the epigenesists complained that they could not find this image in the eggs or sperm of various animals, the preformationists replied that the preformed limbs were transparent!

At any rate, after the triumph of parthenogenesis in aphids, which, by the way, was received with enthusiasm by a certain theologian,

Abbé Pierquin, as supporting the Christmas dogma of the Virgin birth, poor Bonnet was struck all of a heap by a question asked by Abraham Trembley. Who knows, said this colleague, whether one fertilization could last for several generations? "It seemed to me that these two words completely annihilated all I had done." Bonnet determined to raise many successive parthenogenetic generations. He worked desperately, raised three, nine, then got up to thirty, and breathed a sigh of relief. By this laborious operation he felt his thesis was reasonably secure.

A no less careful experimenter was his friend, Lazzaro Spallanzani, of Scandiano, a small village near Modena. The son of a lawyer, he was born in 1729, was educated at the Jesuit school in Reggio, and studied law at the University of Bologna. He was not attracted to law but did well in languages. Although he took minor orders, it does not seem positive that he was ever ordained a priest. For ten years he taught physics and mathematics in the College of San Carlo in Modena. At this period of his life, he wrote literary essays. His father finally allowed him to drop law and turn to science. His first studies dealt with infusoria, microscopic animals and plants. In 1748 an English priest, John Needham, had more or less repeated Redi's experiments but in relation to infusoria. He boiled mutton broth and placed it in a corked vial well closed with gum. When the flask was opened, in a few days it swarmed with animalcules. This, he felt, was proof of spontaneous generation. Spallanzani took up the problem more creatively. He first tried a short period of boiling and discovered that some animalcules died but other species thrived. He then tried leaving a flask open while a similar flask, sealed by melting the neck, was then heated. If boiling temperature was maintained for three-quarters of an hour, no infusoria developed. In the open flask, they appeared. He felt that this was a blow to spontaneous generation. His opponents said he had "spoiled" the air by heating it; the unheated normal air would still cause spontaneous generation.

In 1770 Spallanzani was invited to the University of Pavia by the Empress Maria Theresa. There he continued his work on infusoria, subjecting worms, rotifers and another tiny creeping animal, the tardigrade or water bear, to extreme desiccation. He wrote:

> The solid parts contract and become disfigured, the fluids evaporate, and the animal's body is reduced to an atom of desiccated, hardened matter, which when pierced with a needle breaks into several pieces like a particle of salt.

He then left them alone for as long as three years or even four, after which he applied moisture and a little heat. After three years, 10 percent came to life; after four, just a few individuals. His results, published in 1775, created a sensation. A few other experimenters, including Needham, had done something similar with other small animals; one of them, a more pious abbé than Spallanzani, had been afraid to publish his results for fear of excommunication! Now Spallanzani showed that the animals which, he was convinced, were dead could in their dried state resist extremes of temperature. Moreover, the experiment could be successfully repeated several times. Theological speculations followed. If the animal was really dead and resurrected, what happened to its soul? Was it reborn at the moment of resurrection?

Spallanzani seems to have had particular respect for Voltaire, for he once asked Bonnet to pass one of his books on to the poet and critic. Bonnet, who disliked Voltaire, wrote back:

> You must realize he is neither a philosopher nor a naturalist. His ridiculous *Singularities of Nature* should prove that to you once and for all. . . . He is always airborne on his Pegasus and gets only a bird's-eye view of objects.

Now, Spallanzani communicated the resurrection experiment to the great skeptic. Voltaire avoided involvement by taking refuge in frivolity: "I admit I should like to know why the great Being, author of everything, who gave us life and death hasn't bestowed the faculty of resurrection upon us as he has to the cotifero (*sic*) and the *tardigrado*."

Recent experiments showing that minute animals can come to life after six years in a sealed test tube seem to indicate that they actually die. A spark of life maintained without metabolism or respiration is hard to conceive. Thus the whole subject remains as much a mystery as it was in the eighteenth century.

These experiments gave rise to certain speculations on the part of Réaumur and others. Could large animals and even human beings be subjected to extreme cold and kept in deep freeze for far longer than the usual expectation of life?

The English physiologist John Hunter tried some experiments on carp which failed, but now, with new techniques, the same idea is seriously discussed in our own time.

Spallanzani, who was a convinced ovist and preformationist,

turned his attention to copulation of frogs. Naturally, he set out to prove his ovist philosophy. He first made sure that the eggs were removed from the ovary before fertilization. He knew that fertilization took place outside the body and assumed he could carry it out artificially if he could obtain frog sperm. He finally tried to solve the problem by putting little breeches on frogs which were ready to mount the female. The first model was kicked off by the frog, but he finally achieved little overalls which collected a few drops of a clear liquid. He bathed virgin eggs in it and to his great enthusiasm saw them develop into tadpoles.

This was the first effective artificial insemination in the laboratory, although Arab horse breeders had been using the technique for centuries.

Spallanzani saw a great future for his discovery in breeding various animals, for he extended his technique to dogs. He continued his work on frogs, however, and even filtered the frog semen through layers of blotting paper, after which it proved ineffective. There he effectively destroyed his own ovist theory, for what he did was filter out the spermatozoa. Yet so set was he in his ovism, he did not realize he had proved that spermatozoa were necessary for fertilization. Actually, he hovered on the edge of artificial parthenogenesis, for he tried unsuccessfully to stimulate the virgin egg with electricity and chemicals. Because he was able to discern the growth of the egg within the frog's ovary, he decided this proved preformation and that the egg was actually a disguised tadpole.

Bonnet agreed enthusiastically with Spallanzani's ovist conclusions. Albrecht von Haller, with whom the Italian scientist also corresponded, leaned toward the animalculists but agreed with both on preformation.

Another set of rather brutal experiments carried out by Spallanzani dealt with the instinct that drove frogs to reproduce. He cut off the limbs of copulating males and found that they continued the sex act, even performing it when he cut off their heads. He tells us that he repeated this gruesome vivisection at "the delightful villa" of Bonnet at Genthod, with John Trembley present:

> I was asked whether this pertinacity of the male was the effect of stupidity, insensibility or amorous ardor. Though this appears to be one of those questions of which the determination requires that one should enter the mind of a frog without at the same time becoming one, yet I did not hesitate to say that I thought, as I still do, that this perseverance was less the effect of obtuseness of feeling than

> vehemence of passion, which, as we have seen, renders them insensible to the call of hunger and careless of their safety.

This was, therefore, a rather crude attempt to examine the phenomena of animal psychology. Spallanzani's most important book is *Dissertations Relative to the Natural History of Animals,* published in 1768, the first volume dealing with digestion and reestablishing the function of the gastric juice which he extracted and whose operation he watched outside the animal body. The second volume deals with his work on reproduction. He died in Pavia in 1799.

The scientist who is credited with laying the basis of modern physiology and establishing this study as a separate science is Albrecht von Haller. He is an interesting example of a scientist who was also a literary man. Born in 1708 in Bern, he grew up in a village in the German Alps. His ancestors, one of whom fell at the battle of Kappel, seem to have been mostly clergymen. His father was a well-to-do lawyer. A child prodigy, Albrecht knew Greek, Hebrew and French by the age of twelve, at which time he lost his father, who left instructions in his will destining him for theology. These were evidently not carried out, for by 1723 he was botanizing and preparing to study medicine at the University of Tübingen. From there he went to study under the famous and kindhearted Boerhaave in Leiden. After getting his doctorate at nineteen, he spent some time in England, France and Holland. He visited hospitals, met doctors and some writers in coffeehouses. He acquired a knowledge of English literature, for he was to become an admirer of Alexander Pope. He noted that mortality in French hospitals was higher than that in England and Dutch medical establishments were superior to both in cleanliness.

He returned to teach in Basel and there frequented a literary group which admired English poetry. Haller was temperamentally a romantic, but in his writing he was a split man. A great deal of his verse is in moralizing heroic couplets like those of Pope, but in love lyrics, which abound in tears and tender emotions, he displayed the romantic sensibility.

He moved on to Bern to practice medicine and in 1729 married Marianne Wyss, who gave him three children. By 1736 he was well known as a botanist and a poet. Because he was stereotyped as a writer, the medical men of Basel were antagonistic to him and managed to keep him out of the local hospitals. An offer of a professorship at the University of Göttingen seemed a solution, but the death of his wife, soon after he located in Göttingen, deepened the melancholy of his disposition. Troubled by freethinking, so characteristic of his time, he wrote pamphlets in defense of Christianity. Yet he was an unorthodox Protestant. Unlike Bonnet and Réaumur, who simply kept their religion in a separate compartment, he, on the one hand, favored a mathematical approach to science and, on the other, felt that nature was the medium through which God was disclosed to man. Homesick in Göttingen, where he found some of his colleagues unsympathetic, he got himself elected to a minor municipal office while on a visit to Bern, a sufficient excuse to resign from Göttingen. As a botanist, he tried to set up an opposition system of classification to that of Linnaeus, a system which failed, and the failure embittered him still more. He died in 1777.

Haller's poetry has not stood the test of time, but his work in physiology is important for two reasons: his definition of irritability and his textbook of physiology. Irritability is a property of muscles and organs which contract on being touched or stimulated; other organs are sensitive in that the stimulus is recorded in the brain; some do not react at all. Haller investigated these properties by means of vivisection, concerning which his religious questionings were to fill him with guilt in later life. Haller finally wrote an *Anfangsgründe der Physiologie des Menschen* (Basic Beginnings in Human Physiology), in 1759 which was a model description of the operation of the body. It was used well up into the nineteenth century.

As we have seen, Haller was a preformationist but did not contribute anything to substantiate this point of view. The most important epigenesist was a German, born in Berlin in 1733, whose work was practically unnoticed in his own time. The son of a master tailor, Kaspar Friedrich Wolff attended the College of Medicine of Berlin and then went to Halle where he studied philosophy and got his degree in 1759, his thesis, *Theoria generationis* (Theory of Generation) being his most important contribution to the great debate. Although he was predisposed toward epigenesis as a result of the theoretical ideas of his teacher, Christian Wolff, he drew some concrete conclusions from the development of plants. He showed that the organs of plants, leaves, roots, and even flower buds, were developed from undifferentiated tissue at the tip of a growing shoot or root. The leaves and parts of the flowers developed gradually from tiny bumps which emerged from the plant stem. Thus something appeared which was not there before, which, by definition, is epigenesis. Why this process took place, Wolff explained by reference to an "inner force," a rather vague concept, as the preformationists were quick to point out. Wolff outlined this thesis a second time in terms of the development of the chick. He showed that the intestines, like portions of the plant, also developed from undifferentiated tissue which at first showed no trace of becoming an organ. He also described the folds of the embryo as leaves or layers analogical to the growth points of plants. This is an insight which was forerunner of the discovery of layers of cells which gradually differentiate the organs of the embryo.

Wolff sent this thesis to Haller, who could not accept it on religious grounds. He held firmly to the dogma that all creation went back to Adam. Wolff replied:

> There is nothing contrary to the existence of a Divine Being in the case of an organic body developing through the forces of nature and from natural causes, for this force and these causes themselves, even nature itself, posit a first cause just as much as does the organic body.

He went on to say, "What we seek is truth." If his epigenesis was correct, then Haller should have been enough of a scientist to accept it.

In 1760, Wolff's work was interrupted by the invasion of Berlin by the Russians. During Friedrich's war, he taught anatomy to young army surgeons in Breslau. After peace was made, he came back to Berlin to find a cabal against him in the Medical College. When he gave a course of private lectures, the opposition of the Berlin doctors grew more intense. The two groups of students came to blows. Thanks to the intercession of a friend, Catherine II invited him to come and work for the Petersburg Academy. He married a poor but pretty girl in 1767 and left for Russia to become the forgotten man. His work on the embryology of the chicken appeared in a publication of the Petersburg Academy and went unnoticed. He died, after twenty-seven years in Russia, in 1794. In 1806, Lorenz Oken, also writing on the embryology of the chicken, did not even mention Wolff's important treatise.

The whole controversy over sexual reproduction is instructive, for it shows the forces and the involved and complicated variables in human culture which affect the development and acceptance of new insights. All the preformationists were religiously oriented, and sometimes consciously, as in the case of Haller, perhaps in other cases unconsciously, they wished to preserve the Biblical account of the

Creation. Wolff, with his "What we seek is truth," was espousing an unpopular position and hence it was easy to ignore him.

F. J. Cole, a student of sexual theory, pointed out that microscopic technique was far enough advanced for some researcher who was a more determined observer to have investigated the mechanics of fertilization. The tendency to speculate abstractly, however, was too strong. Even Spallanzani, after filtering out the spermatozoa and proving the residue was useless for fertilization, was so involved in the ovist dogma that he did not draw the obvious conclusions or pursue this line of research further.

Another element in the situation was, of course, the fact that the cell as a biological unit had not been discovered and the cell doctrine had not been formulated. Eighteenth-century thinkers were attempting to go from A to C without the intervening B. Neither preformationists nor epigenesists, in the light of modern developments, were wholly right or wholly wrong. Wolff came nearest to understanding what happened after fertilization, although he did not understand fertilization itself. The preformationists, although they were wrong in their concept of concrete detail, did evolve a metaphor which related to the modern understanding of the function of genes. If the new being is not preformed, it is *predestined,* as a result of operation of the genes, to assume a specific form.

The eighteenth century, as we have seen, was both a period of advance in classification techniques and a period of intellectual activity in the area of theory. It was also in this period that the first breakthrough took place between chemistry and zoology.

15

The Breath of Life

ALL terrestrial forms of life above the single cell need air. Strangely enough, it was not until the end of the eighteenth century that the mystery of air in its relation to animals and plants was solved. This represented a landmark in scientific history for several reasons. Not only did this discovery lay the foundation of modern chemistry, but it also brought chemistry into biology. The conjunction of the two was necessary in order to open up new possibilities for the investigation of life.

All during the scientific revolution, chemistry had lagged. As has been pointed out, the Middle Ages contributed something in terms of alchemical techniques and industrial know-how. With the theories of the appropriately named Bombastus von Hohenheim (Paracelsus) (1493-1541), an alchemist who believed in magic and wrote voluminously and obscurely, doctors turned their attention to chemical remedies. Paracelsus was at least a rebel against Aristotle and Galen, and on the side of novelty. He believed that violent countermeasures should be used against virulent diseases. He and his followers fed their patients such poisons as vitriol, mercury and antimony, and oddly enough, mercury actually proved useful in the treatment of syphilis.

One stumbling block in the way of chemical progress had been a preoccupation with metals and acids combining with them and a lack of awareness of gases. The failure to draw a clear line between the animate and the inanimate was also an obstacle. Even Robert Boyle (1627-1691), who unknowingly liberated oxygen from mercury monoxide, believed that such minerals as saltpeter grew in the earth. Although he was in some ways a thoroughly advanced thinker, he failed to understand combustion. Even the clever Robert Hooke (1635-1703), who was interested in everything, thought that combustible bodies were dissolved by a certain substance in the atmosphere called saltpeter. Flames were a mixture of air and the volatile parts of otherwise indissoluble bodies. This did not add much clarity. The worst obfuscation came from the first real theory of chemical reaction, invented by a wild adventurist, half medical man and half alchemist, who became a doctor by virtue of marrying a privy councillor's daughter and who actually extracted a large sum of money from the Dutch government for a project to make gold from the sand of Holland's beaches. Johann Becher (1635-82), when he was not launching fraudulent West India Companies, took time off to publish his *Subterranean Physics* in 1669. That work stated that he had solved the problem of burning, calcination, and respiration by the concept of phlogiston. It was a chemical of an earthy nature, dry and adapted to combination with solids. When a substance burned, its phlogiston was violently given off in the form of the flame. Weigh a burned body, he explained, and you will find it has lost weight in the process. Other chemists dutifully weighed burned metals and found that the powders had *increased* in weight.

In the case of Becher, the hand was quicker than the eye. At one point, he claimed he had discovered the philosophers' stone, at another perpetual motion. He also rose as high as medical doctor, privy and commercial adviser to His Majesty the Holy Roman Emperor, Leopold I of Austria. It is true that he died in poverty, but it was a good racket while it lasted. To the critics of his phlogiston concept he blandly explained that at times the mysterious substance had no weight; it was in a class with heat and electricity. Besides, he went on glibly, something minus another thing which weighs less than nothing weighs more than the original something!

Nevertheless, the phlogiston theory was one of those conceptions which allowed for some chemical advance and by its vagueness was highly adaptable and stimulated some experiments. At the same time it did not really explain anything and became a deadweight which eventually clogged chemical thinking.

The word "air," however, was beginning to take on new meanings. There was inflammable air and air that extinguished flame. Stephen Hales (1677-1761) did some tentative experiments on the absorption of air by plants and, above all, devised the technique of collecting gases by bubbling them into a jar filled with water.

Then a new "air" entered the picture. Joseph Black in 1756 wrote that something he labeled "fixed air" was emitted with effervescence when lime was dissolved in acid and became quicklime. He proved this by adding a mild alkali to the quicklime, which then precipitated out the original weight of lime while the alkali became caustic. He said that "fixed air" had combined with the lime again in the second reaction because it had a greater affinity for lime than alkali. He further decided that both water and common air contained fixed air which was not the same as common air. Later he discovered he could create his new gas by passing common air over burning charcoal. Black was, of course, by his experiments defining carbon dioxide without benefit of phlogiston.

But in 1703 an otherwise competent German scientist, Georg Stahl, had republished Becher's book and made phlogiston into a classic concept which was to last until the end of the century.

Other minds were working in new directions. Joseph Priestley, conservative enough to believe in phlogiston all his life and radical enough in religion and politics to be persecuted, performed some experiments that began to unite biology and chemistry. This slender, mildmannered clerical soul was born in 1733, in Fieldhead near Leeds. His family were artisans and dissenters. He was educated in a college of his sect and became the pastor of a small chapel. Although he knew French, Italian, German, Arabic, Syriac and Chaldean, in order to live he taught school and also gave private lessons. Eventually, while teaching at an academy at Warrington, he studied some anatomy and chemistry.

When he was thirty-four, he became the pastor of a struggling dissenting chapel in Leeds. During one of his trips to London, he met Ben Franklin, who stimulated his interest in electricity. Franklin even offered him the material to write a history of the subject.

It was chemical experiment, however, that actually got him started in science. There was a public brewery adjoining his home. Gas bubbled continually from the huge vats. This phenomenon caught Priestley's attention. He discovered that this gas had the property of extinguishing burning chips of wood. He was reminded of Joseph Black's "fixed air." Black had been on the trail of a secret remedy to cure the Prime Minister, Robert Walpole, of the gout. Priestley, a purer scientist, extended his curiosity further. He succeeded in making the gas, and when he tried to mix it with water, it fizzed. He had created soda water, hardly to be distinguished from that which was found in certain springs. Priestley demonstrated his discovery before the Royal Society and the College of Physicians. The Society gave him a gold medal, and the physicians wondered if it might be a cure for scurvy. Priestley continued to experiment with gases, using Hales' method of bubbling the gas into an inverted bottle of water when the gas was nonsoluble, and using mercury for those gases which were soluble in water. About this time he, too, was affected by the exploration fever. He was invited to accompany Captain Cook on his second journey. The idea was fascinating, but opposition developed because of his nonconformist views.

Priestley finally obtained oxygen from mercury monoxide but, alas, decided it was "dephlogisticated air." Nevertheless, he examined its properties. A glowing coal burst into flame when placed in it, a red-hot wire glowed and blazed. He was still confused by its nature. He thought that the phlogiston which was added to the air by burning bodies was taken up by plants, thus keeping the atmosphere pure. Finally, he put live mice into dephlogisticated air, floating them up on rafts in the water, and found that they breathed better while in it and survived longer than in the same volume of common air. He had already discovered that when he put a candle in this particular gas, burned it for a while until it went out, and then put a plant in the gas, the "polluted" gas was refreshed by the plant. He now allowed his mouse to die in a limited volume of air and put a plant in the jar. He then put back a mouse and found that the mouse flourished. He wrote: "Plants, instead of affecting the air in the same manner as animal respiration, reverse the effect of breathing and tend to keep the atmosphere sweet and wholesome." He thus had all the clues in his hands but was unable to make the final analysis because he could never rid himself of the intervening phlogiston.

For eight years, Priestley had a patron in Lord Shelburne, and during this period he went to Paris and met Antoine Lavoisier. When he told the Frenchman about his experiments, the latter drank it all in. He immediately set about checking Priestley's work. The meeting was to have important results.

Priestley did not go much further. He tried breathing oxygen and discovered his breath felt light and easy. He even suggested that it might be good for the lungs in cases of illness.

Eventually, Priestley's unconventional political thinking got him into trouble. On Bastille Day in 1791 he and other liberals gathered to celebrate in the meeting house of Birmingham. The agents of the king whipped up a mob by reading an inflammatory document which made such statements as: "The Presbyterians intend to rise. They are planning to burn down the Church. They will blow up Parliament

. . . The king's head will be cut off and dangled before you!" As usual, there were plenty of lumpen know-nothings who were ready for mob action.

A mob of about a thousand burned down the meeting house. Not satisfied, they also attacked the houses of the better-known participants in the meeting. Priestley had received letters from his friend, Ben Franklin, which he publicized. One stated: "Britain, at the expense of three millions, has killed one hundred and fifty yankees in this campaign which is twenty-thousand pounds a head; and at Bunker's Hill, she gained a mile of ground half of which she lost again by our taking post on Ploughed Hill. During the same time, sixty thousand children have been born in America. From these data your mathematical head will easily calculate the time and expense necessary to kill us all and conquer the whole of our territory." Priestley had also publicly answered Edmund Burke's attack on the French Revolution.

The mob rushed to his house, demolished his library, destroyed his apparatus, and burned his manuscripts. From then on, he was a marked man. The gentle, stuttering clergyman fled to London, while for three days George III's riot continued. Priestley had been the center of a brilliant liberal intellectual circle, the Lunar Society in Birmingham, which included Charles Darwin's grandfather, Erasmus. In London, the well-established members of the Royal Society shunned poor Priestley. The natural philosophers, including Sir Joseph Banks and Captain Cook, met at Jacob's Coffeehouse. Priestley was not welcome. He resigned from the Royal Society and emigrated to America in 1794. The great defector was received with adulation. Even the Tammany Society sent a committee to congratulate him. He was offered a Unitarian parish, a professorship at the University of Pennsylvania, a speaking tour. He refused all honors, settled in a small town in Pennsylvania, Northumberland, where he continued to work in his laboratory, discovering carbon monoxide, taking time out now and then to have tea with Washington or attend meetings of the American Philosophical Society. But to the time of his death in 1804 he never gave up phlogiston.

It was Lavoisier, fascinated by Priestley's work, who picked up the torch and carried it onward. Lavoisier made the mistake of exposing Jean Paul Marat's deficiencies, when the latter, a very poor chemist, tried to gain election to the Royal Academy of Sciences. Marat never forgot.

Antoine Lavoisier was born in 1743. His father was a well-to-do lawyer who sent his son to the Collège Mazarin to follow in his footsteps, but the son turned to science. He was much influenced by the demonstrator on chemistry who held forth at the Jardin du Roi. He also met Linnaeus who made a profound impression upon him. While still young, he was elected to the Academy of Sciences. Then he made the second mistake which was to contribute to his death. No doubt influenced by the solid bourgeois desire for security, he bought into the company which collected taxes for the king, the Ferme Générale. Some of the leading lights of the company were important people; Pierre Simon de Laplace, the astronomer, moved in these circles, as did Condorcet, the mathematician and humanist, and even Franklin, when he was in France. Lavoisier also found his wife when he was entertained in the home of the aristocrat and amateur scientist Jacques Paulze de Chastenolles. He married the latter's clever fourteen-year-old daughter, Marie Anne, who had a talent for drawing, which she was to use in illustrating his scientific work. As his wife, she also graciously presided over an intellectual salon.

All along, Lavoisier found himself unable to swallow phlogiston. He substituted for it the term "caloric," which was scarcely more than a synonym for heat. He frankly admitted that he was still in the dark as to its real nature.

He did not give up on the problem of combustion. By 1772 in a sealed memoir to the Academy of Sciences he indicated that something in the air was responsible for the gain in weight during combustion. Priestley's visit two years later gave him important clues. He burned mercury and produced a gas or "air" (in which animals suffocated) and a red powder. The gas was nitrogen, and by heating the red powder, he obtained what we now know was oxygen. And it was Lavoisier who gave it its name. Not a very original experimenter, like Harvey and Newton he possessed the analytical and synthesizing ability to pull together the known facts over which so many scientists had puzzled and to come up with a satisfactory explanation which was to hold up for future generations.

Burning, said Lavoisier, was the union of the burning substance with oxygen. A scrupulous measurer, Lavoisier had the best balances in Europe. Moreover, he stated the axiom upon which all modern chemistry is founded: "One may take it for granted that in every reaction there is an equal quantity of matter before and after the operation." With accurate scales and careful collection of gases it was possible to analyze exactly what went on in a chemical reaction. In addition, he contributed greatly to ease of communication by giving simple and appropriate names to elements.

We have now reached the point where chemistry and biology were once and for all joined—the process of respiration. It was Lavoisier who, in collaboration with Pierre Laplace, made the classic experiments in 1780. They determined to measure animal heat. Devising a three-chambered vessel with a drain at the bottom, they put ice on the outside to keep the temperature constant, as far as the outer atmosphere was concerned. The middle chamber was first filled with crushed ice, which was allowed to drain enough to let the experimental object be placed in it. When the object was cooled, the water melted by it from the inner chamber could be measured. They first placed an ounce of glowing coals in the inner chamber. It was consumed in three minutes, yielding 6 pounds 2 ounces of water. Then a guinea pig was placed in the inner chamber. This seems to be the first emergence of this traditional experimental animal upon the scientific scene. The beast was tenderly installed in a little basket lined with cotton. Lavoisier noted that it did not appear to suffer. The guinea pig emerged from the apparatus with the same heat at which it entered, although the temperature of the inner chamber was zero. Its vital functions had restored its heat, which had melted 13 ounces of ice in ten hours.

Using a bell jar, Lavoisier also showed that in the combustion of carbon, 1 ounce of ordinary air, altered in the process, resulted in the melting of almost 30 ounces of ice, while the production of 1 ounce of "fixed air" (carbon dioxide) melted 26 ounces of ice.

Finally, he put the guinea pig into the bell jar with 248 cubic inches of ordinary air and kept it there for an hour and a quarter. He found that the volume of air was diminished at almost the same rate as by the combustion of carbon. Various other carefully measured experiments followed, leading to the conclusion:

> Respiration is therefore a combustion, very slow, to be sure, but perfectly similar to that of carbon. It occurs in the interior of the lungs without the liberation of any perceptible light because the fire, as far as it is freed, is absorbed by the humidity of these organs. . . . Thus the air which we breathe serves two purposes equally necessary for our preservation: it removes from the blood the base of fixed air [CO_2], an excess of which would be injurious; and the heat which this combination releases in the lungs replaces the

constant loss of heat into the atmosphere and surrounding bodies to which we are subject.

Simple enough, but it took thousands of years of trial and error to arrive at what we now accept as commonplace.

It remains to sketch briefly the reward society bestowed upon the real father of chemistry. During the Revolution, Lavoisier went on working. The vindictive Marat, however, remembered and denounced him publicly. Moreover, his financial relationship to the hated Ferme Générale, which had squeezed the people with excess taxes, was enough to damn him.

In 1794 he was dictating notes to his wife while Seguin, his associate, sat in a varnished silk bag, rendered perfectly airtight, except for a slit carefully cemented to his lips, through which he could breathe. Everything emitted from his body, except the air from his mouth, was being measured. The emissaries of the Revolutionary tribunal burst in and dragged Lavoisier away. He was guillotined in May of that year.

A fitting end to the oxygen story is the scene which was solemnly staged in 1782, the centennial of Becher's death. Madame Lavoisier, robed as a priestess, surrounded by the scientific notables of her salon, burned Becher's books while her friends chanted a requiem for the demise of phlogiston.

16

Revolution and Evolution

◆§ THE Jardin du Roi, which Buffon had made into a leading center of natural science, was now to become the breeding place of evolutionary theory. Three men succeeded Buffon. Between two of them, Baron Georges Cuvier and Jean Baptiste Pierre Antoine de Monet, Chevalier de Lamarck, existed a conflict in theory of primary importance. The third important figure in zoology was Étienne Geoffroy Saint-Hilaire, often at odds with Cuvier but not wholly a partisan of Lamarck's.

Geoffroy Saint-Hilaire, after hearing stories of his distinguished ancestors from his grandmother, said: "I should like to become famous but how to become so?" "By willing it strongly" was the firm reply. Born in 1772 at Étampes, he was sent to the College of Navarre, where the only subject to which he sparked was physics. Finally, he was sent to the Jesuit College of Cardinal Lemoine in Paris, where he discovered mineralogy. In 1792 the Revolution burst, scattering the establishment far and wide and endangering not only aristocrats but clerics and all those suspected of sympathy with the past. Most of Geoffroy's former teachers were imprisoned. The young man flew loyally to the aid of his friends. Indeed, he seems all his life to have been of an impetuous emotional disposition. Some of the faculty he helped to escape, but his mineralogy teacher, Father René Haüy, refused to be rescued in this manner. Undaunted, Geoffroy rushed to members of the Academy, particularly Louis Daubenton, Buffon's former collaborator and now director of the museum, and persuaded them to petition the cleric's release. The old gentleman, however, could not immediately leave his prison; he first had to put his mineral collection, which he had brought with him and which had become disturbed, in order. Haüy told Daubenton: "Love and adopt my young deliverer."

Thus it was that the famous anatomist became Geoffroy Saint-Hilaire's patron. Through his influence the young man became superintendent of the cabinet of zoology in the Jardin du Roi, although he was only twenty-one years old.

He was able to pay his debt to the old zoologist, for during the following year Daubenton reprimanded one of the workmen in the Jardin. In revenge, the man spied and reported bits of conversation he had overheard to the tribunal of the section, denouncing Daubenton as an enemy of the state. Once again, Geoffroy rushed to the defense. It was discovered that Daubenton had once written a monograph on sheep. Suspect as a museum director, as a shepherd he was given a clean bill of health. With the dictatorships of our time in mind, we can sympathize with the scholars who were trying to keep science alive in the midst of the tensions and dangers of civil upheaval, menaced by a political mentality which was to sanction the death of Lavoisier.

Jean Baptiste Pierre Antoine de Monet, Chevalier de Lamarck, was born in 1744, in a small town in Picardy, to a family which maintained a military tradition. How he became Chevalier is something of a mystery since he was the youngest of eleven children. We assume all his older brothers must have died early. Jean Baptiste was destined for the church and put into a Jesuit seminary. As soon as his father died in 1761, he bought an old horse and set out for the Seven Years' War. Immediately after his enlistment, he found himself in a battle near Fissinghausen. His company was enfiladed, and when all the officers were killed, the seventeen-year-old hero took charge and held the position until help arrived. He was immediately awarded a lieutenant's commission.

Despite this brilliant beginning, he was not destined for a military career. The young lieutenant was transferred to Monaco where, in a playful scuffle, a brother officer lifted him by his head, with the result that the glands of his neck became inflamed. Because an operation was not completely successful, he finally resigned from the army on a minute pension of 400 francs a year and went to Paris to recover. There he starved in a garret, began to study medicine, and worked as a bank clerk.

It so happened that he acquired a penchant for going on long walks in the country outside Paris. On one of these, he made the acquaintance of Jean Jacques Rousseau, who liked to botanize. It seems likely that Rousseau may have stimulated his interest in botany, for he dropped medicine, began the study of French flora, and nine years later, in 1778, published his book on the botany of France, which made his reputation. Botany also brought him to the attention of Buffon who saw to it that he was admitted to the Academy of Sciences and engaged him as tutor for his son (destined to die on the guillotine). Buffon became Jean Baptite's patron, getting the young man appointed royal botanist and collector for the Jardin du Roi. He spent the year 1781 traveling with young Buffon in Germany, Holland and Hungary, collecting for the Jardin.

Of Lamarck's personal life we know next to nothing. Gentle, fine-featured, with a kind of philosophic reserve (he never defended himself from attack), always impecunious, he managed to marry four times and produce seven children. We do not know the names of his wives, and only two daughters, Rosalie and Cornelie, seem to have survived him.

He continued to work quietly at botany until the Revolution. During that period of terror, he and the other scientists realized that the name Jardin du Roi was asking for trouble. The denunciation of Daubenton may have influenced them. They changed the name to Jardin des Plantes. When the revolutionary government took notice of the institution, it decided to modify it, and in 1795 it became a museum to which twelve professorships were attached. Since no zoologists were available, the mineralogist Geoffroy and the botanist Lamarck were engaged to divide the territory, the first taking the quadrupeds and higher animals while Lamarck was left with the rest, as they were listed in a museum publication, "insects, or worms and microscopic animals."

It was at this point that the third actor entered the scene. Georges Cuvier, the son of a Swiss Protestant army officer, was born in Montbeliard in 1769, then a part of the Duchy of Württemberg. As a boy, he was considered delicate by a doting mother. She seems to have overwhelmed him, for all his life he had a strong emotional reaction to red stock, her favorite flower. She taught him to read at the age of four and saw to it that he studied Latin, drawing and religion.

By the time he was sent to high school, or gymnasium, he had been fascinated by a copy of Gesner's zoology and he was familiar with Buffon's many volumes. Fluency in Latin (he composed an ode on the duke's birthday) brought him to the attention of the ruler of Württemberg. The latter looked at his drawings, was duly impressed, and became his patron. Duke Alexander and his secretary swept down upon the household, picked up the gifted fourteen-year-old, and carried him off to the University of Stuttgart. Cuvier sat silent and overawed while the two men conversed in German, a language of which he knew not a word. He seems to have made fantastic progress in the language, for after only four months at the university he received a prize for his German studies. He read law and mathematics but found time to become familiar with the achievements of Linnaeus and to botanize. He also continued to develop his gift for drawing, keeping a record of the birds and insects he encountered in his nature studies. The biology teacher Karl Friedrich Kielmeyer (1765-1844) was an imaginative natural scientist and proponent of comparative anatomy. From him Cuvier got a good grounding in biology. The young man's years in the German university seem to have had much to do with forming his character, for there was always something slightly rigid and authoritarian in his scientific personality.

Having obtained his degree at the age of eighteen, he returned home when political changes deprived him of his patron. Although it had been hoped he would go into government service, he had no money and no influence. He therefore took a position as tutor for the children of an aristocratic Protestant family in Caen, Normandy. There he continued his natural science at the seashore, dissecting cuttlefish and mollusks. He also began to think of reclassifying Linnaeus' catchall class of *Vermes* (worms) into which most of the invertebrates were dumped.

Then a certain Abbé Alexandre-Henri Tessier arrived at Caen, apparently under an assumed name, for he was hiding from the Terror. He was not able to refrain from making contacts with the local intelligentsia and encountered Cuvier, who recognized him as the author of some scientific articles on agriculture. The abbé was terrified of being exposed, but Cuvier calmed his fears and became his friend. It so happened that Tessier was a good friend of Geoffroy Saint-Hilaire; in fact, the young man looked up to him as one of his masters. Tessier was keeping up his scientific contacts and in the course of his correspondence praised Cuvier highly, proclaiming, with a Parisian scorn of everything outside the capital, "that he had found a pearl in the dung heap of Normandy."

The impetuous Geoffroy could not wait to acquire the new genius. He wrote to Cuvier: "Come and fulfill among us the part of a Linnaeus—of another lawgiver of natural history." History was to make this statement somewhat ironical.

At any rate, with the concurrence of Lamarck a place was made as assistant to the professor of anatomy, and since this did not pay much, a position was also found for him as a teacher in a private school. At this point, one curious fact is recorded: Cuvier was friendly with the Prince of Monaco and the ruler offered him a room in his mansion in Paris. How or where Cuvier made this connection remains a mystery, but he always had a genius for gaining the favor of people in high places.

Thus in 1795, the three men who were to be the leaders in French zoology became associated in the Jardin des Plantes. Geoffroy busied himself with the zoo. So far, no money had been appropriated to buy animals, and thus one day, when the police arrived with a leopard, a white bear, several mandrills and a panther which they had refused to allow private exhibitors to display to the public, Geoffroy rushed up and down, raised money to indemnify the owners, and set up temporary quarters for the cages. Lamarck, in this period, was working on his reclassification of the grouping under *Vermes* and petitioning for one of the grants now set up for citizens outstanding in the arts. He listed his services to the state, his books, his share in organizing the museum, ending with the statement, "He is the father of six children, most of them under age, and absolutely without fortune." He received a grant of 3,000 francs.

Meanwhile, Cuvier, who had found a very inadequate anatomical collection, set about enlarging it. He worked together with Geoffroy and they became very close. Soon Cuvier moved into the same house. Looking back, Cuvier said, they never breakfasted without first having made a discovery. One evening the worldly-wise old Daubenton directed Geoffroy's attention to Fontaine's fable of the bitch who sought shelter with a friend and then took over the house. Concerning such friends:

> Yield but a foot to their appeal
> And find they have usurped four feet.

As it happened, Cuvier's climb to fame was meteoric and uninterrupted. Good looking in a vivid and overpowering way, he was used to being the handsomest man in every gathering. In addition to turning out brilliant scientific work, he thrived upon government tasks, which did not seem to interfere with his productivity as a scientist. In the end, he dominated the scene and simply by the force of his personality and the universal recognition of his achievements overshadowed his colleagues.

The Revolution ran its course and gave way to counterrevolution in the form of the Directory and Napoleon. Alongside his bloody ambitions, the dictator-to-be cherished a desire to appear as a patron of the arts and sciences. In 1798 both Cuvier and Geoffroy were approached concerning a top-secret project and invited to accompany Napoleon to an unspecified destination. Cuvier declined, but the adventurous Geoffroy accepted. It turned out he was participating in the disastrous Egyptian campaign. Napoleon, encouraging a romanticism of the Orient, was bringing with him one hundred seventy scientists, artists and literary men who were to make a study of the mysterious land of the pyramids. When the fleet stopped at Malta, the zoologist, who was on board the frigate *Alcestis,* caught and dissected a shark and also captured its pilot fish.

Despite losses from lack of provisions and water, Alexandria and Cairo were taken, the cavalry of Mamaluke princes being no match for the disciplined French. Geoffroy spent his time in Alexandria, going on trips (with a military escort) on which he collected birds of the delta and prepared skins and skeletons for the museum. When he moved on to Cairo, the commander put him on a commission consisting of seven scholars to form the Cairo Scientific Institute. Struck by the archaeology of Egypt, he and most of the other scientists visited ruins and pyramids.

In August, the English fleet under Nelson destroyed the French ships in the Battle of the Nile, leaving the army cut off from France. Napoleon still remained hopeful. Ships continued to run the blockade, thus keeping up some communication with France. Things were going badly for France, but for a year Napoleon stayed in Egypt, expecting a change in his luck. Geoffroy, aside from collecting specimens, which included mummies of animals from tombs, was helping to make a map of Egypt. He traveled up the Nile and acquainted himself with the ichthyology of the great river. After a disastrous defeat in Syria, Napoleon learned that the French were being attacked by both Austrians and Russians. In August, 1799, with a handful of chosen men, he ran the blockade and returned to France, abandoning his unfortunate army.

Shortly before leaving, he made one of his statements meant for history. As he and Geoffroy Saint-Hilaire were viewing the pyramid of Gizeh, he announced gravely: "I find myself conquering Egypt, as did Alexander; it would have been more to my taste to follow in the footsteps of Newton."

The group of French intellectuals, comprising engineers, physicists, mechanics, surveyors, mathematicians, astronomers, cartographers, architects, chemists, economists, archaeologists, Geoffroy, the zoologist, as well as artists, musicians and poets, were lodged in a number of fine houses in the garden district of Cairo. They had set up their studios, laboratories and workshops and were preparing a vast report on Egypt. They lived well and worked happily under General Jean Baptiste Kléber, who was left in command of the country. Kléber hung on until he was assassinated and replaced by the inept General Jacques François de Menou in June, 1800.

By now the unhappy intellectuals were anxious to return to France with their data and collections. Since Cairo was ravaged by disease, they moved back to Alexandria. Menou agreed to negotiate a safe conduct but tried to deprive them of all their documents and collections, for he stupidly seemed to think these would betray the hopeless military situation.

When this conflict over censorship was fought out, they set out in the brig *Oiseau* but were shelled by the British admiral. When they appealed to Menou, he sent a message that he, too, would sink the brig if they did not return. The harassed scholars returned to share in the misery of a siege that ended in plague and starvation.

Finally, when the British shelled the city and it capitulated in 1801 (through all of this Geoffroy was studying electric rays), Menou signed an agreement by which all the scientists' precious material would be turned over to the British, who had sent a representative to snatch it for their museums. Geoffroy and a committee went to the British General Hutchinson. When he proved adamant, the fiery Geoffroy Saint-Hilaire cried out that they would burn their collections and documents and the British would become as infamous as those who destroyed the library of Alexandria.

The threat worked and the scholars returned home.

Meanwhile, Cuvier had been consolidating his position. He had brought out a book on anatomy and was now professor of that study in the Collège de France. In two years, he had become famous, and in 1802 Napoleon appointed him one of six inspectors to set up a new public school system in France. He married the widow of a general executed during the terror and eventually had four children.

In 1802, Lamarck, who had quietly been developing his ideas all along, published his *Recherches sur l'organisation des corps vivants* (Studies of the Organization of Living Bodies), which contained his ideas on classification and a first statement of his theory of evolution or "transformism," as it was to be called in France. Despite the importance of his insights, they seem to have gone practically unnoticed.

Geoffroy, too, was developing some ideas of his own. As early as

1795, he had written: "All the most essential differences which affect each family included in the same class arise merely from different arrangements, from complications or modifications of the same organs." He had been influenced by Buffon's belief in the arbitrariness of Linnaean species and seems to have rejected his own work on mammals (which was published in 1803 and then withdrawn from sale to the public).

In 1808 he was sent on a new mission by Napoleon "to explore the scientific riches of Portugal." Napoleon, having made himself dictator by a coup on the strength of his so-called Egyptian victory, was marching roughshod over the rest of Europe. French culture was proclaimed supreme, and foreign museums were rifled of their best works to stock those of France. Although Geoffroy was armed with a document giving him the power to confiscate whatever he liked, his son reported that he took with him crates of zoological material which he planned to exchange for specimens from Brazil. Unfortunately, he fell into a hornet's nest. The dictator had used Spain to maneuver against Portugal, and now he was playing Ferdinand, the Crown Prince, against his own father, Charles IV of Spain. Meanwhile, Napoleon had poured troops into that country. The result was a bloody, spontaneous popular uprising which was both leaderless and without specific aims, except hatred of the French.

Spain was seething when Geoffroy arrived in Madrid. He immediately set out for Portugal, only to be arrested at the border. Mobs screaming "Death to the French!" collected outside the jail and even tried to burn it down. Fortunately, Geoffroy and his friends had helped a Spanish lady, whose coach had overturned, by taking her into their own carriage. She happened to be the niece of the governor of Estremadura, who released them and allowed them to proceed to Portugal. As a result, the governor was subsequently murdered. At first, the arrival of the French zoologist created a similar uproar in Portugal, where it was thought he would plunder the museum. Luckily, according to his son's account, he managed to convince everyone of his intent to trade specimens and he even put the collections of the Museum of Lisbon into proper scientific order. The Peninsular War, however, was also heading for disaster. With the arrival of Wellington, the French had to evacuate and Geoffroy was once more faced with the loss of his collection. This time he negotiated, giving up four crates of material, the rest being released as his personal property.

On his return to France, he was appointed to a professorship in the science faculty of the University of Paris, which Lamarck had turned down on the basis of his age. Lamarck was only fifty-three, but he seems to have always been of a retiring disposition. Cuvier had gone on setting up school systems in various conquered countries.

Once more, in 1814, France was shaken apart politically, this time by the fall of Napoleon. Cuvier had just been made a Councillor of State, but he was confirmed in this office after the Restoration by Louis XVIII, even though he was a Protestant. Geoffroy also was elected to the Chamber of Deputies, but retired after less than a year. From this time on, Cuvier's honors and responsibility increased. In 1817 he visited England, stayed with Sir Joseph Banks at his country estate, and familiarized himself with various natural science collections, including Hunt's. On his return home, he was made a member of the Council of State and also awarded the Legion of Honor. He published many volumes on ichthyology and continued his investigations of the fossils in the environs of Paris which he had begun in 1800. His *Le règne animal* (The Animal Kingdom), published in 1817, developed his plan of classification in which he opposed both Lamarck and Geoffroy. After another visit to England, he returned to find that France had once more suffered a political upheaval. So firmly fixed was he in public life that Louis Philippe not only retained him on the Council but made him a peer. His daily routine reminds us of Buffon. He rose at seven and read the newspapers. Then he looked over and passed on school books in his role of Commissioner of Education. He then set the tasks for his assistants in the Jardin for the day. He inspected the menagerie, then went off to fulfill his governmental role in the Council. At night, he placed a lamp in his carriage so that he could read and write. Before dinner he joined his family in his wife's sitting room. After dinner he spent an hour in the drawing room with friends, then went back to work. His favorite relaxation was to lie on the sofa while his wife or daughter read literary works in various languages to him.

Cuvier never learned English. When in that country, he depended upon his daughter to interpret. Indeed, he was always indignant when he encountered Englishmen who could not speak French. Although he generally maintained an Olympian calm, at home he was inclined to explode over petty frustrations.

Cuvier maintained his dominant position under Charles X and succumbed to an epidemic of cholera in 1832.

Lamarck, after going blind and being forced to give up teaching, died in 1829. He was buried in a pauper's grave. His two important works, *Philosophie zoologique* (Zoological Philosophy) and *Histoire naturelle des animaux sans vertebres* (Natural History of the Invertebrates), were published in 1809 and 1819-22. After his death, Cuvier wrote a eulogy which is a decided blemish on the excessively noble public image created by the paleontologist's admirers. Cuvier waited two years to write it and died before it was delivered (by a colleague) to the Academy of Sciences. Its monumental contempt was so extreme that it was censored by the Academy before it was published. By deliberately misinterpreting Lamarck's ideas, Cuvier set going a prejudice against his scientific colleague which tended to obscure Lamarck's importance for decades. "A system resting upon such bases," he wrote with authoritarian finality, "may entertain the imagination of a poet but it cannot for a moment support the examination of anyone who has dissected a hand, a vital organ or a mere feather."

Cuvier was indifferent to general theories and stuck to a tradition in comparative anatomy, created by such men as Daubenton and Camper, but he differed from them in placing more emphasis upon the animal kingdom than upon man. He was particularly concerned with the correlation between the organs in the body of the individual. A carnivorous animal, for instance, must possess good vision, efficient locomotive organs, claws for clutching, sharp teeth for tearing and a digestive system adapted to handling protein. It never had hooves or flat molars, for these characteristics are correlated with a herbivorous diet. He therefore divided the animal kingdom into *Vertebrata, Mollusca, Articulata* and *Radiata*. Each had a specific ground plan, modified in different ways, in the families included under the group. Between the four groups absolutely no comparison was possible. To arrange all animals in a series was absolutely indefensible. Species were immutable.

Although Cuvier's work in fossils led him to the discovery of many extinct pachyderms and ruminants, not to mention a species of gigantic extinct reptiles of the Cretaceous age, his reaction was to develop his "catastrophe theory," which became a dogma. Cuvier was a pious Lutheran, vice-president of the Bible Society, and it is clear that unconsciously he wished to save something of the Genesis account of the Creation. Cuvier had clearly seen the stratification of the geological deposits which he had studied near Paris. His catastrophes were produced "by diverse overflowings and retiring of the sea." For instance, the preservation of entire mammoth corpses in the far north was due to the sudden flooding of the icy sea. Cuvier did not suggest special creation of new species after each catastrophe (this was the contribution of the pious-minded who seized upon his ideas with approbation). Instead he suggested that if Australia were flooded, all the monotremes, such as kangaroos and koalas would disappear, since they existed nowhere else. Then, if a land bridge were created with Asia by a falling sea level, a whole new group of animals would eventually populate the continent. He took refuge in negative arguments to refute Lamarck's transformism:

> If species have gradually changed, we must find traces of the gradual modifications that existed between the paleotheria and the present species. . . . Why have not the bowels of the earth preserved the monuments of so remarkable a geology?

The answer is, of course, that since his time, such intermediates have been found, proving that it is dangerous to generalize on the theory that everything is known. He stated flatly that there were no human fossil bones and all those found in the ground were simply the remains of men who had fallen into crevasses or been buried in mines.

Finally, he drew upon the mummies of dogs, ibis, birds of prey, monkeys, crocodiles and bulls that Geoffroy Saint-Hilaire had brought back from Egypt. These were identical with living species, which proved there was no modification. This, of course, betrayed Cuvier's concept of geologic time as being influenced by Christian theology, which reckoned the age of the earth at about six thousand years.

On the positive side, Cuvier's book on fossil bones, *Recherches sur les ossements fossiles,* published in 1812, really founded the separate study of paleontology, or animals of the past. He became such an expert from his study of correlations that he was credited with being able to reconstruct a whole animal from a single bone. He was often called upon to unmask frauds. One of these was the alleged skeleton of an ancient German king Teutobochus Rex, found in a mythical thirty-foot sepulcher. Teutobochus was composed of elephant bones. A giant extinct salamander was the basis of a skeleton, the remains of "the accursed race" drowned in the biblical Flood. All in all, Cuvier discovered one hundred sixty-eight fossil vertebrates, belonging to fifty genera.

Both Lamarck and Geoffroy have been accused of fantastic speculation, but in both cases their flights of imagination were united with valuable insights, insights more progressive than Cuvier's despite the value of his analytical and descriptive work. The temper of the times was changing. Romanticism was in full flower in Germany and philosophy was once more invading the sciences. Philosophy has always tended to set up systems, which on later critical inspection are found to have little solid basis. If system building and overambitious synthesizing were the weakness of the new zoological theorists, nevertheless, in the case of Lamarck the basic viewpoint of transformism marks the beginning of modern zoology. The dogma of fixity of species was symptomatic of a static point of view toward the world. Whether acknowledged or not, it stemmed from the old Christian synthesis, which was basically static, hooked on absolute and literal interpretation of the Bible.

Even though Lamarck often mentioned the Creator, he was one of those who kept his religious sentiments apart from his concept of dynamic change. And it was just this belief in change and modification of living entities that constituted a first statement of the modern view of the evolutionary history of the world.

There had been hints of this in earlier writers. Bonnet had toyed with a connected series of animal groups; Erasmus Darwin had suggested that animals were affected by environmental changes and that these modifications were passed on to their descendants. Buffon believed that one species might give rise to another. It was Lamarck, however, who boldly insisted that all living forms could be arranged in a developmental hierarchy and that the world had not survived catastrophes but had, throughout its history, undergone gradual change.

Lamarck's independence of mind was demonstrated in his book on geology, printed in 1802 at his own expense and ignored by the scientific world. Albert Carozzi who translated and reprinted it in the United States in 1964 points out that by his conception of millions of years of geologic time the author was fifty years ahead of his generation. He wrote: "Oh, how very ancient the earth is! And how ridiculously small the ideas of those who consider the earth's age to be 6,000 odd years!"

It was Lamarck, the taxonomist, who gave us the broad subdivisions of the animal kingdom, vertebrate and invertebrate. He had a tendency toward two-part classification. The concept of vertebrate uniting many phyla by a positive characteristic of a backbone is thoroughly sound; invertebrate, although it is still used, fails to be "natural" in that it is negative and unites such disparate beings as earthworms, giant squids, and bees.

As his ideas developed, he added more classes to his subdivisions of the invertebrates, ten in all by 1807. His method was to start from the top and work downward as organs become simplified and finally disappear. Reptiles came lower than the mammals because of their cold blood and incompletely formed heart and lungs. The duckbill platypus (just discovered) seemed to him a link between birds and mammals. The mollusks stood higher than the true worms because they possessed a brain, nerves and a single-chambered heart.

His whole scheme went as follows:

Mammals
Birds
Reptiles (including amphibians)
Fishes
Mollusks
Cirripedes (barnacles and the like)
Annelids (worms living in water)
Crustaceans
Arachnids (spiders)
Insects
Worms (including parasites)
Radiates (starfish and their relatives)
Polyps (sponges and hydra, etc.)
Infusoria

Although some groups have been ranked higher or lower since his time and many more phyla have been created, Lamarck introduced some new and well-founded distinctions by separating insects, crustaceans and spiders, for instance.

Lamarck is best remembered, of course, for his theories concerning the origin of species which have given rise to controversies still continuing today. Fundamentally, said Lamarck, all life is motion, which underlines the difference between his attitude and that of Cuvier.

The basis of his theory of the modification of species was the belief that organisms reacted to their environment. He maintained that frequent and sustained use of an organ makes it larger, size is inherited, and also that the reverse is true. He could point to the near disappearance of the whale's hind legs, which are only a few inches long, and also to the small useless wings of flightless birds. He also believed "that nature preserves by heredity those forms" which arise from use and disuse. The production of a new or modified organ arose from a need (not *wish,* as Cuvier had distorted it in the elegy). In other words, "if an animal is urged to a particular action . . . the organs which carry out this are immediately stimulated."

An example of this would be ancestors of the duck getting into the water, stretching their toes and attempting to paddle. As a result, skin somehow grew between their toes. Since Lamarck knew nothing of cell theory or chromosomes, he could not explain the mechanism of physical change in any further detail. Although he believed that the better adapted animal would be more likely to survive, his theory differed from random change (from which the best examples are preserved by natural selection) in that he felt something took place in response to a direct stimulus. The fact that certain species of animals grow thicker coats of fur when they live in colder regions was another detail that fitted in with his ideas.

As we shall see, the problem of transformism, or the modification of species and the creation of new ones, has become one of the great questions of biology since Lamarck's time. More and more evidence from paleontology has convinced us of organic dynamism in nature, but the how and the why of natural selection are still matters of controversy.

In Lamarck's own day, the theological notion of the fixity of species was so strong that he was simply considered slightly mad, and it was easy for Cuvier to destroy him with majestic contempt. The fact that he was interested in meteorology and printed a bulletin of weather prediction, which he carried out by observing clouds from his top-floor apartment, was also held against him. Yet there again, he was on the right track, for he tried to get reports from different distant areas, a project which, to be practical, had to wait for the telegraph and other modern forms of rapid communication.

Lamarck had his partisans. His loyal daughter, Cornelie, used to say: "Posterity will honor you and avenge you!" And Geoffroy Saint-Hilaire, always his friend, cried at his funeral: "Attacked on all sides, injured likewise by odious ridicule, Lamarck, too indignant to answer these cutting epigrams, submitted to the indignity with a sorrowful patience. Lamarck lived a long time, poor, blind, and forsaken, but not by me. I shall ever love and honor him!"

But what was Geoffroy Saint-Hilaire's position in the realm of biological theory? Actually he had difficulty accepting the idea of an evolutionary scale such as Lamarck's. Instead he looked for archetypes out of which all other organic forms had developed. This led him to find *analogs*—for instance, the gill plate of fishes and the mammalian ear. Or he tried to interpret the exoskeletons of insects as analogs to external vertebrae of higher animals, likening them to the backbone and sternum of turtles. While such comparisons were superficial and not based on a strict investigation of anatomy, his insistence on relating the great groups of animals (in defiance of Cuvier's inalienable separations) and finding essentially similar patterns of structure was in line with evolutionary thinking that would be supported by further discoveries in paleontology. He was anxious to relate birds to mammals and persuaded himself that he had found teeth in the jaws of a young parrot. While this was no doubt self-delusion, in 1861 a fossil toothed bird, the *Archaeopteryx,* was actually discovered, which formed a link between birds and not mammals but reptiles. Once again, theory was running ahead of itself and, while intuitively right in its general direction, had to wait for more data to correct the details.

Geoffroy Saint-Hilaire's search for archetypes seems to have run parallel to some of Johann Wolfgang von Goethe's researches in anatomy in 1784. Of course, Geoffroy also had Pierre Belon and Buffon to draw on as ancestors. Goethe was an admirer of the Frenchman's work, and as a result, there was probably an exchange of influence. In 1830, the antagonism between Geoffroy and Cuvier broke out in the Academy of Sciences. At one point, the impetuous Geoffroy is said to have leaped to his feet, crying that a single plan ran through all animal structure from simple to complex. All men would recognize this if a certain person in the room did not use his great authority to obscure truth. Both men published monographs refuting each other. When Goethe heard of the controversy, he cried enthusiastically, "The volcano has burst forth," and described it as "a revolution in human intellect." He sent a message of encouragement to Geoffroy. The latter, with Cuvier's death in 1832, was left master of the scene. His ideas were carried on by his son, Isidore, when he died in 1844. Their general significance, however, continued to be developed in Germany and gave rise to a school of biological thinking that is known as nature philosophy.

17

Contemplation of the Universe

☙ IN 1774, young men all over Europe began wearing blue coats, yellow waistcoats and boots. This was the result of a literary success, *The Sorrows of Young Werther,* which Goethe wrote at the age of twenty-five, when he was entirely a creature of undisciplined feeling. Werther worshiped nature, devoted himself to an orgy of sensibility, and finally, when frustrated in love, committed suicide.

The Werther syndrome was highly symptomatic of a change in life-style which had been going on in the latter part of the eighteenth century. The deification of feeling came as a reaction to seventeenth- and eighteenth-century rationalism and, in science, particularly against Cartesian mechanistic approaches to the organic world. The excesses of romanticism were also a reaction to a renewed notion that the Greek and Roman classical civilizations had never been surpassed and that this culture of the past embraced all that was worth knowing.

Although it may seem that science is an intellectual discipline, which goes its austere way, disregarding the fashions and irrationalities of culture—and this is indeed the way scientists would like to see themselves—the truth is that all human culture interacts. In order to modify the myth of scientific aloofness, the writer has all along been trying to trace biological thinking in relation to the very human situation of those who developed it. Now the movement, generally called romanticism, produced a romantic biology that was affected by both the virtues and defects of this new attitude toward the world.

The key to much of what was going on lies in the word "nature." For the emotional generation it had as many meanings as the word "God," which it often replaced. It was attached to vague feelings of awe, enthusiasm or ecstasy which might be aroused by anything from a landscape to a woman's naked body. When it was considered philosophically (and in Germany, romanticism and philosophy went hand in hand), it generally led to a search for some cosmic unity, some grand and all-embracing synthesis.

Since nature could easily lead to natural science, and science was linked to increase of knowledge and progressive attitudes in general, when Goethe was seeking for a foundation for his *Weltanschauung* after his trip to Italy, he became involved more and more with botany. After the great emotional release of Italian sunshine and Italian art, he maintained that the transition to cosmic thinking was easy:

> My primitive plant is becoming the weirdest creature in the world, for which nature herself shall envy me. With this for a model, and the key to it, it will then be possible to invent an endless series of plants which will be consistent, *i.e.*, which, even if they do not actually exist, yet could exist and are not just the shadows and fancies of painters and poets but have an interior truth and inevitability. The same law will apply to all other living things.

By "primitive plant" Goethe meant his *archetype*, a generalized form which would reflect nature's basic plan and serve as the key to all living organisms. It is noteworthy that at the end of this passage (from a letter from Italy in 1787) he applied the same idea to the zoological kingdom. This was seven years before Geoffroy Saint-Hilaire's remarks about modifications of basic forms, which indicates that they were thinking along the same lines. In 1790 he published his paper *Versuch, die Metamorphose der Pflanzen zu erklären* (An Attempt to Clarify the Metamorphoses of Plants), in which he stated that plant tissue metamorphosed into cotyledons, stem leaves and flower leaves (petals, stamens, and pistils). Actually, these speculations had been anticipated by the neglected K. F. Wolff in much more scholarly form in 1759, when he pointed out that leaves, roots and flower heads all developed from undifferentiated root tissue.

The chief interest in Goethe's thinking lies in the fact that he was rejecting preformation and fashioning a dynamic and evolutionary image of nature. A vitalist, like Wolff who spoke of an "inner force" which caused growth, Goethe employed the more cliché term "spiritual force." In 1795, he wrote an *Erster Entwurf einer Einleitung in die vergleichende Anatomie* (First Sketch of an Introduction to Comparative Anatomy) in which he again talked of an ideal type to which all animal forms were to be compared. Curiously enough, Platonic ideas which were considered to be absolutes and which contributed to the Christian contention of special creation and fixity of species bear a remote resemblance to the basic archetypes of the nature philosophers. The new archetype, however, being capable of change and evolution, was a useful concept in destroying the old. It applied in a limited sense to the structure of the genera within one family. Nevertheless, some of Goethe's typically poetic notions about anatomy are fanciful enough and even comic, as when he suggested that mammals' tails "can be seen as an indication of the endlessness of organic existence."

On the whole, Goethe's specific contributions to biology were generally wrong or else had been anticipated long before. One of these was his rediscovery of the intermaxillary bone in human beings. This bone was clearly present in the orangutan and, as we have pointed out, was discussed in the controversy over man's relation to the great ape. Goethe, who was in favor of a unity throughout nature, took time out to prove that the intermaxillary bone existed in man and also in several other animals. Unfortunately for Goethe's originality, the bone had been known from the time of Vesalius. It is clearly separated in the embryo, but the sutures disappear in the adult. On the other hand, they remain clearly visible in adult orangutans. Camper, to whom Goethe sent the paper, was polite but unimpressed.

The second contribution to zoological anatomy, involving a theory of the structure of the archetypal vertebrate, brings us to Goethe's involvement with Lorenz Oken, a professional zoologist, one of the best known of the nature philosophers and, in some ways, one of the most fantastic.

His early history is one of homespun virtue and poverty. Born in 1779 to a peasant family, he went barefoot in summer and wore wooden shoes in winter. An industrious scholar in the local Franciscan school, he so impressed the monks that they paid his expenses at the Baden-Baden Seminary. He wrote that because the boys distorted his real name, which was Ockenfuss, to Ochsenfuss (oxfoot), he cut off his "foot" early in life. At any rate, he continued to be an outstanding student and came under the influence of Anton Maier, the professor of mathematics and nature study. In rags and with worn-out shoes, he set out on foot to Freiburg in 1800, where he entered the university, registering in the medical faculty. There he lived and worked in poverty, existing on some minuscule scholarship. Several of his professors took an interest in him and invited him to their homes, where the uncouth country boy was gradually civilized by the refined faculty wives. His subsequent history indicates, however, that certain rough edges were never smoothed away. He formed an attachment for Charlotte Ittner, the daughter of the University chancellor, but later this affair broke up. He maintained that she failed to mend a shirt she washed for him and that she was inefficient in the kitchen. She blamed their incompatibility on his stubbornness and stormy disposition.

Oken read the work of such predecessors as Swammerdam and Bonnet. In his early writings, he took the position that empiricism and philosophy must go hand in hand and that philosophy would give form to empirical knowledge. Unfortunately, he also read the works of Friedrich Wilhelm Joseph von Schelling, an unsettling experience, for Schelling was perhaps the wildest of the nature philosophers. Having started out with the problem that philosophy had invented for itself with Descartes—the cleavage between the subjective and the objective, the physical body and the spirit—he built a verbal bypass around it:

> Nature must be the visible spirit, spirit the invisible Nature. Here, too, in the absolute identity between the spirit in us and Nature outside of us, we must solve the problem how Nature can exist outside of us.

Intoxicated with the magic of the word, like so many philosophers, "absolute identity" took care of everything: "The absolute identity and true inner homogeneity of all matter in every possible differentiation of form is the only real kernel and central point of all material phenomena."

After much talk of electricity and galvanism, about which the nineteenth century knew so little, Schelling arrived at such obscurities as the following:

> The most positive pole of the earth is the brain of the animal and next to this, that of man. For the law of metamorphosis, not only from the point of view of the whole organism but also from the point of view of the individual, means that the animal is but the positive pole (Nitrogen) of metamorphosis of the most fulfilled, that is, the most powerful pole.

By a flight of ideas this led to the remarkable conclusion that the sex organ was the animal's root and the blossom the plant's brain.

Oken got himself involved in Pythagoras' mystique of numbers. All this was to bear fruit in his later writing. Yet in spite of the phantasmagoria of romanticism which boiled in his blood, he made one of the first statements of the cell theory. "The basis of the organic is pure droplets of jelly [*Schleim*] which can be independently established by means of the microscope. They are called cells and the whole complex, cell network." The concept of *Schleim* is, of course, what was to be eventually named protoplasm, and this stands out as one of the insights of the imaginative mind which shows that Oken had a touch of genius, even though he mixed the fanciful with the factual.

The teacher of natural science at Freiburg was not sympathetic to Oken's thinking, and consequently, after obtaining his doctorate, he set off for Würzburg in the hope that Schelling would be able to do something for him. Unfortunately, the philosopher's disposition was as temperamental and arrogant as that of his disciple. An expert at making enemies, Schelling was already out of favor with the powers that ruled at Würzburg.

Oken was finally sent to Johann Friedrich Blumenbach at Göttingen, where he obtained an assistantship to this pioneer of comparative anatomy in Germany. Blumenbach, by his comparison of the skulls of different ethnic types, was laying the basis of physical anthropology (Camper had begun with the facial angle). Unfortunately, there Oken had his first run-in with his superiors. He quarreled with Blumenbach over the intermaxillary bone, which Goethe had rediscovered in 1784. Oken had read Goethe and had seen the bone in embryos, but Blumenbach (no doubt upholding the dignity of man) denied it existed in humans.

In 1805 Oken retired to the remote island of Wangerooge in the North Sea, where for a year he worked as a general medical practitioner and continued to study biology in terms of such sea animals as mollusks and holothurians. It is from this period that his other important insight dates, a concept to be quoted by Goethe in *Faust*, Part II: "Primal sea jelly out of which all life arises. . . ." Or as Oken explained it in his *Die Zeugung* (Generation) in 1805:

> All organic beings originate from and consist of droplets of cells. These, when detached and regarded in their original process of production, are the infusorial mass or protojelly whence all larger organisms fashion themselves or are evolved. The production, therefore, is nothing else but a regular agglomeration of infusoria—not, of course, of species already elaborated or perfect, but of droplets of jelly which, by their union or combination, first form themselves into particular species.

Although the details are confused, the basic insight is remarkable. Another work by Oken during this period was a study of the development of the alimentary canal in which he did not mention K. F. Wolff's pioneer study which, however, he had read.

Because of Goethe's influence Oken was able to escape from his intellectual expatriation to the University of Jena, where he became an assistant professor. At first, he and Goethe got along splendidly. They worked together on anatomy, and when Oken published his theory of the vertebral origin of the skull, Goethe supported him. As time went on, however, their personalities clashed. Oken complained that the poet did not want to treat him as an equal. On the other hand, Oken seems to have pushed his relationship to the duke, Karl August, who gave him the run of the ducal museum and library. This did not please Goethe, who wanted to remain his patron. According to Oken, "they were in a fluctuating state toward each other." This changed to dislike on the part of Goethe. Oken published a *Beitrag zur Optik* (Contribution to Optics), which Goethe felt infringed on his own ideas (as expressed to friends and to which he had witnesses). Oken, however, had published first. By 1810 the great poet said he wanted to have nothing further to do with his former friend. The quarrel was heightened by the fact that Goethe subsequently laid claim to priority in formulating the vertebral skull theory. Oken published a retrospective account of its origins in his journal, *Isis*. During his residence on Wangerooge, while resting from a walk, he happened to contemplate the skull of a deer, when all of a sudden it flashed through his mind like lightning that the skull must be formed from vertebrae. The mammal, in his view, was developed from a prototype, a segmented creature with appendages on each segment. Apparently he talked much of this theory, saying, "Man is a backbone." His colleagues used to greet him with "Good day, Mr. Backbone."

Goethe, too, went on record. In the collected works there is a statement that "in the year 1790, in Venice the origin of the skull from vertebrae was revealed to me." The irony of the whole tempest in a teacup consisted, of course, of the fact that further studies in embryology were to reveal that the mammalian skull does *not* develop out of vertebrae.

Oken's temperamental disposition and liberal political sentiments finally got him into trouble with the University of Jena administration during the wave of reaction after the fall of Napoleon. It was in the important natural science journal *Isis,* which he founded in 1816, that he expressed his sentiments and went so far as to criticize the university administration. Goethe, who had withdrawn his support, wrote unkindly of the periodical,

> It rushes rashly everywhere,
> No broom could be unrulier,
> And wants to be prophetic, too,
> A creature most peculiar.

The journal was prosecuted, the editor sentenced to six weeks in jail and costs, but he appealed the decision and won. Eventually, he had to publish in Leipzig, and by 1819, although the faculty supported him, he lost his professorship.

After a trip to Paris, where he met Cuvier, he secured a position in the University of Munich, where he lasted two years and again got into a squabble over the use of the Royal Library and Royal Museum. When King Ludwig I of Bavaria decided to ease the situation by transferring him to Würzburg, Oken wrote:

> Your Majesty, a German professor is not transferred, he is invited.
>
> OKEN

At fifty-three, he was once more without a job. He hoped to go to Freiburg, but the Minister of State for Baden knew his reputation and did not wish to become involved. Finally, in 1833, he was called to Zurich, where he became a Swiss citizen and had an honorable career until his death in 1851. Alexander von Humboldt praised him for founding the German Society of Natural Scientists and Physicians. Louis Agassiz spoke of him as a great teacher.

A glance at the work that first sketches his nature philosophy, *Lehrbuch der Naturphilosophie* (Textbook of Nature Philosophy), 1809, reveals a certain amount of Schelling's influence, particularly in such statements as: "Galvanism is the principle of life. There is no other life force than galvanic polarity." Oken's is, however, a very different literary personality. While Schelling wrote in cumbersome, turgid, abstract language, Oken's writing has an almost biblical cadence. The sentences are short and clear, the images concrete, even when they are most extravagant. In short, a curious kind of poetry plays in and out of Oken's account of the natural world: "The universe is a ball. If God is to be real, he must appear in the form of a sphere." And again: "Space is an idea like time, a form of God, like time; it is the possible form, the all-embracing $0 = +0-$." Having established that life comes from the sea, "The sea perceives light and it lives."

Oken went on with his myth: if there is light at the point when the protojelly arises in water and air, a solar organism, an animal, is formed; if there is darkness, an earthly organism, a plant, is created:

> An animal is a whole solar system, a plant is only a planet. Originally the plant is a droplet or cell network. In seeds, stem and root there is cell network which is called parenchyma.

Most charming of all is his aphorism: "Animal is blossom without a stem." Aside from being imaginative metaphors, many of his statements carried insights which pointed toward later discoveries:

> The animal cell is an organized cell, an organ, no longer a particle in an anatomical system. . . . All hydra and similar creatures consist of molecules of sensitivity. Higher animals possess nerve molecules.

Quaintly descriptive is: "The highest terminal or head of the gut is the tongue. This is the organ of the gut-sense." Male supremacy peeps out in: "The protoanimal is female. The male is a higher development of the female, not a previously and individually developed animal." Oken's anatomical remarks display the same poetic insouciance: "The eye is a whole body, a whole animal. Milk is a vegetable product of animals."

When he ran through the various classes of creatures, he had more lyrical things to say. The snail excited him:

> When a man looks at a snail, he could believe he is looking at the most exalted goddess sitting on a tripod. What majesty in a creeping snail, what deliberation, what seriousness, what shyness and yet, what absolute trust! Indeed, a snail is a noble symbol of the deeply slumbering spirit within.

Birds came in for their share of metaphorical attention:

> A bird is an insect with fleshy limbs, an insect with a face.
> The capacity to recognize an image of things, I call imagination.
> The bird has imagination and no doubt about it.
> That is why birds can dream.

When Oken came to the human race, he displayed the usual racism, dividing it into Moors and white men: "The Moor is the apeman who cannot blush while the apogee of all, the White Man, can blush." He did admit, however, that the American Indian was an "improved" Moor.

In histories of biology, Oken is generally deplored as an example of undisciplined fantasy. The mechanistic and mathematical approach to scholarship, however, has begun to reveal a certain bleakness, and the gap between humanism and so-called exact sciences has more than once been considered a weakness in our culture. Oken's blend of poetry and biological insight was unique and not without a certain charm.

A third German scientist, also a friend of Goethe's, who emerged from nature philosophy circles was Alexander von Humboldt. He is known, however, less for any specific advances in zoology than for his achievements in geography. He took the whole field of nature for his own and eventually produced an ambitious book, *Kosmos*. Humboldt, like Captain Cook, was attracted by far-off exotic lands and became one of the great travelers of the nineteenth century.

His father was a major, who had served Frederick the Great, and also a baron; his mother was a French Huguenot. Alexander was born in 1769. His youth was spent in the Castle of Tegel, situated on an arm of the Havel River, not far from Berlin. He and his brother, William, who was two years older, were educated by a tutor of liberal views, who indoctrinated them with the romanticism of Rousseau. He was followed by another tutor who excelled in languages. William took to literature and was outstanding in botany. Alexander, who was considered delicate and sickly, did not distinguish himself as a scholar.

In view of the robustness with which Alexander was later to endure the hardships of travel in the jungle, it seems likely that his early ills were psychogenic and the result of a rather pampered childhood. While William went through a Wertherian period in college, Alexander attended the University of Göttingen and in 1788 came under the influence of Blumenbach. Most important of all, he became a close friend of Johann Georg Forster, who had sailed with Cook. In 1790 he went on a trip with Forster through the Rhine area and, finally, to Holland and England. By this time Alexander was seriously studying mineralogy. When his brother went into government service, Alexander prepared himself for commercial pursuits, studying in the Mining Academy at Freiburg in 1791. For the next four years, he was employed as a superintendent of the Prussian mines in Bayreuth.

Nevertheless, his friendship with Forster had stimulated a desire to see the world and explore foreign parts. Since his father had died in 1777, the death of his mother in 1796 meant that the brothers came into money. Alexander was now in Jena, studying the effect of electricity on muscular tissue, in close contact with Goethe. The poet wrote: "I have spent some time with Humboldt agreeably and usefully; my natural history studies have been roused from their winter sleep by his presence." This is testimony to Alexander Humboldt's mature interests. Having sold his share of the inheritance, an estate in Neumark, he was now free to travel and needed only a goal. For a time, history conspired to frustrate him. A group of French scientists were planning to go to South America, but the Napoleonic campaigns put a stop to the project. Humboldt's next plan, from his Parisian base, was to join the expeditionary force in Egypt, but before he could get there, he was thwarted by Nelson's victory and the isolation of the French troops.

Finally, in 1799, together with a French scientist, Aimé Bonpland, he went to Madrid, which was, of course, the base for reaching the southern continent of the New World. He was presented to the king, who graciously issued a royal permit to visit all the Spanish dominions across the sea. The next problem was how to get there.

At the port of Corunna, a corvette, the *Pizarro,* was due to sail for Mexico, but unfortunately the port was blockaded by the British. Humboldt and his friend, Bonpland, had to control their impatience while English gunboats patrolled the harbor. A storm at last came to their rescue by driving the warships out to sea. Full of excitement, Humboldt and Bonpland watched the last lights of Corunna disappear as the corvette successfully ran the blockade.

Humboldt was captivated, like Sir Joseph Banks, by phosphorescent medusae and, at Tenerife, climbed the peak. From this point on, he began to keep careful notes concerning the geographic diffusion of plants and animals. As we shall see, his mountain climbing was extraordinary; indeed, he became an amateur of volcanoes.

An outbreak of fever on board the *Pizarro* caused the captain to change course and run to the nearest port. By the time he reached Cumaná, Venezuela (the voyage took forty-one days), a young Spaniard had died. Since as yet medical knowledge was primitive, epidemic disease was mortally dangerous and it is indeed amazing that Humboldt survived the infections with which he was continually threatened.

Cumaná was a sleepy port whose inhabitants set their chairs in the water, at the edge of the Manzanares River, and smoked cigars as they watched the moon rise. Around the main square were stone arcades, and this was where the slave market was located. Every morning, young blacks from fifteen to twenty years of age were exposed for sale after their skins had been rubbed with coconut oil to make them shine. "The persons who came to purchase examined the teeth of these slaves to judge of their age and health, forcing open their mouths as we do those of horses in a market."

Humboldt pointed out the irony of the fact that these slaves had been brought by a Danish ship, and Denmark was the first country to abolish the slave trade. In Venezuela, there were sixty thousand black captives.

Humboldt took a trip to a ravine containing a cave full of birds similar to the goatsucker:

> It would be difficult to form an idea of the horrible noise occasioned by thousands of these birds in the dark part of the cavern. Their shrill and piercing cries strike upon the vault of the rocks and are repeated by subterranean echoes. The Indians showed us the nests of the *guácharos* by fixing a torch to the end of a long pole. The nests were 50 or 60 feet high above our heads, in holes in the shape of funnels with which the roof of the grotto is pierced like a sieve.

The Indians killed thousands of the young birds, which they took from their nests, for their fat was a principal source of oil. The interior of the cave was supposedly inhabited by the Indians' ancestral dead.

Humboldt pointed out that the Indians were not being wiped out as in North America, but rather a gradual mestization was taking place. First, missions pushed into the interior, then colonists followed. He said that the native tribes fought each other and sometimes practiced cannibalism. The monks tried to increase their little mission villages by taking advantage of native dissensions, and the military lived in a state of hostility to the monks whom they were supposed to protect. Although he admitted that Indian languages which he studied were capable of many delicate distinctions and were rich in vocabulary, he was not a very enlightened anthropologist. He reacted against the custom of painting the body red (which we now know has a religious significance), considered the natives to be dirty and stupid, and remarked that he would prefer to think of them as degenerate groups rather than as primitive ancestors.

Humboldt traveled down the Manzanares, palm-lined and crocodile-infested, in a thirty-foot boat with a three-foot freeboard, no deck, and a huge triangular sail, steered by the Pole Star, until the vessel reached the sea. The port of La Guayra was full of yellow fever, which Humboldt escaped by immediately pushing on to Caracas, in a mountain valley, where he stayed two months. He climbed a volcano, never before attempted, accompanied by sixteen Venezuelans, most of whom dropped out before the summit. He visited sugar and cotton plantations and finally set out for the Orinoco and Apure rivers which he wished to explore.

From Caracas to the interior the terrain consisted of llanos or treeless plains, naked sands or grass over which the winds whirled:

> These sand winds augment the suffocating heat of the air. Every grain of quartz, hotter than the surrounding air, radiates heat in every direction . . . all around us the plains seem to ascend to the sky, and the vast profound solitude appeared like an ocean covered with seaweed . . . the earth was confounded with the sky. Through the dry mist and strata of vapor, the trunks of palm trees were seen from afar, stripped of their foliage and their verdant summits, and looking like the mast of a ship described upon the horizon.

Since the sections of the Orinoco, the Apure and their tributaries ran through these savannas, in the rainy season, when the waterways rose, the plains were covered with twelve or fourteen feet of water. The wild horses, swimming in these drowned plains to reach the tall grass, often became the prey of crocodiles.

These same wild horses, which seemed to be expendable, were used by the Indians to capture electric eels (*Gymnotus*). The horses were driven into a stream whereupon the five-foot fish attacked them, discharging their paralyzing currents. Some horses, so weakened, fell and were drowned. The eels also were finally so weakened that they could be captured and tied with very dry cords.

Humboldt accidentally stepped on an eel and had violent pains in his knees and every other joint for a day. Since he had already made some studies of animals and electricity, he was fascinated by the power of these sinuous fish. He discovered that if two people held an eel, sometimes one and sometimes the other received the discharge. Up to this time, such animals had been compared to batteries, which discharged automatically. Humboldt established the fact that the current could be controlled and directed.

Along the banks of the Apure, he saw jaguars, peccaries, peacock pheasants and bands of tapirs which lived in troops of about fifty. As large as pigs, they swam better than they ran. Often they showed no fear and wiggled their upper lip, as a rabbit wiggles its nose, at the explorer.

Entering the Orinoco, in a sailing piragua with a crew of four Indians, the party found this majestic stream extending like a lake, lined with vast parched beaches. Now there were no birds, none of the herons, flamingos and spoonbills which inhabited the Apure. When a heavy gust of wind nearly capsized the boat, Humboldt's books and manuscripts were deluged. Fortunately, the sheet broke and the boat was righted.

Humboldt encountered certain Indians who used curare to poison their hunting arrows. He promptly studied the effects of the poison. Like most nineteenth-century travelers he had no ethnological interest in Indian dancing and the music of panpipes, which he found monotonous when he attended a feast the Indians staged after a successful hunting trip.

At one point, he found inscriptions and primitive sculpture carved on rocks. He suffered from mosquito bites, saw great waterfalls, stopped at missions, listened to the sad songs of his Indian crew, watched howler monkeys hang by their prehensile tails and swing from branch to branch, and, altogether, covered 375 miles on five great rivers of the interior of South America.

He sent much of his collection off to Havana in a boat which was eventually lost, while he, with his faithful Bonpland, set off for the other side of the southern continent, in the mistaken belief that a friend of his was sailing around the southern straits. Arriving at the Magdalena River in Colombia, Humboldt did some botanizing and the party pushed on to Bogotá on mules. There he stayed until September, 1801, and climbed all the snow-topped volcanoes he could find. He traveled on to Quito, where he spent nine months and climbed Chimborazo and Cotopaxi, the last being the highest volcano in the area, by his reckoning some 18,216 feet. He noted that he could scarcely breathe when he reached the top and blood flowed from his eyes, lips and gums. Although he now learned that his friend had sailed to New Zealand, he decided to go to Lima to see the planet Mercury traverse the disk of the sun. In southern Peru, he admired the Inca viaducts which led from Cuzco to the Páraino de Assuay. He visited mines and ruins near Cajamarca and journeyed up the coast to Lima, where he spent some months investigating the cold currents and gave his name to the one which results in Lima's being covered with a blanket of clouds all winter.

Back in Ecuador, he would have liked to have revisited Cotopaxi, which was in eruption, but he had to secure a passage on a boat from Guayaquil to Acapulco, where he stayed a couple of months, avoiding another epidemic of yellow fever. He visited the mines of Guanajuato and then took a look at a flaming newborn volcano near Michoacán. From Mexico City he, of course, climbed Popocatepetl and Ixtacihuatl and on the way to Veracruz did not omit the peak of Orizaba.

He took a boat from Veracruz, once more escaping yellow fever, and reached Havana. From Cuba he sailed to New York and spent two months in the States studying politics.

The intrepid and indestructible traveler reached Paris in 1804, where he was hailed as a second Columbus, for he had been repeatedly reported dead. He spent his whole fortune on his expedition by the time his great report was published. Various experts worked on different types of material, Cuvier handling the zoology. Bonpland took care of botany (he had collected more than six thousand specimens). The work, in three great folio volumes, with charts, maps and illustrations, took forty years to complete and cost more than Napoleon's famous report on Egypt.

From 1808 to 1823, Humboldt lived mostly in Paris, while his companion, Bonpland, after the fall of Napoleon returned to South America to become a professor of natural history in Buenos Aires.

Humboldt was invited back to Berlin by the King of Prussia in 1823. In Germany, he became a highly popular lecturer on cosmography. It was these lectures which were the germ of his last book,

In 1828, Czar Nicholas of Russia offered to back an expedition to the Urals. Taking a group of scientists with him, Humboldt traveled 2,500 miles and brought back important geological data.

Although Humboldt always refused university appointments, the King of Prussia appointed him Chamberlain and Privy Councillor. He spent the latter part of his life working on his book, *Kosmos,* part of which was published after his death in 1859.

Like most nature philosophers, Humboldt took all the natural world for his own. In his youth, he published a mythological account of life, in terms of chemical affinities, embodying a good deal of fantasy. His pre-South American studies on the effect of electricity on animal bodies, however, were already fairly sober. His impressive years of travel, during which he scrutinized firsthand all of nature, turned him in the direction of collecting factual data. Even though some of his enterprises were truly titanic in concept, in line with the romanticism of his period, as a scientist-explorer he opened up a new world.

His material on zoological and botanical geography, on climate and terrestrial magnetism was invaluable. In his ambitious book, *Kosmos,* "the history of the contemplation of the universe," he devoted his first volume to astronomy, the early history of the earth, climate and both zoological and botanical geography. His second volume contained historical material, covering landscape and nature study, the history of natural science and geographical discovery. In Volume III, he returned to astronomy in detail and continued this to Volume IV. Volume V was devoted to geology. His studies of igneous rock formation were, of course, outstanding, for he had spent untold energy in examining mountains firsthand. He established the fact that volcanoes were grouped in ranges along cracks in the earth's crust.

Not the least interesting is his approach to the relation between the climate, soil and plants in various latitudes. Interested in the shape of plants which dominated specific landscapes, he distinguished sixteen landscape-forming vegetable types. Although his approach was romantic and artistic, such a departure from dry Linnaean classification and such an interest in interrelationships were laying the basis for ecological study to be developed in the distant future.

18

Everything Alive Has Cellular Origin

☙ THE lenses of early microscopes suffered from the defect of chromatic aberration—that is, a colorless object seen under magnification shimmered with the colors of the rainbow. As a result, incorrect interpretations often arose. Leeuwenhoek had somehow partially corrected for this defect by reducing the aperture of the microscope and using only a tiny central portion of the lens. Progress in microzoology had to wait until microscopes were improved. It was not until the early nineteenth century that achromatic lenses were developed, and the improved microscopes were put on the market by an Italian firm in 1827. Consequently, in this period a more accurate analysis of organic tissue began to be possible.

From now on, in anatomical areas of zoology the reductionist process went on rapidly. By this we mean the reduction of matter and processes to their simplest elements and generally the identification of smaller and smaller units. It is a process that generally goes hand in hand with the mechanist point of view. While it has made possible great advances in the understanding of life processes and of microorganisms and much progress in medicine, it must be admitted that the synthetic view of the world has suffered. It is precisely the opposite direction in science from that cultivated by the nature philosophers of the nineteenth century. It has resulted in the rise of hordes of specialized disciplines and specialists who are scarcely able to communicate with each other. Indeed, the reductive process must bear some of the blame for the unchecked exploitation of nature which now alarms us. From the nineteenth century science moves on like a triumphant procession led by the blind.

Yet such developments arise imperceptibly. The pioneers had their immediate problems and worked, as usual, under difficulties. One of the most outstanding investigators to make use of the new microscope started out as a monk. Jan Evangelista Purkinje (born in 1787) was the son of a Czech estate supervisor living in northern Bohemia. The boy did well in school and was sent to the gymnasium in Mikulov, Moravia. This was made possible because he obtained a scholarship for singing in the choir. By 1804, impressed by his Piarist teachers, he decided to become a Piarist father. His education at this point was in German, but he was helped in his work by an excellent knowledge of Latin. Purkinje came to admire the German romantics, particularly Schiller, and was later to translate some of his poems into Czech. The Czechs were a suppressed minority in the Austrian Empire and thus had to accept German culture if they wished to rise in the world. In fact, Jan Evangelista's brother became so Germanized that he was no longer able to speak Czech.

Being a poor boy, Jan Evangelista believed the idea of solving his economic future by settling into the Piarist Order was a good one and it pleased his parents. After two years of teaching during his novitiate, however, he discovered that he wanted more scientific education, that he was overloaded with teaching duties, and that he was pretty much a slave to the officials of the order. He left it and the school and set off for Prague on foot, wearing a pair of shoes that did not fit. He finally hung them over his shoulder and went his way with bleeding feet. In Prague, the first thing he did was to attend a performance of one of Schiller's plays.

Back home, in Libochovice, he found his mother ill and only added to her miseries by his insistence that he could manage without the Piarist Order. Somehow he got to the University of Prague, where he supported himself by teaching and studied philosophy, physics and biology. In this period he was affected by the newly organized Czech nationalist movement and began writing poetry in Czech.

Nature philosophy and Professor Emmanuel Pohl, who taught botany, led Jan Evangelista to the decision that he wished to become a naturalist. Fortunately, he found a patron in Baron Franz Hildeprandt, whose son he tutored. Purkinje had become interested in new theories of education, especially those of Johann Heinrich Pestalozzi and the baron, who seems to have been both generous and progressive, even offering him a building in which he could set up a progressive school. Just at this time, 1813, Austria became involved in the Napoleonic wars and Purkinje's pupil volunteered to fight. His father, however, continued his interest in the young scientist and financed his further education which resulted in a medical degree.

Purkinje became the protégé of Professor F. I. Fritz, lecturer in anatomy, who had connections in Berlin. Meanwhile, the young student was experimenting on his own and wrote his dissertation on the subjective phenomena of vision. Baron Hildeprandt offered him the post of his personal physician, but Purkinje preferred to teach in the Prague medical school and try for appointments in universities. He was rejected by several of these, probably for chauvinistic reasons, for Germans despised Austrians and ranked the Czechs even lower as uncultivated barbarians. Finally, through Fritz's connection with Johann Rust (also a Czech) of the University of Berlin, he became a candidate for the chair of physiology at Breslau. There were many competitors, including Carl G. Carus, a well-known nature philosopher, while Purkinje was facing the usual discrimination against Czechs. Nevertheless, he went to Berlin and made a point of meeting useful people. Fortunately, he had sent his dissertation on vision to Goethe who liked it and used his influence. Humboldt also wrote a letter. The Prussian Minister of Education finally appointed Purkinje over the vote of the faculty.

As a result, for many years Purkinje had to face the hostility of the professor of anatomy and also the curator of the university, a military Prussian bureaucrat.

Nevertheless, by sheer strength of character and ability, Purkinje made his way. He married the daughter of Professor Karl Asmund Rudolphi of Berlin University. Mary bore him four children, but his personal life was destined to be tragic. In 1832, the two girls died; in 1834, his old mother died; and in 1835, his wife died of typhoid fever. Purkinje was left with two young sons.

In 1832, he persuaded the university to buy one of the new achromatic microscopes. Never a theorist, Purkinje from now on embarked on a series of important individual discoveries and made many technical advances in microscopy. He wrote on the development of the hen's egg and identified the nucleus in the primary cell. Bones and teeth had formerly been softened before sectioning for microscopic investigation. Purkinje invented a knife which was the forerunner of the microtome and which could section these hard substances without resorting to the distortion of softening techniques. He was the first to seal in preparations with Canada balsam between two slides. He studied the work of Louis Daguerre, learned his technique, and made the first photographs of microscopic material. He discovered moving cilia in the mucous membranes of mammalian oviducts.

One of his interesting technical devices was borrowed from the primitive experiments with moving pictures at this time going on in

Paris. His kinesiscope consisted of a moving drum on which a sequence of pictures were drawn, which, when projected on a screen, created movemnt. He used this to demonstrate the action of the valves of the heart.

In 1837, in a paper read before the Congress of Science in Prague, he made the statement that there were three fundamental elements of animal organisms: fluids such as blood, plasma, and lymph; fibers in loose connection, such as tendons and fasciae; and granules [*Körner*], by which he meant cells. He also compared the nucleus in the hen's egg with the nucleus in other cells.

Purkinje thus formulated a slightly more precise definition of the cell doctrine than Oken's and did so two years before Theodor Schwann published his famous statement and a year before Matthias Schleiden's work appeared in print.

The reactionaries in the university could not bear his progressive methods of teaching, which involved experiment, laboratory work and prepared demonstrations. They even attempted to force him to teach from a physiology handbook. They also blocked his efforts to obtain space and equipment for a physiology institute. As a result, he did much of his teaching in his home. Every room was filled with bottles, instruments, and preparations. Many brilliant young students wrote their dissertations with the aid of his home laboratory, which has been called the "cradle of histology," histology being the study of organic tissue.

Finally, in 1836, a new Minister of Education came to power. Purkinje hastened to send in the blueprint for his institute. Two years later, he got a small dark building which at least had space for a laboratory, a microscopy room, and a place to keep experimental animals. At first, he was denied funds to pay an assistant, but at last this, too, was forthcoming.

Most of his publications consisted of short papers, but after the creation of the institute, he had little time for his own work. Although his experimental work was carried out from a strictly mechanistic point of view, touches of nature philosophy appeared in his writing: "Life with all its power draws into its substance the physical forces of light and electricity and uses them for its own activities. Then it brings these forces into its deeper center, consciousness, where they are reflected in a higher level than in the sensory organs. . . ."

Purkinje was finally called to the chair of philosophy at the University of Prague. There, he felt, it was his duty to work for Czech nationalist aims. The emperor Franz Joseph had made Austria a police state, and thus on the one hand, the government wished to exploit the scientist's fame, but on the other, the police considered anything Czech subversive. The latter part of Purkinje's life was spent fighting for equality of the Czech language. Before his death in 1869 he started a journal in Czech and founded a Czech medical society.

Although German natural science was affected by nature philosophy, its practitioners were able to do precise descriptive work and, at the same time, hold romantic ideas in the sphere of theory. Oken's *Isis* was an important outlet for their papers and his Society of Natural Scientists and Physicians was a forum for new discoveries. Indeed, its annual congress of 1828 marked one of the important steps forward in the investigation of embryology. To this meeting, which was opened by Humboldt, came Purkinje and Johannes Müller. The paper, which we now know was a landmark, was presented by Karl Ernst von Baer. Up to this time, certain follicles identified by Graaf in the seventeenth century had been accepted as the eggs of mammals, and Graaf gave the ovary its name. Now, Baer announced:

> While I was looking at the ovary of a bitch solely with the aim of comparing the structure of this organ with the ovaries of other animals examined by me, entirely without hope of finding the egg, upon which I counted all the less since the animal was pregnant, I saw quite casually and with unaided eyes, a little yellow body in each Graafian vesiculum in this ovary, which turned out to be an ovulum under the microscope.

To Baer's disappointment, his paper was received in silence. The atmosphere of the meeting must be imagined in all its stuffy formality, a gathering of tall hats, frock coats, beards and whiskers. Baer had no real conception of the ovum as a cell, and the importance of the cell had not yet crystallized. Consequently, the assembled savants did not realize that von Baer had laid the factual basis for the detailed study of mammalian embryology.

Curiously enough, many of the scientists of this period who became eminent seem to have suffered from psychic and emotional difficulties. Indeed, today most of them would probably have been in analysis. Baer considered himself Russian, having been born in Estonia in 1792. He was educated in the provincial University of Dorpat, where scholars were still teaching the Aristotelian notion that the ovum was formed by crystallization. Later, he went to the universities of Berlin and Würzburg. He taught in the University of Königsberg for sixteen years, at first hoping to study the embryology of all animals to prove Oken's archetypal notions. He found, however, that fetuses from different classes differed. He published a treatise on embryology in three parts and had a disagreement with his patron, Karl Friedrich Burdach, who had brought him to Königsberg. He became a prey to depression, could not work, and once shut himself up in his house for a year. When he came out and saw the ripening rye, he threw himself on the ground and wept. In this period, even nervous breakdowns followed the romantic pattern. He somehow felt that travel would cure him, but nothing was done about it until his older brother died and he was called home to manage his father's estate in Estonia. Obtaining a position with the Academy of Sciences of St. Petersburg, he left Königsberg in 1834, even though he had built an anatomy museum from "one moth-eaten stuffed bird, a nest of a marsupial mouse and a cassowary's egg," and become director of the Anatomical Institute in the Russian university.

In St. Petersburg, he had no library, no laboratory, and no assistant, but he was made director of the Zoological Museum. He continually suffered from depression, but now he was able to persuade the Russian government to send him on trips to study the fishing industry, trips which seemed to help maintain his mental balance.

Eventually, he taught at the University of Dorpat and died in 1876 a confirmed anti-Darwinian because he felt that the structure of the organism was shaped by inner forces. He showed that from the ovum, layers of tissues are developed, which give rise to the various organs. This is called the germ layer theory. On the other hand, although he saw the frog's egg subdivide, he still did not recognize the ovum as a single cell.

The cell doctrine was to be developed by a pupil of another great neurotic, Johannes Peter Müller (1801-1858), who had been destined by his father (a master shoemaker of Koblenz) to become a saddlemaker. By the time Johannes was seventeen he had decided to become a scientist. While he was studying at the University of Bonn, he met his wife, Nanny Zeiler, to whom he was engaged for eight years. One of his earliest papers, on the movements of insects, was published in *Isis* and became his dissertation. It was full of Okenism and so fantastic that he repudiated it in later years, even destroying any copies he could lay hands on. Cuvier, K. F. Wolff, Oken and Goethe were his heroes in this period. Wishing to study in Paris with Cuvier, he applied to the grand duke's minister for a fellowship and was awarded one—to Berlin. There he came under the influence of Purkinje's father-in-law, Karl Asmund Rudolphi, professor of anatomy and a more exact type of scientist, who became his "fatherly friend." While still partly a nature philosopher, he produced a book on fantasy, taking off from Goethe's optical studies, which were purely subjective. Müller seems to have had a tendency to hallucinate. In his paper, he clearly stated that sickness of the nerves resulted in religious imagery of all types, gods, demons, and the belief in spirits. The work was of value to further investigation of sense physiology, but it was disastrous for Müller. He had worked on himself as a guinea pig, observing his fantasies which either led to or were symptomatic of a serious nervous breakdown, which took place just as he was to be married to the incredibly patient Nanny.

She must have been a woman of great stability of character, for she married him anyway. One of his colleagues at Bonn, where Müller had a professorship, succeeded in getting him a travel fellowship from travel with the ever-ministering Nanny, he was able to pull himself to-the Minister of Education. By now he was suicidal; yet after a year of gether and teach once more.

The influence of Rudolphi and also the work of the great materialist and biochemist Jöns Berzelius (who visited Bonn) steered him in the direction of exact science, to which he made many contributions, although he always remained a vitalist. He worked on glands, discovered lymph hearts in frogs, and described the development of sex organs in men and animals. He finally wrote an important handbook of physiology.

All his life he suffered from insomnia, and like Baer, he traveled to cure his depressions, writing long, tender letters to the faithful Nanny.

Müller did studies on lampreys and sharks, the lowest vertebrates, and on the amphioxus, a curious fishlike animal, which is now ranked just below the vertebrates. He also worked on echinoderms and mollusks. Thus he was a great pioneer in marine biology.

The unsuccessful revolution of 1848 in Germany was another period of crisis for poor Müller. The students demonstrated, demanding an end to examinations, free education, and crying "Down with Eichhorn!" who was Minister of Culture. Müller, temperamentally a conservative, was made rector the year after the students were quelled (intimidated by displays of the military) and had to endure a year of minor sabotage and hostility. Once more, he was obliged to travel. The fact that he would work ten hours without a break shows that he did not really take care of himself.

A great teacher, Muller trained many of the most important natural scientists of the next generation. In 1833, he was called to a professorship in Berlin, where he fought unsuccessfully for a new anatomical institute. It was there that his influence extended on the two men who were to take the next step forward and, once and for all, crystallize the cell doctrine which had been touched on by numerous scholars but never seen as a universal principle. The two men, whose names are generally linked, are Matthias Jakob Schleiden and Theodor Schwann. Both conform to the pattern of neurosis and brilliance in physiology.

Schleiden born in 1804, the son of a famous Hamburg doctor, studied law and became a lawyer in his native town. Since he already showed a tendency to depression, lack of clients and little success in his profession brought on a suicidal attempt in 1831. He shot himself in the head, but fortunately his aim was poor and he recovered, abandoning law, which he had come to hate. He studied botany and medicine, first in Göttingen and then in Berlin. During his period in Berlin, he published in a periodical edited by Müller, *Archiv fur Anatomie und Physiologie,* an essay, *Beitrag zur Phytogenesis* (Contribution to the Genesis of Plants). Robert Brown, a Scottish botanist, had rediscovered the nucleus of the plant cell in 1831. Leeuwenhoek had seen the nucleus in red blood cells and also in epithelial cells of plant tissue long before, but he arrived at no cell doctrine and his work seems to have been forgotten. Brown went on to state that the nucleus was an essential component of all plant cells. Schleiden picked up Brown's insight and developed it further, saying that the plant is a community of cells. Although he conceived cells as little boxes, an image which goes back to Hooke, who saw only the dead walls of cork cells, he did write: "Cells are organisms and entire animals and plants are aggregates of these organisms, arranged according to definite laws."

Shortly after this, Schleiden departed to take a degree in philosophy at Jena, where he became a professor of botany. After thirteen years at Jena, he took a chair of botany in Dorpat, but quarreled with religious authorities over his natural science point of view. After this, he wandered from city to city, engaging in private practice as a doctor and writing on botany. He went from Dresden to Frankfurt, to Darmstadt and finally to Wiesbaden, pausing at the last city to celebrate the fiftieth anniversary of his law degree, a rather odd affair, since he had not practiced law in forty-five years. He died in Frankfurt in 1881.

While in Berlin, Schleiden made the acquaintance of Theodor Schwann, one of Müller's favorite pupils and collaborators. Schwann (born in 1810 at Neuss-am-Rhein) was of a timid, introspective, pious disposition, whose early studies at the Jesuit College of Cologne brought him in touch with Georg Simon Ohm (who taught mathematics and physics). Unfortunately, the influence of the professor of theology seems to have intensified what was almost religious mania. He desired to become a saint and explained his timidity as a congenital weakness in the spinal cord! This physical weakness was the form which original sin took in him. He moved on to the University of Bonn where, one day out walking, he met Müller. They became friends and collaborators in experiments on the spinal nerves of the frog. Schwann seems to have followed Müller to Berlin, fascinated and stimulated by the older man, and got his degree in medicine in 1834. The following year, Müller made him his assistant in the anatomical museum. Although Schwann wrote on the need for oxygen for the proper development of the chicken embryo (a beginning in chemical embryology), it was probably contact with Schleiden, in 1838, which turned his attention to the cell.

In 1839, after talking with Schleiden, he wrote, "I have found, with the aid of my microscope, that these varied forms of the elementary portions of animal tissue are nothing but transformed cells whose uniformity of texture is found in the animal kingdom and, in consequence, everything alive has cellular origin." His paper, published in 1839, *Mikroskopische Untersuchungen über die Übereinstimmung in der Struktur und dem Wachstum der Tiere und Pflanzen* (Microscopic Investigations Concerning the Similarity Between the Structure and Growth of Animals and Plants) further developed Schleiden's belief in the independent properties of cells. He pointed out that cells were "possessed of an independent life, so to speak; in other words, the molecules in each separate, elementary part are so combined as to set free a power by which the part is capable of attracting new molecules and so increasing. . . ." This really meant that cells were able to assimilate nutritive material—in other words, that they possessed "metabolic power."

Of course, Lamarck in 1809 had written, "But no one, so far as I know, has yet perceived that cellular tissue is the general matrix of all organization and that without this tissue no living body would be able to exist, nor could it have been formed." In those days, however, no one was willing to listen.

After Schleiden and Schwann, other scientists took up the study of cells in both plants and animals, for the cell doctrine had effectually united botany and zoology and from now on we cannot discuss zoology without occasional reference to the science of plants. From 1835 to 1839, Hugo von Mohl worked on cell division, by mitosis or splitting, and Carl von Siebold by 1845 had recognized protozoans as one-celled animals. And finally in the same year Rudolf von Kölliker showed that the spermatozoan and the ovum are cellular products. From now on, it was established that the activities of an organism were the sum of the activities of the cell units and that the cell was the key to the understanding of many aspects of the phenomenon of life.

The effect of the cell doctrine upon Schwann is extremely curious. Under the stiffening influence of Müller, with whom he did his best work, he had become more of a rationalist than his master. Müller believed in "an organic creative force" and that consciousness was "rational creative force," found only in the higher animals. Schwann wrote that the organized body "is not produced by a fundamental power which is guided in its operation by a definite idea, but is developed according to blind laws of necessity. . . ." Apparently he brooded over these points and finally went into another religious crisis in which he had trouble reconciling this point of view with his Catholicism. He wrote in his autobiography:

> If suddenly God should endow bees with the principle of spiritual freedom, this principle would think rationally: my ancestors constructed hexagonal cells, but I can prove mathematically that a cylindrical cell is better and that its walls have a greater capacity. If the bee should follow this reasoning and consider its stubborn opinion superior to its instinct, we can see that that would be the end of the species. This thought made me abandon rationalism.

He also abandoned all specific scientific work, teaching in Louvain from 1839 to 1848 and Liege from 1848 to his death in 1882. The last years of his life were dedicated to a long work which would reconcile all the manifestations of life, including the cell theory, with Catholic dogma. He never finished it.

19

The Triumph of the Atom

☙ ALTHOUGH the microscopists were observing tinier and tinier organic particles until they arrived at the cell, finally agreeing that the cell consisted of something Purkinje baptized as protoplasm, they eventually came up against the problem of the nature of protoplasm. They could go no further without the aid of another branch of natural science, chemistry.

During the nineteenth century, this science advanced by leaps and bounds. We have already seen how such primal gases as oxygen and carbon dioxide were finally isolated. It was an important beginning but no more than a beginning. What was the nature of elements and compounds? How did chemical processes come about? Alchemy and phlogiston had been chased off the stage; new ideas must replace them.

An English Quaker, son of a poor weaver, was to introduce a most revolutionary concept which was to make possible all modern chemistry. John Dalton (1766-1844) was one of those charming English amateurs, practically self-taught, whose mild and amiable eccentricities put him in decided contrast to the string of tortured, emotionally troubled college professors with whom we have been dealing.

By the age of twelve he was teaching school to help out in a household with five other children. His hobby was the study of air. He made his own primitive thermometers and barometers. In 1793, he became a tutor in mathematics and natural philosophy at New College, Manchester. After discovering that he was a poor lecturer, he finally resigned and made his living as a private tutor. Curiously afraid of women (he never married), he nevertheless liked to take tea with pretty girls. Apparently his only recreation was bowling. Everything was written down in his notebooks, from weather conditions (for forty-six years) to his bowling scores. Unlike many such harmless pedants, he happened to be a genius.

It was now known that air was composed of oxygen, nitrogen, carbon dioxide and water vapor. How did these hold together? How were they mixed? Samples of air had been taken from the tops of mountains, in cities and in various other locations, but the gases seemed to be present in the same quantities.

Oddly enough, in this period when the emphasis began to be placed more and more on accurate experimentation, Dalton was a rather clumsy laboratory technician. He set out to solve his problems by merely thinking. He was enough of a classicist to know that in the early history of philosophy the word "atom" had been bandied about. We have earlier mentioned Democritus, the antivitalist who ascribed everything to necessity. Around 500 B.C., he taught that all matter was made up of atoms, invisible particles separated by space through which they traveled. He believed that the perceptible qualities of matter were due to the different types of atoms. Those of water were smooth and could glide over each other.

We have also cited the great poem of Lucretius which put the atomic theory into artistic form. Then there was Newton, who also held a type of atomic theory, writing that matter was formed of impenetrable, movable particles.

Thus an atomic theory had appealed to certain minds, but nothing practical had ever been done about making it more concrete. Dalton decided, without any experimental proof, that chemical compounds were composed of invisible atoms which were indivisible throughout chemical change but which caused change by uniting in various ways. Mercuric oxide, for instance, consisted of one atom of mercury, a metallic liquid, and one atom of the gas, oxygen, which in the presence of heat combined to form a new product, a red powder.

This was a perfectly acceptable scheme for teaching. But further ideas came to him. After some brisk controversy, chemists had shown that when chemical elements combined to form specific compounds, the elements united in definite proportion by weight. Dalton decided that all the atoms of one element would be alike but those of different elements might differ in shape or size. Thus they would have different weights. There was no way of putting invisible particles on a scale, but it would be possible to work out the relative weights.

He took hydrogen gas, the lightest element known, and gave it the value of one. Oxygen and hydrogen united in the ratio of one to seven. Since he believed water consisted of one atom of each element, oxygen had the value of seven. He made out a table of fourteen elements on this basis. He also found that the same weight of one element combined with different weights of another to form compounds with different properties. Although Dalton's actual weights were often wrong, the principle was correct and was to be validated by other scholars. Thus, all at once, the atom, which had been a clever and poetic concept in classical times, began to assume a new kind of reality because it had become an entity susceptible to measurement.

For a time, Dalton's theory was opposed, but gradually opposition faded away as it was realized that in any case, it was a hypothesis of importance as a working tool. Now, of course, we know that it was much more, a breakthrough to a fundamental concept as basic as the cell theory in biology.

From England the torch of chemistry passed to a brilliant Swede, Jöns Jakob Berzelius, born in 1779 in Väversunda, the son of a poor clergyman, who died young. His stepfather was also a clergyman who got him tutors and even tutored him himself, encouraging him to botanize, saying he might become another Linnaeus. When his mother died and the stepfather married again, the stepmother wanted him out of the house. She sent him to board with an uncle. Unfortunately, his aunt was an alcoholic, who continually complained that she wasn't paid enough for his keep. His relations with his cousins were ambivalent. Eventually, early in life he became a tutor. His first job, on a large estate, saddled him with another female alcoholic, the wife of his employer. The employer was a miser who made him sleep in the room where potatoes were stored. Fortunately for the young scientist, potatoes need a certain amount of warmth or they will freeze. His pupils were his age and older and told him flatly they intended to learn nothing.

Berzelius finally worked as an apothecary. One of his tutors, who kept in touch with him, persuaded him to study medicine rather than theology. In the University of Uppsala, he learned French, English and German. Johann Afzelius, the professor of chemistry, who was a legalistic and devious character, refused to open the small laboratory for him more than the official once a week, but Berzelius bribed a janitor and got in. Afzelius eventually caught him, but confronted with a fait accompli, grudgingly let him work.

Berzelius had heard of Galvani's experiments with electricity. He made his own voltaic battery and showed that various compounds could be broken down by passing an electric current through them.

He also propounded the theory that metals always move toward the negative pole, nonmetals to the positive.

Eventually he became professor of chemistry, biology and medicine at the University of Stockholm, where he continued to make important contributions to chemistry. Up to his time, in order to describe chemical processes and elements, a series of cabalistic signs were used, which looked as esoteric as Egyptian hieroglyphics. Indeed, some of them went back hundreds of years, having been inherited from alchemy. Berzelius instituted a revolution by using the initial or first two letters of the Latin name of the element. At first, he used figures to represent the number of atoms in the molecule, above the letters; later they were placed below, giving us, for example, H_2O for water. His system has the great virtue of using type elements already available to the printer and of easily being arranged in mathematical equations. Amusingly enough, the great pioneer Dalton found the new symbols horrifying and obstinately stuck to the old ones.

Berzelius spent months of grueling work measuring the atomic weight of the fifty known elements, many of which he had discovered. He worked them out so accurately that the weights now used, which have been checked with the advantage of modern equipment, only vary from his by a few decimal points.

Known as the greatest chemist in the world, he became a friend of Goethe's, was presented to King Louis Philippe of France, numbered the Crown Prince of Sweden among his pupils, and was visited by the Czar of Russia.

At fifty-six, the great scholar, overworked, suffering from bronchial infections and gout, fell into a depression. For two years, he had no impulse to work and became convinced that he was unable to. The thought occurred to him that marriage might do him good, although twenty-three years earlier a brother scientist had advised against it (his friend had far too many children). Now he asked the advice of Count Trolle Wachtmeister who told him it was just what he needed. Berzelius perked up, married a woman of twenty-four, the daughter of a town councillor, took a new lease on life and lived happily until 1848.

The Swedish scientist contributed to all areas of chemistry, but we are concerned with his contribution to the study of animal bodies and their functions which in his time he called animal chemistry. *Lectures on Animal Chemistry,* 1806, and later his *A View of the Progress and State of Animal Chemistry,* 1813, cover all that was known of what we should now call organic chemistry. He made a forthright statement concerning vitalism. After pointing out that "the constituents of the living body are the same as found in inorganic matter," he went on, "We may consider the whole animal body as an instrument which, from the nourishment it receives, collects material for continual chemical processes, of which the chief object is its own support." Berzelius felt that the cause of many of these processes was so profoundly hidden that it might never be found. "We call this hidden cause vital power . . . we make use of the word to which we can fix no idea. . . . We know, after this explanation, as little as we knew before."

This is an important formulation of a point of view, for it is in essence a restatement, in cautious form, of Cartesianism, the body as machine and the machine powered by chemical processes.

Berzelius did go a step further in analyzing his chemical processes. Rejecting electricity, which the nature philosophers happily labeled a vital agent, he attributed everything to the action of the nerves. He had performed some vivisectional experiments by cutting nerves near the esophagus after which he believed that the blood in the adjacent arteries became venous. To an extent he was right, for the brain activates chemical processes and is, in turn, activated by chemical processes.

The enumeration of organ and body processes that follows shows how little work had as yet been done and how little technique was available for analyzing organic matter. It was known that gastric juice decomposed organic compounds and coagulated milk. Aside from the study of urine and urea through which physicians had learned something about diagnosis of disease, the functions of such body products as saliva, bile, pancreatic juice and intestinal fluid were unknown.

Berzelius worked before the formulation of the cell doctrine, and hence there was no overall structural point of view, from which to start. He spoke of "cellular matter" which filled the space between the organs. On the whole, a few acids and fats had been found in animal tissues, and bones were known to be composed largely of calcium. How the fetus was nourished was unknown because the blood vessels were not connected to those of the mother. He summed up: "I have endeavored to unite the pursuit of one common object, in order to thus give to the investigation of the Animal Chemist a determined and scientific tendency, and to his efforts a physiological view."

It was one of his pupils who took an important specific step forward in relating organic and inorganic chemistry.

In 1823, a young German chemist knocked on the door of Berzelius' house. The latter, himself, came to open it. Friedrich Wöhler had been expected sooner, but a particularly stormy sea trip had delayed him. Now he looked at the "soberly dressed, stately man with a florid complexion" with awe. The famous Swede, however, was kindness itself. He took young Wöhler immediately to see his laboratory where "it was as if I were dreaming. . . ."

Wöhler was the son of a veterinary who had beaten a petty prince, who insulted him, with a riding whip and dashed off over the border to Eschersheim, near Frankfurt, where Friedrich was born in 1800. The elder Wöhler became manager of the Duke of Meiningen's famous private theater and later president of the Frankfurt Institute for the Advancement of Art. The boy was turned over to tutors who developed his weak muscles and his clever mind. While attending the Frankfurt gymnasium, which was purely humanist, he studied chemistry as an extracurricular activity. Since his mother was the daughter of a professor of philosophy and his father was educated in the field of medicine, he could count on support for a scientific career. Even though he was already working busily at chemical experiments in his gymnasium days, he planned to become an obstetrician when he obtained his degree in medicine at the University of Heidelberg. Professor Leopold Gmelin made the suggestion that he should become a chemist and urged him to write to Berzelius with whom he ought to study.

The famous savant's reply went in part as follows:

> Anyone who has studied chemistry with Professor Leopold Gmelin will find he has little to learn from me. But I shall not deny myself the agreeable opportunity of making your acquaintance personally and will therefore be delighted to take you on as my collaborator [*Arbeitskamerad*]. . . .

So it was that Wöhler went to Stockholm for a year with the master, his parents paying his expenses. The association with the great Swede developed into a warm friendship which lasted until the death of Berzelius. Wöhler described him as telling jokes and laughing heartily as they worked when he was in good health. At times, his periodic migraine headaches descended upon him, whereupon he shut himself up for days, eating nothing and seeing no one.

Wöhler became his constant companion and his assistant when, in the great hall of the Academy of Sciences, he performed experiments before an audience of the crown prince and his wife, the queen and a group of courtiers, Berzelius talking in French as he worked.

When important foreigners came to visit, including the famous chemist Humphrey Davy, Wöhler met them and often traveled around Sweden with his master on geological jaunts. When after a year it was time to say good-bye, the great man walked down to the ship with him, silent and depressed. It is easy to see from the fragments of his biography that Berzelius suffered from loneliness. After they parted, the two men kept up a lively correspondence.

Wöhler was teaching in an industrial school in Berlin, where he heard Humboldt's famous *Kosmos* lectures, when in 1828 he wrote exultingly to Berzelius: "I can't, so to speak, hold my chemical water, and must tell you that I can make urea, without kidneys, or even an animal, human or dog, being at all necessary."

Even the antivitalist Berzelius did not believe that organic compounds could be synthetically produced. Thus for the first time, a truly experimental proof had been reached to show that the products of living beings were essentially the same as those of inorganic chemistry and that no essential "vital force" was needed to produce them.

From this brief sketch of developments in chemistry, it will be seen that the study was destined to play an important role, affecting the sister science of biology. And indeed all the French physiologists of the next generation were involved with chemistry.

20

A Living Laboratory

☙ THE best-known French physiologist of the nineteenth century, Claude Bernard, revealed his relation to the romantic movement by starting out as a would-be playwright. If *Werther* was the symbol of the movement in Germany, Victor Hugo's *Hernani* was the revolutionary battle cry in France. Claude Bernard, born in 1813, was the son of a wine grower, who also served as a schoolteacher. The boy attended a couple of small second-rate colleges, and finally, since the family fortunes were at low ebb, he was forced to take a job as a pharmacist in a suburb of Lyons. He hated such menial work as washing bottles but enjoyed visits to a veterinary school where he went to deliver drugs. The theater, however, was his real love. On his evening off, one night a month, he went to the Lyons theater where he saw operettas, vaudevilles and romantic historical dramas. He even wrote *La Rose du Rhône,* a vaudeville, which was performed and netted him one hundred francs.

Meanwhile, as he mixed chemicals, his first practical task was making shoe polish. Pharmacists of the period stocked a cure-all remedy, called thériaque, which was supposedly composed of sixty drugs and was mixed with Canary wine. Whenever young Claude spoiled some compound he was preparing, it was saved and tossed into the thériaque.

Unfortunately, Bernard was so theater-struck that he spent more time composing a five-act drama than tending to business and was dismissed. Undaunted, he finished his historical drama, *Arthur de Bretagne*. In this work, the influence of *Hernani* was tempered with borrowings from Shakespeare's *King John*. Somehow, through family connections, he obtained a letter to Louis Philippe's librarian. Bernard dashed off to Paris, where his letter got him passed on to the professor of literature at the Sorbonne. This worthy read his play and said to him: "You have done some pharmaceutical work. Study medicine. You have not the temperament of a dramatist." Judging from one song quoted by his biographer, the play was probably no worse than most second-rate dramas being successfully acted in this period. Yet on such accidents do careers hang. The verdict of a stuffy academic teacher of literature devastated Claude Bernard who was after all an innocent youth from the provinces. It transformed him from a minor playwright into a great physiologist.

He enrolled in the medical school of Paris, but the lectures bored him, although he excelled at anatomy and dissection. In addition to his studies Bernard taught natural history in a girls' school, but finally fell under the influence of a professor at the Collège de France, a famous lecturer and a ruthless vivisectionist, François Magendie. This crusty, arrogant and scornful man was a hardheaded antivitalist and made many enemies. Yet his record in fighting a cholera epidemic and ministering to the poorest part of the population commanded respect, as did his brilliant experimental work in which he used his knowledge of both physics and chemistry. He was the antithesis of nature philosophy and romanticism, and the effect of his strong personality on Claude Bernard was destined to cure him of any lingering leanings toward literature. At first, Bernard attended Magendie's lectures and finally obtained the post of preparationist. After he had done three or four preparations to be used in lectures, Magendie said gruffly, "You're better than I am," and walked out of the room. He continued to be rude and difficult, however, despite his respect for the young man's work. Although Bernard was often discouraged, he finally learned how to get along with his mentor; he was older than most students and more mature. Moving in the direction of organic chemistry, he wrote his thesis on the operation of gastric juice on cane sugar and obtained his degree in 1843.

From then on, he wrote a number of papers in collaboration with a young chemist and also worked with old Magendie. The latter was president of the War Department's Commission on Equine Hygiene, which meant that the two scholars had the opportunity of studying horses and using them for experiments.

Bernard failed to obtain an appointment on the faculty of the college but got several prizes for various papers (Magendie was often on the juries). In 1845, he contracted an "arranged" marriage with Marie Françoise Martin, the daughter of a doctor. She brought him a little money (he was hard up) but was pious and hated vivisection. Although they reared two daughters, their private life was misery.

In 1849 Bernard published an important work on the pancreas in which he showed that that organ emulsified fat, and for this he was awarded the ribbon of the Legion of Honor. Unfortunately, when this was reported in the press, the story referred to the "musical" instead of "medical" properties of the organ. Bernard had little sense of humor, and when teased by his friends for being a great composer, he wrote letters of indignation to the paper.

His greatest achievement dealt with the liver, a study in the tradition of Wöhler and Lavoisier. Up to the time of his investigation, it was thought that any sugar found in the blood of a carnivore must somehow be introduced through the diet. Bernard proved that there was absolutely no sugar in the diet of the dogs upon which he worked. A further hypothesis was that the gland itself supplied nothing, that its tissue supposedly had a catalytic action upon the elements of the blood that passed through it, causing sugar to form directly in the blood. Bernard took the liver of a dog which had been fed on a meat diet and washed it until it was emptied of blood. The washed-out blood contained no trace of sugar. He then left the liver at room temperature for twenty-four hours. "I found that this organ, washed entirely bloodless, which I had left the night before, completely free of sugar, now contained sugar in abundance." He said there was a substance in the liver tissue "which gradually changed to sugar by a kind of fermentation . . ."

He showed that by boiling the liver tissue, this process was stopped. He made pulp of such tissue after washing and dried it on filter paper. This was free of sugar, but when he added water he once more found sugar after a few hours. After being soaked in alcohol and dried, the liver pulp still retained its properties. He finally concluded that it was the task of chemists to isolate this substance, which was the forerunner of sugar. In 1857 he did this himself and called it glycogen.

This was, of course, a model of the kind of careful experiment which Berzelius had called for. It showed that the animal body was, as Bernard put it, a living laboratory. He later showed that the cutting of the vagus nerve stopped the secretion of blood by the liver, thus proving Berzelius' speculation concerning the function of the nerves.

Although in 1867 Bernard was made perpetual president of the Société de Biologie which he helped found, it took three applications for him to get into the Academy of Sciences, which was now limited to sixty-six members. When Magendie died in 1855, Bernard was called to his chair in the Collège de France. In a eulogy of his old teacher he pointed out that Magendie had said, "When I experiment I am all eyes and ears. I have no brain."

Bernard was described by a visitor to his laboratory at the college as working "in a narrow, damp corridor, where Bernard stood before his animal table, his tall hat on, his long gray hair hanging down, a muffler about his neck, his fingers in the abdomen of a large dog which was howling mournfully."

The Empress Eugénie liked to invite groups of distinguished people to the royal country estate at Compiègne. Artists gave theatrical representations, poets recited, and scientists performed experiments. On one occasion, when Bernard had been invited, the emperor, Napoleon III, who enjoyed private discussions with his guests, asked him questions about his work. Carried away, the physiologist talked fluently, to the scandal of the court, for a full two hours. The emperor, however, was fascinated and said, "I want to do something to please you." He told Bernard he could have anything he wanted. The scientist chose an assistant.

In the cholera epidemic of 1865, Bernard worked with Louis Pasteur, endeavoring to find the germ responsible. Bernard actually did some tentative experiments that led to those of Pasteur which, once and for all, put an end to the perennial topic of spontaneous generation. Boiling a sugar solution and sealing the neck of the retort together by melting showed that no organisms appeared in the solution. It was Pasteur who substantiated Schwann's belief that fermentation was caused by microorganisms. His opponents tried to show that the organisms were generated *from* fermentation.

Bernard's public life continued to be brilliant. He was made a senator in 1869, although from 1865 to 1867 he showed that his nervous makeup was not unlike that of the group of physiologists we have previously encountered. His weak point was his stomach; gastritis became so severe that for those two years he had to retire to his old home at Saint-Julien, where he gardened, transplanting bushes and grafting trees. The acute source of his problems was obviously his wife. Fanatically religious and bigoted, she hated his work, and when he brought home miserable experimental dogs which had to be nursed back to health, she ostentatiously contributed to the humane society. The quarrels between them became so intense that he came home only for meals. Finally in 1869, the year his health improved, they agreed to live apart. He managed to keep up relations with his daughters and for female companionship maintained a long friendship with a Russian intellectual, Marie Raffalovich.

His reaction to the war of 1870 was another attack of gastritis which sent him back to Saint-Julien. He believed that France was defeated because of lack of "scientific sense" in its people. After the social upheavals were over, he returned to Paris, and continued to work until his death in 1888.

If Claude Bernard tied physiology all the more closely to chemistry, another Frenchman, Jean Baptiste Joseph Dieudonné Boussingault (1802-1878), strengthened the links between the animal and vegetable world. A native of Paris, he had a friend who worked in a large commercial chemical laboratory. Even while in secondary school, he visited this institution and learned basic techniques. After the fall of Napoleon, a school of mines was set up in Saint-Étienne at which he studied. He then worked briefly as director of the mines at Lobsann in Alsace. What really resulted in a turning point in his life was his friendship with Humboldt. "I stood to him as a disciple does to a master and soon learned the difficult art of listening." The famous traveler had not come back from South America unscarred. At fifty-three, he was white-haired, pockmarked from smallpox contracted at Cartagena, and his right arm was partially paralyzed from rheumatism, attributed to the months he had spent on the Orinoco. At night, he slept on a moist blanket. When he wrote or shook hands, he had to lift his right arm with his left. He always dressed in Directoire style, a blue coat, yellow waistcoat, striped white trousers, turned-down boots (the only pair in Paris in 1822) and a white tie. He worked on an unpainted pine table, writing figures on the surface until it was wholly covered and then planing it clean.

In 1822 Simón Bolívar, who was campaigning for independence in South America, decided that he wanted to encourage science in the newly emergent republics. He sent an emissary to Humboldt, asking for capable people to help set up an institute. Boussingault jumped at this chance. Since it was not possible to sail from France because Louis XVIII was a reactionary and was opposed to any dealings with the new Latin republics, Boussingault had to sail from Belgium. Shortly after the brig *New York* left from Antwerp, it met another ship, took on some cannons, pulled down the Stars and Stripes and raised the Colombian flag to loud vivas from all on board. After a brush with a Spanish schooner, the brig safely reached La Guaira.

Boussingault was to spend ten years in Latin America in Venezuela, Colombia and Ecuador. He climbed many of the same volcanoes conquered by Humboldt, studied earthquakes, which were common, and made notes on mineral deposits. Some of his comments on life in

Colombia are quaint. According to him, the women were seductive and although they could read and write, never opened a book. They all smoked cigars and spat neatly over the heads of visitors. When experimenting with some poison, which the natives used for fishing, he got the material in his eyes, which became painfully inflamed. Friends saw to it that he got the right treatments—a lactating servant girl squirted her milk in his eyes.

Back in Paris in 1832, he published some papers on his observations in the New World. He had no degrees, however, and needed a job. Somehow, through his connections, he obtained the position of professor of chemistry at the university of Lyons. In 1838, he married the daughter of an agronomist. His father-in-law's professional ties seem to have helped him and indeed channeled his interests, for between 1860 and 1874, he produced a five-volume work on agricultural chemistry and physiology. After the war of 1870, he accepted a professorship at the National Institute of Agronomy.

Boussingault's significant contribution was in the area of plant and animal nutrition. Up until his time, it was thought that plants absorbed nitrogen from the air. He proved by experiments in which he grew plants in sand totally devoid of nitrogen that most plants could not thrive and died, unable to obtain it from the air. The exceptions were legumes and red clover. Nitrogen salts were obtained by the majority of plants from the earth as a result of the decay of plant or animal material. His table of comparisons went as follows:

Animals	Plants
Mobile organisms	Immobile organisms
Oxidating organisms	Reducing organisms
Exhale carbon dioxide, water, ammonia, nitrogen	Fix carbon dioxide, water, ammonia, nitrogen
Consume oxygen, neutral nitrogen compounds, grease, starch sugar, gums	Produce oxygen, neutral nitrogen compounds, grease, starch sugar, gums
Produce heat	Absorb heat
Reintegrate their elements in air and earth	Derive their elements from air and earth
Transform organic matter into inorganic	Transform inorganic matter into organic

The interdependence of plant and animal life had never been established so definitely, and what has since been called the nitrogen cycle was, for the first time, defined. It is well to remember that this basic ecological relationship was known from the middle of the nineteenth century, but it is only today that we are suddenly discovering how heedlessly man has tampered with it.

As nineteenth-century physiology became more and more chemically oriented, the vitalist point of view was less and less acceptable to leading scholars. The most uncompromising statement was made by Emil Du Bois-Reymond (1818-96), a pupil of Johannes Müller, who studied electricity in live organisms and discovered that muscles and nerves produced electricity which could be measured with the apparatus used in physics.

He maintained therefore that knowledge of the organic world "is to go back to the atom." In other words, organic and inorganic chemistry were essentially the same and if organic matter behaves differently, it is merely because different relationships are involved. "Law and chance are merely other names for mechanical necessity." On the other hand, he maintained that consciousness and emotional values were an insoluble problem:

> With the first quiver of enjoyment or pain experienced by the simplest being at the beginning of animal life on earth, or with the first awareness of a quality an unbridgeable gulf opened up in the world which became twofold and incomprehensible.

He pointed out that Descartes had bridged the gap by using the word "God." For him that was merely a confession of ignorance. Indeed, the same criticism applied to the concept "vital force" which was used by the vitalists to cover what was yet unexplained by reduction. Other still more uncompromising materialists were to insist that there was nothing material or spiritual which could not be explained mechanistically. Du Bois-Reymond's position, however, is that of many scholars today. And as we shall see, the chemico-physical explanations of the phenomena of life are still being pushed further and further.

21

"It's Dogged as Does It"

☙ THE shape of Charles Darwin's nose almost changed the intellectual history of the nineteenth century. When, as an agreeable young sportsman of twenty-two, he was interviewed by Captain Robert Fitzroy of the *Beagle* in 1831 for the post of naturalist on a voyage around South America, the latter studied his nose. Fitzroy was an adherent of the system of Johann Lavater, a Swiss mystic, who had worked out a method of character analysis based on physiognomy, or the shape of the face. Fitzroy came close to rejecting Darwin because his nose did not show "sufficient energy and determination."

Darwin, born in 1809, was the son of a successful, freethinking doctor, who had married a member of the wealthy Wedgwood pottery family. He was a dropout from the Edinburgh Medical School, because he hated dissection. He was then sent to Cambridge in the hope that, all else failing, he could become a clergyman. At Cambridge, he not only felt no call to the Church but got into a shooting set and, according to his own story, drank and sang his way through college. This was a slight exaggeration based on his dislike of the classical tradition in education. He did acquire a good grounding in geology with Adam Sedgwick and also began to collect beetles as a result of his friendship with John Henslow, pious fundamentalist and professor of botany, who was also competent in entomology, chemistry and geology. It was Henslow who heard of the opening on the *Beagle* and saw to it that it was offered to young Charles. The experience of this famous voyage was to make Darwin a key figure in the history of zoology and in the history of human culture.

Darwin was very much a child of his industrial society. If the effect of the Industrial Revolution was not so dramatic as the change in human culture which took place in the Renaissance, it was nevertheless far-reaching. First, the fact that industry and capitalism could make use of science tended to make it highly respectable. Instead of being regarded as eccentric intellectuals, its practitioners began to seem like magicians. We have seen that Boussingault wrote a practical guide for farmers. Many early geologists, who called themselves mineralogists, were concerned with finding veins of minerals and ore containing useful metals. The early American anthropologist Henry Schoolcraft, for instance, in his mineralogist phase was taken on government expeditions in the hope of finding copper. The age of the steam engine was also the age of iron and coal. With the expansion of industry there was a new need for raw material, producing a new wave of exploration, which extended the colonial system to remote parts of the world. This expansion carried both the missionary and the scientist with it. Along with all this practical profit seeking went a reasonable amount of pure science. It was in the nineteenth century that the tacit bargain was made between science and society; scholars were allowed freedom to pursue whatever lines of inquiry interested them, provided that a percentage of their work paid off in profits. We begin to see that a certain abdication of control was slowly woven into this relationship. The scientists did not bother to check the probable results of their labors against humanistic values. They were content to render unto Caesar.

We have already seen how chemistry was being integrated into the life sciences. In Darwin's time, geology was to play a leading role, affecting the direction of investigation. The link between Darwin and geology, aside from his university teachers, is Charles Lyell (1797-1875), a Scottish scientist of independent means and originality of insight. Before him, various one-sided concepts had prevailed concerning the earth's history. One of Humboldt's teachers in the Mining Academy of Freiburg, Abraham Gottlob Werner, had taught a neptunist theory, stating that all rocks, including basalt, had been laid down by water. Werner simply argued from his own backyard; he was a scientific provincial. Humboldt's activity in volcanoes, on the other side of the globe, was corrective to this distortion by clarifying the importance of "vulcanism," or fire-molding of rock. Through his extended travels he broadened the field of geological thinking.

Cuvier subscribed to a catastrophe theory which posited great floods and changes of climate that would eliminate all organic life in certain areas which later became repopulated by new creations or diffusion. In this way, he was able to account for extinct species and preserve his final Flood for Genesis. Although he sometimes hedged on new creations and used the argument of redistribution, those who followed him created "progressionism." This was a last attempt to preserve the Great Chain of Being, which had lingered on from the Middle Ages. Keeping the ladderlike image in mind, a supernatural intelligence had worked out a design by which, instead of a literal creation of all species in one week, successive creations took place with man as the final and crowning product. The desperate efforts of British fundamentalism, evidenced in the work of conservative geologists, to preserve some remnant of the biblical myth and to distort geology in its image are blamed on a reaction to the freethinking of the French Revolution. It is clear that public prejudices of this sort were sufficiently strong to affect Lyell, who was cautious and sensitive to social pressure. Although he already had most of the facts in his hands before Darwin, he never made the final statement concerning evolution.

What was his contribution? Having been trained as a lawyer, he was particularly good at marshaling arguments and presenting evidence. He also made several voyages in order to examine the evidence of the rocks in various parts of the world. He came to the conclusion that the same forces were operating on the crust of the earth today as in the past. Water was still laying down beds of sedimentary rock, volcanoes were still pouring forth lava, which would become igneous rock. The surface of the earth, as the core cooled, would wrinkle like a dried apple; formations would be twisted and cracked; seas would shift; and thus the arrangement of and relationships between rocks formed in the past would continually shift and change.

This point of view necessitated a rejection of the limited notion of geologic time which the pious had expanded from a week in Genesis to six thousand years.

Lyell's work was detailed instead of speculative. From fossil trilobite eyes he drew the conclusion that the light of the sun, far in the past, must have been similar to what it is today. He even investigated fossil prints of raindrops, which also gave a clue to a similarity of atmosphere in the distant past.

He drew a picture of a succession of species. The introduction of new ones, multiplying widely and requiring large amounts of food, might cause the death of those already existing. He was obliged to admit that new types seemed to have arisen, but there he stopped and remained noncommittal as to their formation.

The first volume of his book, *The Principles of Geology,* appeared in 1830, and Darwin took it with him on the *Beagle*. The second volume was sent to Darwin in 1832. Another book, we now note as a constant stimulus to nineteenth-century naturalists, was Humboldt's *Personal Narrative*. Darwin read it and reread it, and he was destined to follow in the footsteps of the famous traveler-scholar.

The *Beagle,* under Robert Fitzroy, had made other voyages to survey coastlines and map harbors. His activities, like Captain Cook's projects, are an example of the marriage between capitalism, the Industrial Revolution and a certain amount of pure science. Fitzroy was really concerned with missionaries, and Darwin's post was something of an afterthought. Fitzroy, the grandson of a duke and the nephew of an earl, a fundamentalist really wanted someone who would, by his investigations, support Genesis. The irony of history was felt bitterly by this overemotional, though capable, seaman whose life ended in suicide.

It was only through Darwin's tact that the two men managed to survive the trip amicably. At one point, when they were visiting Brazil, Fitzroy recounted how, when a slaveowner had called in slaves for his inspection, he had asked them if they were content. All had answered yes. Darwin ventured to suggest that this was not the best way to get a sincere answer. The captain was so angry that he banished Darwin from his table, but later invited him back.

Darwin, who was humanely antislavery, was scarred by what he saw of the institution in South America. In later life, he used to mention screams he would hear at night, which he felt sure were those of tortured slaves.

The *Beagle* sailed down the east coast of South America, Darwin leaving the vessel now and then and sometimes spending weeks or months on land. He was particularly struck by the gaucho culture of the pampas. The Argentine cowboys were rough, bloodthirsty but hospitable and, in their way, honorable. The Indians had adopted the horse much as North American Plains Indians had in the United States. Darwin called them "a tall, fine race." He recorded that the Indians made sacrifices and left offerings under certain trees. The rich left meat, bread, cigars and poured liquor into holes at the base of the trunk. Poor Indians, having nothing else, pulled a thread from their ponchos and tied it to a branch.

On his trip to the Colorado River, Darwin made the acquaintance of General Juan Rosas, encamped in a square of wagons, artillery and straw huts, on the bank of the river. Rosas, the son of a wealthy man, was a gaucho who owned 74 square leagues of land. A self-made general, he was pursuing a policy of extermination of the Indians, who still resisted the expansion of the Europeans. Ironically enough, he had enlisted some tribes to fight against their own countrymen. The soldiers, Darwin wrote, "were of a mixed breed, between Negro, Indian and Spaniard. I know not the reason, but men of such origin seldom have a good expression of countenance." Darwin decided Rosas was a man of extraordinary character who would use his influence for the prosperity and advancement of the country. Darwin added a note in 1845: "This prophecy has turned out entirely and miserably wrong." The year after Darwin left, Rosas had become an oppressive dictator, one of those tyrants from which Latin America has suffered during all its history.

Darwin made notes on the great flightless bird of the pampas, the rhea, which he saw captured with bolas. The male, he was told, hatched the eggs. Deer were plentiful, as were the wild species of llama, the guanaco. The agouti, a twenty-five pound rodent, was as common as the rabbit of the northern continent. Sometimes it dug its own burrows, sometimes it lived in those of the viscacha, another rodent of the plains. A second squatter in empty viscacha burrows was a small burrowing owl.

The viscacha's most endearing habit was collecting anything portable to decorate the entrance to its burrow. When small bits of property were lost, their owners searched the viscacha holes, and such objects as pipes or watches had in this way been recovered.

The largest rodent of all was the capybara; one which Darwin shot weighed ninety-eight pounds. It was characterized by a short stump-like tail and a deep heavy-toothed jaw. At a distance, groups of capybara looked somewhat like pigs, but when they sat up, they betrayed their rodent connections. These animals were active swimmers and fed on water plants. Since jaguars had been exterminated, they were very tame and allowed the naturalist to come within arm's length before they grunted and dashed into the water.

On the subject of Argentine public life, Darwin had comments to make which still hold good in Latin America: "Nearly every public officer can be bribed. The head man in the post office sold forged government franks. The governor and the prime minister openly combined to plunder the state. Justice, where gold came into play, was hardly expected by anyone."

Darwin made copious notes on geology as well as zoology and was again and again aware of the rising of rock stratas, many of which had formerly been under the sea. He also collected fossils of giant extinct sloths and giant extinct armadillos which, nevertheless, were related to existing species. "The great number of these extinct quadrupeds lived at a late period, and were contemporaries of most of the existing seashells." What happened to them? "The mind at first is irresistibly hurried into the belief of some great catastrophe; but . . . an examination of the geology of La Plata and Patagonia, leads to the belief that all the features of the land result from slow and gradual changes." Thus he was testing out the catastrophe theory in terms of his own observation and finding it wanting.

Condors, the largest of the carrion birds, could be seen high in the air, wheeling in graceful circles. "The Chilean countrymen tell you that they are watching a dying animal or the puma devouring its prey." They also attacked live kids and lambs. The Chileans captured them by enticing them into an enclosure of sticks and when they were gorged with the bait, closing the door. The condors were trapped, for there was not space enough for them to get the running start necessary to lift their heavy bodies. Darwin, unaware of Friedrich II's experiments hundreds of years before, cited some new ones to prove that these carrion eaters did not have a keen sense of smell:

> Except when rising from the ground, I do not recollect ever having seen one of these birds flap its wings . . . it is truly wonderful and beautiful to see so great a bird, hour after hour, without any apparent exertion, wheeling and gliding over mountain and river.

By February, 1832, the *Beagle* reached Tierra del Fuego and stopped at the same mountain where Sir Joseph Banks nearly froze to death. The party experienced the usual stormy weather which characterizes this bleak tip of the continent. Captain Fitzroy was returning two natives to their own people. One of them, Jemmy Button, had been brought to England on a previous voyage, taught a little English, and dressed in white man's clothing. After three years in Europe, he had almost forgotten his own language. Within a few weeks he was stripped of his frock coat and naked as his kindred. Matthews, the missionary whom Fitzroy had brought so far to save the soul of the naked Fuegians, had, according to Darwin, "little energy of character." After a week with the natives, he lost heart and returned to the *Beagle*. The Alakolufs had tried to strip him naked and also seemed to want to depilate his face and body.

Jemmy Button got himself a wife and seemed readjusted to his people. Darwin hoped that some day he or his descendants might protect some shipwrecked sailor. His experience in England seemed to have a negative effect, for it was he who organized the massacre of six missionaries who tried to found a colony in 1859. The confused lack of communication with and general contempt for the natives, who lived on a very low material level, is summed up in Darwin's quaint nineteenth-century capitalist comment:

> At present even a piece of cloth is torn to shreds and distributed; and no one individual becomes richer than another. On the other hand, it is difficult to understand how a chief can arise till there is property of some sort by which he might manifest his superiority and increase his power.

After the ship had sailed up the west coast of South America, Darwin found various formations in the vicinity of Peru which clearly showed the rising of this relatively young continent. At Valparaiso "in the 220 years before our visit, the elevation cannot have exceeded nineteen feet, yet . . . there has been a rise, partly insensible and partly by a start during the shock of 1822 of ten or eleven feet." He was finding proof that in this continent of continual earthquakes, the great molding forces were still at work as in the past, substantiating Lyell's doctrines.

The Galápagos Islands, those arid volcanic dots which now belong to Ecuador, represented for Darwin a challenging problem. Distinctly different species of the same genus inhabited islands only 30 to 60 miles apart. The tortoises differed, as did a genus of finch which could be separated into several species by the size of the beak. Darwin said he was late in noticing this important phenomenon, "but I ought perhaps to be thankful that I obtained sufficient material to establish this most remarkable fact in the distribution of organic beings." When the botanist Joseph Dalton Hooker came to examine the plants brought back from the islands, the same differentiation was found to hold true. All species were related to South American flora and fauna.

The ship sailed on to New Zealand and Australia. On King George Sound, Darwin watched uncomprehendingly a native dance, corroboree—or "corrobery" as the whites mispronounced it—in which two totemic groups performed their ceremonies. "It was a most rude barbarous scene, and, to our ideas, without any sort of meaning. . . ." Yet native women and children watched with the greatest pleasure. "Perhaps these dances originally represented actions, such as wars and victories; there was one called the Emu dance, in which each man extended his arm in a bent manner, like the neck of that bird."

Actually this was a dance to increase the emus and was part of a complicated world of mythology and ritual that was to preoccupy anthropologists for a good fifty years. But in the eighteen thirties, the science of anthropology was not yet invented, and to Darwin anything which deviated from the smug European norm was rather shocking. "The group of nearly naked figures viewed by the light of the blazing fires, all moving in hideous harmony, formed a perfect display of a festival among the lowest barbarians."

When the *Beagle* returned to London in 1836, Darwin was no longer the amiable dilettante of five years before. He had become a capable naturalist and collector and was already mulling over certain theoretical ideas. He turned his collection over to specialists who classified and wrote reports on zoology, botany and paleontology.

In 1838, he married his cousin, Emma Wedgwood, who brought him enough Wedgwood money so that he was not forced to become a wage earner. Since it happened that his illness began about this time, his wife rapidly became the careful ministrant to his health, nursing him throughout his retired life in the country during which he so carefully husbanded his forces in order to continue his work. Opinions differ as to the illness. He said he could forget it when he worked, and some biographers have decided that he suffered from psychosomatic problems. The symptoms seem to have been palpitations, vomiting and headaches. He was unable to endure large groups of people and, consequently, seldom attended meetings of scientific societies and limited his entertaining to inviting one or two persons down for the weekend. Now and then, he went to an establishment which gave "hydropathic treatment" consisting of cold baths, which he felt were beneficial. Since this was another example of a quack nineteenth-century fad, it argues on the side of a psychogenic cause for his illness. In 1959, however, an expert on tropical diseases pointed out that Darwin's symptoms were close to those of Chagas' disease, an ailment produced by a parasite similar to that causing sleeping sickness and carried by "the great black bug of the Pampas" which Darwin recorded attacked him while he was in the Argentine province of Mendoza: "It is most disgusting to feel soft wingless insects, about an inch long, crawling over one's body. Before sucking they are quite thin, but afterwards they become round and bloated with blood. . . ." It was perhaps this king-size bedbug which was responsible for the great scientist's forty years of ill health.

From 1839 on, Darwin published his *Journal of the Voyage of the Beagle,* his work on South American geology, and his theory of the formation of coral reefs. He suggested that the basis of such an island was a volcano which had sunk below the surface and on which, around the edge of the crater, the corals had built up to the surface or above, so flora and fauna were able to gain a foothold. Subsequent study has confirmed Darwin's analysis in many cases.

It was in 1846 that he began his eight-year exhaustive work on barnacles. Before he was finished, he was heartily tired of it, but it was good training in the drudgery of the zoological profession. He had to clear away duplications. "I find every genus of Cirripedia has half-a-dozen names and not one careful description of one species in any genus." This was the result of the post-Linnaean mania for finding new species and tacking the discoverer's name to them without bothering to find out if they had been previously described. Darwin isolated the cementing apparatus and thanks to his newly purchased compound microscope, discovered, in some cases, minute parasitic males. His book on the subject appeared in two volumes in 1851, and Darwin confessed ruefully that he was probably the origin of Professor Long, in a novel by Bulwer-Lytton, who had written two huge volumes on limpets.

According to his own story, he did not devote his whole time to work on species until 1854, but in 1838 he had read *An Essay on the Principle of Population* by Thomas Robert Malthus, a clergyman who also studied economics. This famous book pointed out that the urge to reproduce in nature, if not opposed by counterchecks, would result in overpopulation and an ensuing struggle for existence in which the natural increase would be thinned out by carnivores and the weaker animals would die from lack of food or other natural causes. In man he saw the same thing happening, with war and disease the only natural checks on overpopulation. He believed that men should refrain from rearing families they could not afford. Curiously enough, despite some hard geographical and historical statistics to prove his point, Malthus' work seems to have been forgotten for decades and only recently the full impact of his warning has begun to be felt.

Darwin was impressed by the concept "struggle for existence." He had already been thinking and making notes on the problem of species, as is indicated in his letters to his most important scientific friends, Joseph Dalton Hooker, the botanist, and Charles Lyell, both of whom he had come to know since his publications had begun to make his name. His ideas on natural selection and variation of species were fairly well developed, but he felt the need of tremendous documentation before publishing. There were probably various reasons for this. The pious climate of Victorian England was more reactionary than that of the continent. Darwin read German with difficulty and does not seem to have been well acquainted with the rather evolutionary thinking of the nature philosophers. He knew and spoke of Lamarck's ideas as "rubbish." On the whole, he seems to have visualized himself as a lone pioneer. Moreover, he was not a controversial personality. When the time came, a champion arose who could and did speak and write polemically, but at this period Darwin did not know that he was to be supported by the vigorous personality of Thomas Huxley. He also felt he had to overcome the stigma of being labeled an amateur, for he did not have a thoroughgoing academic background in any of the natural sciences, except perhaps geology which he had studied under Adam Sedgwick. And Sedgwick was to become one of his bitterest opponents. Full of self-doubts and hesitations, he discussed with his friends and went on piling up material; one of his mottos was: "It's dogged as does it."

Meanwhile, across the world in the Malay Archipelago another natural scientist was also reading Malthus. Darwin's codiscoverer, Alfred Russel Wallace (1823-1913), was an amateur who had made himself into a scientist. He had, however, less formal schooling than the naturalist of the *Beagle*.

His father was a dabbler in various arts who had enough to live on until he married. Then his efforts to increase his income by "entrusting" money to various advisers only resulted in losses. The family retired to a simple cottage in Wales near the Usk River, where the fishermen still used hide-covered coracles. Wallace enjoyed his country childhood until, at the age of fourteen, he left home to work with his brother, a surveyor. In 1844, the year he came of age, his brother had no more work for him. His next job was that of a schoolteacher in a private school in Leicester, where he taught drawing, reading, writing and mathematics. He felt embarrassed by his lack of classical languages and disliked being under the thumb of the head of the school. Wallace, who was a firm believer in phrenology, attributed his impatience with a subordinate position to a poorly developed "organ of veneration." He also experimented with hypnotism, which in his time was mysterious and little understood. In his mind there was some sort of a connection between hypnotism and phrenology. Curiously enough, even Darwin spoke respectfully of phrenology. Neither one of these naturalists had had a thorough training in anatomy, which would have convinced them that the bumps on the skull had nothing to do with the brain underneath. In both there was a touch of amateurism which, in Darwin's best work, was overcome by practical experience and rigorous logic but which devaluated Wallace's thinking toward the end of his life.

Wallace also read Humboldt and William Hickling Prescott's *Conquest of Peru* and began to feel the urge to travel. While a teacher, he became friendly with the entomologist Henry Walter Bates, and he himself began collecting insects.

Back in Wales, he took over the surveying business on his brother, William's, death and worked on the plans for a new railway. He also ventured into architecture and designed a couple of buildings. Darwin's journal of the *Beagle*'s voyage added to Wallace's urge to explore the naturalist's paradise of South America. He and Bates eventually made an arrangement with a dealer to visit the Amazon and collect specimens for sale. Apparently the dealer advanced funds for the trip.

In 1848, the two young men embarked in the *Mischief* which took them to the Brazilian city of Pará. Wallace was sick for six days of the voyage, but once ashore, the New World delighted him. He was thrilled to see oranges and bananas growing on trees, he tasted the meat of turtles and iguanas, and he ascended tropical rivers, looking for rare specimens. The rich variety of tropical butterflies was a challenge to the collector who visited the Guaraná River and the Río Negro. He imported his younger brother to work with him, but his brother showed no real taste for natural history and managed to catch a fever and die on the way back to England. Wallace had dysentery and malaria but possessed the iron constitution necessary to survive as an explorer. In search of a rare umbrella bird, he explored the Uaupés River, a tributary of the Río Negro, which was twenty miles wide and nearly black in color. It was not to be visited again by a European until 1887.

On the way home, he traveled in an unseaworthy ship which nearly sank in a storm and then narrowly escaped destruction from a fire in the cargo. He brought with him about a hundred species of fish and many quadruped skins and skeletons and a large collection of birds, among them the umbrella bird, which sported a kind of parasol of blue feathers on its head. His material brought him to the attention of zoological and entomological societies before which he spoke. His travel book also made him better known and although published at his own expense, brought in a few pounds. He was elected a fellow of the Royal Society and obtained the post of professor of natural history and paleontology in the Royal School of Mines.

The Crystal Palace's International Exposition in 1851 included some groups representing South American Indians. Wallace was called in as a consultant. Italian workmen, who had a classical formula for figures, were making the models. Wallace tried to suggest modifications of the brown Venuses and Apollos which would bring them a little closer to reality. Wallace, with considerable prophetic insight, said that museums should place such groups behind glass, well lighted and with a real habitat background, instead of mounting them on pedestals like sculpture.

The itch to explore and collect was still strong. Wallace tried for a free passage on a Navy ship, having decided upon the Malay Archipelago as his next hunting ground. The first ship, to Singapore, got new orders which shunted it to the Crimea. He finally embarked on a Peninsula and Oriental steamer, the *Bengal.* This got him to Alexandria, where he viewed some of Egypt's ancient monuments. A horse-drawn bus took him to Suez, where he boarded a boat which sailed through the Strait of Malacca to Singapore. He was destined to spend eight years in the Far East.

He first stayed at a French mission near Singapore while he collected and sent home thousands of beetles. Later he moved on to Sarawak, where Sir James Brooke, who had made himself the raja of the island, was exploiting the mines and bringing the blessings of British civilization to the Malays. Brooke had had a bad press with accusations of mistreating natives. Wallace was enthusiastic about the raja who put down conspiracies of native chiefs with ease and ruled over "inferior and superior races." Native ringleaders who tried to promote a strike for shorter hours and more pay for miners were immediately discharged, and native policemen were brought in to act as strikebreakers.

Wallace spent fifteen months collecting orangutan skins. He narrated a perfectly incredible incident. He shot something in a tree and brought down a mother with a child in her arms. Since she moved easily in the branches, Wallace concluded that she was a "wild woman of the wood." He took her skin and skeleton and was rearing the baby with a pap made of boiled rice and water. Wallace left no doubt that it was a human child. Aside from pygmies and Malays, the archipelago was peopled by some tribes of mixed blood who inhabited the woods. Presumably it was one of these woodland people whom Wallace shot. His shocking indifference toward this accidental murder tells us something about the Victorian attitude toward simple people who were often confused with the great apes. Ironically enough, Wallace congratulated the raja for having put a stop to the practice of head hunting and thus establishing "law and order."

By this time the naturalist had read Lyell's book and also Malthus'. He wrote a paper, "On the Law which has regulated the Introduction of New Species," and published it in the *Annals and Magazine of Natural History* in 1855. It attracted no attention. During the next few years, which were spent in Macassar and Amboina, he began to correspond with Darwin. The latter cautiously admitted that he was working on the problem of species. While down with a bout of fever in 1858, Wallace, prompted by the ideas of Malthus, wrote "On the Tendency of Varieties to depart Indefinitely from the Original Type" and sent it to Darwin. The scientist of the *Beagle* was absolutely flabbergasted; the concept of the evolution of species by natural selection was the same as his own! His friends had warned him that he might be forestalled if he did not publish, but he had been mulling over an immense work with hundreds of pages of documentation.

At first, he wished to bend over backward and give up his own book. His friends, Hooker and Lyell, persuaded him to let them read the essay he had written together with a synopsis of his work and a letter from the American botanist Asa Gray, dated 1857, which would establish a priority for Darwin before the Linnaean Society. This was finally done and excited very little reaction. Wallace proved to be amiable and unpretentious. Urged on by his friends, Darwin worked strenuously for a year and published a short book, the first edition of *The Origin of Species,* in 1859. It immediately went into a second edition and soon sold three thousand copies.

The matter of priority presently seemed unimportant, for at least two other amateurs had published brief statements of an evolutionary nature, and as we have seen, naturalists all over the world were beginning to think in evolutionary terms. In addition, another amateur, Robert Chambers (1802-71) had published anonymously, in 1844, *The Vestiges of the Natural History of Creation.* This book, somewhat influenced by progressionism, stated that species were constantly varying and that a struggle went on in which the action of the carnivores was a check on overfertility. Like Lamarck, Chambers believed that a creative life force caused organic growth and could be adapted to environmental demands. Although he accepted spontaneous generation and stated that evolution "was the manner in which the Divine Author has been pleased to work," the book created a furor. It was condemned by the pious as atheistic and immoral with the result that between 1844 and 1860, twenty-four thousand copies were sold. Scientists had no use for it, and Thomas Huxley, not yet an evolutionist, attacked it as a "work of fiction." It is easy to see why the history of this book, which he read carefully, made Darwin cautious about his own. In a sense, however, *Vestiges* had drawn the fire of the opposition and made things a little easier for Darwin.

Darwin's basic statement was that species varied; he used examples of domestic animals such as the dog to prove his point. Domestic animals as a result of the work of the breeder were selected for certain traits and would eventually breed true. Such disparate types as the bulldog and the greyhound also were derived from a common ancestor and although they could interbreed and did not show the sterility of the hybrid, were on the way to becoming separate species. He could also show variation in nature, witness the Galápagos finches which on each island had varied enough to become separate species but which were all seemingly derived from a common ancestor.

Most important was the geological record, which though still imperfect showed that although some species such as the giant sloths and armadillos had died out, smaller species, related to them, still existed in South America. The evidence for the mutability of species was strongly presented and, after the work of Darwin, could no longer be rationally denied. The dynamic view of nature and the organic world, which had been in conflict with the static one for decades, was now firmly established.

How did these changes in species come about? It was there that the ideas of Malthus seemed to provide a solution. Natural causes kept down the population of both animals and humans. In other words, some survived and some did not. Those that survived were stronger or better adapted and thus triumphed in "the struggle for existence."

As Darwin put it: "More individuals are born than can possibly survive. . . . The slightest advantage in certain individuals, at any age or during any season, over those with which they come into competition, or better adaptation in however slight a degree to the surroundings will, in the long run, turn the balance."

Nature, therefore, acts in a parallel fashion to the animal breeder, and the environment causes the best adapted to survive. It will be seen that this position begins to sound like that of Lamarck. Lamarck had envisioned an ideal type toward which each species strove, which was modified by the needs of the environment. Darwin apparently had read a version of Cuvier's vicious eulogy in which Lamarck was made to say that the variation had to do with the animal's will. This misinterpretation seems to have biased Darwin toward his illustrious predecessor. But Darwin himself had to face the problem of inheritance of acquired characteristics, or inheritance of variation. To escape from the problem, he resurrected and modified Buffon's old idea of pangenesis. The sexual products contained atoms or gemmules from all parts of the body which were combined in the embryo. Therefore, if certain parts or organs varied, the gemmules from these organs would transmit the change to the next generation.

Oddly enough, there is a metaphoric half-truth in the concept. Although genes, of which Darwin knew nothing, govern heredity, it has been discovered that the necessary genetic material in plants can be taken from any part of the plant. In other words, the same genes are present in all the body cells and can give rise to a new individual. Dr. J. B Gourdon of Oxford has used an intestinal cell nucleus of a frog grafted in a frog's egg to grow a new individual.

How the environment acts upon the individual to produce inheritable variation is still a matter of controversy and the preponderance of the evidence is against the transmission of acquired characteristics. Although Darwin rejected supernatural design and believed in accidental variation, which turned out to have survival value, he argued with Lamarck on the importance of disuse as causing organs to degenerate and special usage developing them. And indeed, the idea of struggle for existence suggests an activity on the part of the individual, which is not far from Lamarck's need to adapt. For these reasons it must be indicated that Darwin owed something to his predecessor, who has been badly treated by history and his contemporaries. The fact that Darwin maintained that he got nothing from Lamarck is a trifle disingenuous. Darwin, although the most modest of men and most generous toward Wallace, had a strong unconscious desire to remain the proprietor of evolution and tended to deny outside influences on his thinking.

The history of the reception of the *Origin of Species* has often been told. Theologians, as usual (and rightly from their point of view), were antagonistic. The most serious scientific opposition came from Richard Owen (1804-92), a leading anatomist, director of the Hunter Museum and head of the Natural History Department of the British Museum. Although he adhered to some of the nature-philosopher mythology, especially the vertebral theory of the skull, he did important work in descriptive anatomy and paleontology. He was particularly important for having distinguished between homology, the same organ in different animals, regardless of difference in form or function, and analogy, a part or organ of an animal having the same function as a part or organ of a different animal.

The gills of a fish and a mammal's lung were analogous, but the fish's swim bladder and the mammalian lung were homologous.

Owen, who was progressionist, as would be expected with his lingering romanticism, attacked Darwin in anonymous reviews, and it was he who supplied Bishop Wilberforce with his ammunition for the famous episode in which the fatuous cleric was slapped down at Oxford by Darwin's defender, the redoubtable Thomas Huxley, in 1860 at a meeting of the Association for the Advancement of Science. Strangely enough, it was Chambers, who had never dared claim authorship of his *Vestiges* for fear of ruining his publishing business, who persuaded Huxley to attend the meeting.

Despite the weaknesses of many of Darwin's arguments such as sexual selection, in which he overstressed esthetic reactions, the solid labor, the mass of carefully collected material and the overall outlines of his theory of evolution were impressive. Not only did he practice his motto, "It's dogged as does it," but he convinced leading figures such as Joseph Hooker, the botanist, and even the hesitating Lyell. Huxley became his standard-bearer; in the United States Asa Gray defended him against Louis Agassiz. The evolutionists became dedicated champions of the new, very much like the followers of Galileo.

There is no doubt, in looking backward, that Darwin's theory changed the direction of the intellectual currents of the period. From his time on, a developmental approach has been used in history and anthropology, as well as the natural sciences. It was the first time that natural science had caused a cultural revolution equivalent to that of Galileo's revision of astronomical thinking. Actually, the establishment of the cell doctrine was probably more important, but it did not catch the public's attention. The triumphs of the Industrial Revolution had created an image of progress in a world of flux, which had burst the bonds of theology. The bourgeois audience, which had now taken to reading, was able to follow the controversy and tended to side with the new ideas. And yet we are confronted with the curious situation in which a concept that was liberal as far as science went and a stimulus to creative work in zoology, and biology in general, became in the social sphere the rationale for laissez-faire capitalism and the reactionary answer to the demand for economic justice. If the struggle for existence was a natural law, holding good among men as well as animals, then he who triumphed by tooth and claw, proving his fitness by his ability to acquire riches, was in the right. The poor were poor because of their inherent weakness and lack of talent. Thus the most cruel indifference to the condition of the factory worker and most patronizing attitude toward the underprivileged were easily justified. Social Darwinism, as it was called, became one of the commonest weapons against that other great seminal thinker of the age, Karl Marx.

Thanks to his friends and defenders, Darwin was not crushed and cast aside like Lamarck. He continued to work in his country retreat, publishing the *Descent of Man* in 1871. The reception of the book will be discussed in connection with Thomas Huxley. It is sufficient to say that Darwin saw his prestige grow steadily, and at his death in 1882 he was buried in Westminster Abbey.

Wallace, his co-propounder of evolutionary theory, went through a curious metamorphosis. He returned home in 1862, became friendly with Darwin, sold his specimens, wrote a book on the Malay Archipelago, which sold well, and settled down to a literary career. He was made president of the Entomological Society, and after an unsuccessful courtship in which his Victorian sweetheart suddenly rejected him with no explanation, he married William Mitten's eldest daughter.

Wallace drifted away from Darwin, for he insisted that natural selection did not explain man's mental and moral nature. In response to a question by Darwin, he wrote a book on protective mimicry and warning colors, describing cases in which animals seemed to mimic other dangerous animals, an adaptation which supposedly had survival value.

In 1886, Wallace went on an extended lecture tour in America. By this time he had become a Socialist and almost an ecologist as an ardent exponent of nationalization of the land, for he felt that coasts, streams, lakes and mountains should be exploited for the public good and not be hoarded as private property. Wallace lacked the drive toward documentation of his ideas and strict critical sense which made Darwin such a great figure in the history of natural science. Somehow through phrenology and mesmerism he became a convinced spiritualist, wrote *Miracles and Modern Spritualism* in 1875, and was hurt when the physicist John Tyndall and the journalist G. H. Lewes were not impressed by the seances he staged for them. Although in the latter part of his life he ceased to be a scientist, his work in zoography, dealing with the Malay Archipelago, is a permanent contribution.

Once evolutionary theory gained general acceptance, it led to further discoveries and developments in the area of heredity and the descent of man. It is the followers and critics of Darwin we encounter next in the history of zoology.

22

Darwin's Champions

☙ WHEN the news of the famous fracas at Oxford over Darwin's theory came to the ears of the wife of the Bishop of Worcester, she cried, "Descended from the apes! My dear, let us hope it is not true, but if it is, let us pray that it will not become generally known!"

The man who saw to it that her prayers were not answered was Thomas Huxley (1825-95). His family seems to have been marked for tragedy. His father, an unsuccessful schoolmaster, lapsed into early senility. One of his brothers sank into insanity. His mother died while he was still young. Thomas, the seventh of eight children, was one of those strenuous and seemingly optimistic Victorians who nevertheless suffered from an inner canker. Beatrice Webb remarked that when he was not absorbed in work, "he dreams strange dreams: carries on lengthy conversations between unknown persons living within his brain. There is a strain of madness in him."

Huxley was thoroughly unhappy at the school where his father was headmaster. Although the institution had a reputation (both Thackeray and W. S. Gilbert went to it), Huxley as son of the director may have been in a difficult position. The school was obviously the model for Dr. Swishtail's school which Thackeray described in *Vanity Fair* in terms of incredible snobbery on the part of the well-connected and brutality exercised by the older boys toward their fags.

The young scientist-to-be read geology on his own, recorded his youthful philosophizing in his diary, and after two years of formal schooling, studied by himself with such success that he was admitted to the medical school of the University of London. After he took his degree, since he had no money to set himself up in practice, he obtained the position of assistant surgeon in the Royal Navy and was assigned to a leaky, cockroach-infested survey ship, the HMS *Rattlesnake*. The official naturalist was John Macgillivray, who was glad of the young doctor's help. During 1846 the ship visited the Great Barrier Reef and stopped at Sydney, where Huxley immediately became engaged to Nettie Heathorn. Nettie was another patient, nineteenth-century fiancée, for the engagement lasted nine years. Huxley sailed off to visit the Torres Strait, destined to be the scene of the first anthropological expedition in 1898, when Alfred Cort Haddon, on Huxley's advice, went there to study native life. Huxley, himself, did no anthropological work. He was too busy with his nets, collecting and sorting marine invertebrates. When the ship returned to England, Huxley published a paper on the medusae which made his name and got him into the Royal Society. At this point, it was Richard Owen, his antagonist-to-be, who helped him obtain a special service appointment to HMS *Fishgard* with leave of absence which allowed him to work at zoology. He prepared papers on mollusks, echinoderms, rotifers and crustaceans.

Huxley insisted that the Admiralty should publish his work, but the government was not science-minded to the point where it would support a zoologist, even though he had received the Royal Society's gold medal. He was ordered to report to his ship. Huxley's aggressive traits came to the fore. His willingness to fight seems to have been a healthy compensation for his unhappy childhood and the slights he received during the *Rattlesnake*'s voyage, when officers sometimes pulled in his nets against his wishes or threw out his preparations on the pretext that they were in the way. He announced that he would not join his ship unless his papers were published. The Admiralty reacted by dismissing him.

For a time, he had a struggle. He tried for several teaching jobs which he did not get and wrote unhappily to the patient Miss Heathorn that he was finding he could not live by science. Finally, the break came when he got the job of paleontologist in the Royal School of Mines. In 1855 Nettie was sent for, just in time, for she arrived half dead from the ministrations of incompetent Sydney doctors. London physicians told Huxley she would not live. Happily they were wrong; perhaps rejoining Thomas was the stimulus she needed. They reared four children in a family atmosphere which one of the offspring described as "a republic tempered by epigram."

The turning point in Huxley's career came in 1859 when he read Darwin's book and wrote a warm review. The two men became friends. After the clash with Bishop Wilberforce, he carried on publicity for Darwin's ideas, speaking everywhere with eloquence and success. Indeed, Darwin, Wallace and Hooker all said they felt inferior when they listened to him, so rich was his scholarly background.

The conflict with Owen had been brewing for two years before the Oxford incident. Owen insisted that certain lobes of the brain, present in man, were missing in the higher apes. Huxley, by dissection, proved he was wrong. Owen replied, repeating himself. Huxley replied by pointing out that Owen was using an illustration from a German paper which was incorrectly drawn. The battle became notorious. Comic pamphlets were written and *Punch* published a poem which began:

> Then Huxley and Owen,
> With rivalry glowing
> With pen and ink rushed to the scratch:
> 'Tis brain versus brain
> 'Til one of them's slain,
> By Jove! It will be a good match!

Finally, the German authors of the paper with the drawing admitted it was incorrect and Huxley commented: "During two years . . Professor Owen has not ventured forth a single preparation in support of his oft-repeated assertions."

The incident is significant in that it shows to what extent Darwin's work had attracted public attention. Owen was fighting to prove that there was a great gap between men and the apes. Huxley in one of his lectures, in that stronghold of Presbyterian orthodoxy, Edinburgh, announced that he knew an "Australian blackfellow who is as intelligent and as good a man as one half of the British Philistines." Needless to say, he was attacked in the Scottish press.

Interestingly enough, Huxley as a young man accepted most of the lower-middle-class prejudices against blacks and Jews, but he gradually shed such provincialisms. He was near enough to the working class, however, to be able to reach them with his lectures which were clear, logical and humorous.

The Edinburgh lecture was the basis for his most influential book, *Zoological Evidences as to Man's Place in Nature,* published in 1863. Darwin had dodged the issue in *The Origin,* merely saying that his theory would no doubt throw light on man's early history. Huxley courageously linked man with the other higher vertebrates. Huxley had by this time become what he called an "agnostic," a term he invented. Although he liked to bait clerical obscurantists, he was always willing to go halfway to meet liberal clergymen.

In his book, he reproduced drawings of a series of skeletons as a frontispiece: a gibbon, an orangutan, a chimpanzee, a gorilla and a man. He then examined the zoological history of the higher apes, clarifying the name "pongo" which was applied to both chimpanzee and gorilla, while the chimpanzee had also been called both satyr and pygmy. He then brought up the evidence of embryology with illustrations, showing the similarity between the embryos of a dog and a human being in their early stages, both exhibiting gill slits. He summed up: "Indeed it is very long before the body of the young human can be distinguished from that of the young puppy." Such evidence as gill slits contributed to the later theories of the German biologist Ernst Haeckel in which he advanced the belief that ontogeny (development of the individual) reflected phylogeny (development of the group). In other words, the early stages of the mammalian embryo suggest an early historical derivation from the fish.

Huxley's third section compared skulls of men and apes, and he concluded: "Hence it is obvious that greatly as the dentition of the highest apes differs from that of man, it differs more widely from that of the lowest apes." He also compared hands and brains, arriving at the same conclusion.

Finally, he sketched the history of his battle with Owen and then took up the matter of the first and highly controversial Neanderthal skull, found by Jacques Boucher de Perthes in 1856. Since the links between modern man and his presumed apelike ancestor were precisely what was lacking, the antievolutionist scholars called the skull a freak, a fraud, a pathological exhibit. Huxley, who had examined casts of the skull, pronounced on the side of authenticity and antiquity, and his judgment has been sustained. He was cautious about dating it and merely concluded by saying: "If the doctrine of progressive development is correct, we must extend by long epochs the most liberal estimate that has yet been made of the antiquity of man."

Huxley wrote easily; his style had color and reflected his personality. In contrast, Darwin was long-winded and somewhat crabbed. To a large extent, Huxley's book really anticipated *The Descent of Man*. Darwin used some of Huxley's data but added some new material on natural selection and also tried to derive emotions and mental qualities from those of animals—a first, sketchy attempt at animal psychology. There his work suffered from not distinguishing the learned reactions of domestic animals from their basic emotional constitutions. Part of his book also dealt with sexual selection, and his evaluation of the intelligence of females compared with males exhibited male chauvinism.

In 1876, the United States had the opportunity of viewing Huxley's tall, imposing figure on the lecture platform. Unlike Matthew Arnold, the classicist, who did nothing but complain of vulgarity while in the United States, Huxley, a child of his age, was delighted with the atmosphere of bustling industrialism. He told New York *Tribune* reporters, as he viewed the harbor where steamers were being maneuvered here and there: "If I were not a man, I should like to be a tug."

His influence on education, which resulted not only from his popular lectures but also from his positions on various school boards and committees, was always on the side of a scientific curriculum. He pushed for a modernization of Oxford and Cambridge. He acted as rector of Aberdeen University and worked to make it more contemporary. The School of Mines, in which he taught so long, was turned into a College of Science through his efforts.

Made a Privy Councillor by the Crown, he summed up his efforts to educate the working class and speaking as a doctor, said: "The English nation will not take science from above so it must get it from below . . . if we cannot get it to take pills we must administer our remedies par derriere."

The last ten years of his life were a constant struggle with ill health. In 1895 he died writing a paper in answer to Arthur James Balfour's attack on agnosticism.

In France, the scientific world remained oddly indifferent to Darwinism and, when it did return to evolution, tended to revive Lamarckism. But in Germany, the controversy over the change in biological thinking was as intense as in England and perhaps even more so because it involved politics. The storm center, and the figure in Germany equivalent to that of Huxley, was Ernst Haeckel. His lifespan, 1834 to 1919, covers the spread of evolutionary ideas around the world, as well as the contributions deriving from the work of Gregor Mendel and August Weismann, and the rise of genetics, which have since modified evolutionary thinking. Haeckel, however, stuck to one position, a position which defines him as a belated romantic.

His family was conventionally pious and respected literary culture; indeed it was typical of the Prussian middle class, for his father was a government official. His education in Merseburg and in the gymnasium, in particular, was in the old-fashioned classical curriculum which enabled him to read Greek (he continued to enjoy Homer in later life), but looking backward, he proclaimed it full of falsehood instead of the truth of natural science. Nevertheless, he read Humboldt's *Aussichten der Natur* (Appearances of Nature), Darwin's *Voyage of the Beagle,* and Matthias Schleiden's book on general botany. He was also an admirer of Goethe and, at some point, read Oken. When it came to his higher education, his father wanted him to become a doctor. His preference was for botany. Under parental pressure he entered the University of Würzburg. Fortunately for him, Rudolf Albert von Kölliker (1817-1905), a Swiss student of Oken and Müller, was a professor at the school. A great teacher, he made important contributions to cytology, publishing on cell division of the egg of the cephalopods. He contributed also to our knowledge of the cell nucleus and worked on the cell structure of nerves and blood vessels.

Another pupil of Müller, Rudolf Virchow (1821-1902), was also teaching at Würzburg. His important research was in pathology. Virchow was trained in the Friedrich Wilhelm Institute of Berlin, where a promising student could obtain an education if he agreed to become an army doctor. In the Berlin school he also came under the influence of Emil Heinrich Du Bois-Reymond. After obtaining his degree, he became an intern in the Charité Hospital and continued his studies in chemistry and microbiology. He wrote on the need for a mechanistic approach to medicine and insisted that life was the sum of mechanical, physical and chemical laws, in the tradition of Du Bois-Reymond. In 1847, he traveled in Belgium and Holland, returning to Germany just in time for the revolution of 1848. He helped build barricades in Berlin and armed himself with a pistol, but since the government soldiers kept well out of reach of side arms, he was not involved in any intense action. As a result, however, he lost his post as a medical officer in the industrial district of Silesia. It is then that he accepted a position in Würzburg. His *Cellular-pathologie,* published in 1858, set forth the thesis that disease was a sickness of the cells. Influenced by his political viewpoint, he described the body as a democratic collection of cells. Pathology was an undemocratic oligarchy of tissue. He showed concretely that the diseased cells were formerly normal in such illnesses as leukemia, and coined the phrase "all cells arise from cells."

Together with Kölliker, Virchow put the Würzburg Medical School on the map. Haeckel studied profitably with both and eventually moved to Berlin to spend a year with the old master, Müller. The latter gave him permission to work in the museum at the university and took him on a summer trip to study marine biology in Helgoland. The trip was a turning point in his life. The scientists enjoyed rowing along the shore, dragging their nets. Haeckel wrote: "I will never forget the astonishment with which I wondered at the horde of transparent marine animals as Müller dumped them into the glass container." There were medusae (jellyfish), ctenophores (comb jellies), copepods (tiny crustaceans), larvae of worms, echinoderms (sea urchins) and many other invertebrates. "Once you are really caught up by the magic of the marine world, you can never give it up," Müller told him, and he was right.

Haeckel, who was esthetically sensitive, sketched his specimens in watercolor; the radiolaria on which he was to do his best work probably appealed to him because of their shape. They belonged to a class of one-celled animals called the *Rhizopoda* (root-footed). From an inner nucleated portion threadlike projections, pseudopods, extended. These false feet often contained a glassy silicate skeleton, basketlike or star-shaped in many variegated forms.

Haeckel next worked as Virchow's assistant after that scholar had been called to Berlin in 1856. Their early association appears to have been amicable. In the summer of 1859, Haeckel made a trip to Messina where he studied radiolaria and underwent a psychical crisis, for that year he read *The Origin of Species*. He said, "Even at first reading it impressed me deeply." It did more than that; he underwent a conversion. Darwinism became the basis of all his thinking, his *raison d'être*. An emotional man, his early letters to his parents express conventional piety. The events of the revolution were to have a powerful effect on his religious attitude.

Something must be said of this political ferment, for it became related to evolutionary theory. In 1848, the middle-class intellectuals opposed Friedrich Wilhelm IV and set up barricades in the streets of Berlin. The students and other young activists did not ally themselves even temporarily with the working class as the bourgeois revolutionists did elsewhere. At first, the king sent the army against them, then vacillated, declared himself a liberal, and allowed a liberal diet to be elected, representing all the disunited German states. It was an assembly of professors and lawyers. They expected the petty princes, who ruled the states they represented, to undergo a change of heart and become liberals. Marx and Engels wanted a tougher program involving the masses, but the professors remained in a state of romantic utopianism.

At this point, the persecuted Czechs tried to profit from the situation by asserting Czech nationalism. Their Austrian overlords made a deal with the Prussians. The Diet made the mistake of supporting the suppression of the Czech minority, but the army, once let loose, did not stop and became involved in war with Poland. When the smoke had cleared away, the impotent liberals saw the army in the saddle, Friedrich Wilhelm again a reactionary, and the professors victims of their own illusions as all hopes of a unification of progressive German states collapsed.

Virchow and Haeckel were on the side of liberalism. The return of absolutism meant the Catholics were placated by the government and could push for reactionary legislation. It was this state of affairs which roused Haeckel strongly against religious obscurantism. This, together with reading and digesting Darwin's book, undermined his faith. He became an enemy of Christianity and also of Bismarck, who was soon to become the ruling force that was to put an end to hopes of democracy in Germany.

Haeckel wrote a dissertation on crayfish, practiced medicine briefly, and finally, as the result of a recommendation by his friend, Karl Gegenbaur, a zoologist in Jena, was called to a professorship of zoology at that university, where he remained until his retirement.

His work on the *Radiolaria,* published in 1860, was monumental, an important contribution to protozoology. In it he attempted his first application of Darwinism by trying to classify these animals in terms of development from a single type.

When Haeckel went to Jena, he married his cousin, Anna Sethe, but marriage did not make him cautious, for in 1863, at a meeting of the Society of Natural Scientists and Physicians, he bravely delivered a speech summarizing Darwin's views, announced his adherence to the cause of the English scholar, and called him the "Newton of the organic world." This action was significant. Haeckel was in the liberal opposition to Bismarck but was not active in politics. His championing of Darwinism was the beginning of the association of the evolutionary position with political liberalism. (Marx and Engels were to color it with radicalism as they absorbed it into their political thinking.) In the scientific world, however, Haeckel stood alone. Darwin was considered crazy, for Müller had always stood fast on the immutability of species and both Virchow and Kölliker were severe critics of natural selection. Interesting enough, however, Virchow was one of the founders of the Progressive Party and a member of the Diet. This party continued rather impotent criticism of Bismarck, who was unifying the German states by force and promoting the Industrial Revolution. From 1863 on, Kölliker and particularly Virchow (Haeckel's former friends) were his permanent antagonists. In 1866, he visited Darwin, who supported him against his critics. The two men exchanged warm letters up to Darwin's death.

By the time Haeckel had published his *General Morphology* in 1866, in which he analyzed all the natural world in terms of matter, form and energy, assumed a mechanistic viewpoint, and first announced his "monist" philosophy, which excluded existing religions, he came into conflict with the whole religious establishment. Even less tactful than Huxley, who also aroused his share of hostility, Haeckel indeed persisted in sin and published a series of lectures on *The History of Creation* in 1868 and in 1874 *Anthropogenie,* his own descent of man. In the eyes of the reactionaries he had assumed a role close to the Antichrist. A clerical adversary, who had urged the Duke of Weimar to dismiss him from his post in the university because of his immoral writings, was asked by the duke if he thought Haeckel really believed what he wrote. "Of course," said the indignant prelate. "Then he merely does what you do," said the duke cheerfully.

Haeckel dedicated the first of these books to Darwin, Goethe and Lamarck. There was much in his temperament that allied him to the nature philosophers. In his semiphilosophical work he indulged in much uncritical speculation. There is no use scolding him in retrospect as some historians of biology have done. His mistakes are now forgotten, and in his time he undoubtedly performed a useful function.

On the credit side, he elaborated the concept that phylogeny is recapitulated by the individual embryology of higher animals. The presence of gill slits, multiple aortic arches and a two-chambered heart in embryos of mammals are still considered ancestral relics, indicating that gill-breathing vertebrates preceded those who breathed air. In point of time the evolutionary sequence (substantiated by paleontology) was fish, amphibians, reptiles, birds and mammals.

Haeckel also evolved a theory that from the cell division of the egg, a form (the gastrula) evolved with a double layer of cells. From each of the layers another split off to form the mesoderm, the lining of the interior cavity of the body. Haeckel thought this was true of all animals, but subsequent research has shown that his process is true of only the higher animals, the third layer being formed from either of the first two or both. In the lower phyla, up to jellyfish, no third layer forms. Huxley first observed the two-layer gastrula.

Haeckel also put forward some notions concerning energy and spirit as a motivating force activating atoms in the development of organisms, a theory which did not constitute a contribution of any value. It should be noted, however, that his pupil and supporter, Oscar Hertwig (1849-1922), on a trip to Sardinia with Haeckel in 1875 discovered the actual process of fertilization in the sea urchin. He wrote:

> Although spermatozoa attach themselves to the gelatinous envelope of an egg in great numbers—many thousands of them when concentrated seminal fluid is employed—still only one of them is concerned with fertilization, and that is the one by a lashlike motion of its filament first approaches the egg. Here it strikes the surface of the egg with the point of its head. The clear superficial expanse of the egg protoplasm is at once elevated to a small knob that is often drawn out to a fine point, the so-called receptive prominence or cone of attraction. At this place the seminal filament, with pendulous motions of its caudal appendage bores its way into the egg. At the same time a fine membrane detaches itself from the yolk over the whole surface, beginning at the cone and becoming separated from it by an ever-increasing space.

He went on to describe how the egg and sperm nucleus became joined and fused. Thus an old mystery was made somewhat clearer and the work of Hertwig emerged as a triumph for Haeckel's influence.

The warlike Darwinian was beginning to accumulate a following among both scientists and laymen. Du Bois-Reymond was mildly pro-Darwin but anti-Haeckel. Younger naturalists were accepting evolution, and in 1877, when the Prussian government was drafting a new educational law, the intellectual conflict came to a head in a kind of preview (on a much higher level) of the famous "monkey trial" in Tennessee in 1925.

Haeckel made a speech in September, 1877, before the annual meeting of the German Association of Naturalists and Physicians in which he stated that evolution was a philosophical point of view which was destined to become the basis of all natural science in teaching. He concluded:

> How far the principles of the doctrine of universal evolution ought to be at once introduced into our schools, and in what succession its most important branches ought to be taught in different classes

—cosmogony, geology, the phylogenesis of animals and plants and anthropology—this we must leave to practical teachers. But we believe that an extensive reform of instruction in this direction is inevitable and will be crowned by the fairest results.

Four days later, Virchow answered him in an address, *The Freedom of Science in the Modern State,* a sad performance for a former liberal. After congratulating his countrymen on their freedom to teach, he then warned them that they must discipline themselves or they might lose that freedom. Evolution was an unproved hypothesis, and "we must therefore say to the teachers in schools, 'Do not teach it.' " He then tried to smear evolutionary theory with guilt by association, because socialism "had already established a sympathetic relation with it." In other words, evolution was to blame for the Paris Commune which all German bourgeois regarded with horror. Virchow, on the one hand, used the sound scientific argument that one must always remain skeptical and never succumb to dogma and, on the other, insisted that only the "scientific certainty" must be taught. He wound up by revealing his emotional motivation. As far as the relation of man to the animal went, "every positive advance we have made in the province of prehistoric anthropology has actually removed us farther from such a connection." Needless to say, he was one who pronounced all Neanderthal skulls to be pathological.

Haeckel answered him, pointing out all the inconsistencies and specious arguments. It was to no avail. The reactionary clerics and politicians triumphantly pointed to Virchow as their justification and the Prussian Minister of Education prohibited not only Darwinism but also the teaching of biology in the secondary schools. Since evolutionary principles were steadily gaining credence throughout the world, the whole episode is one more example of the sins against humanity committed by politicians.

Haeckel wrote one more book, *The Riddle of the Universe,* published in 1899, which had an extraordinary distribution, being translated into many languages and hailed by the radicals and freethinkers all over the world. Regardless of the fact that it contains some of his most muddled thinking which distresses both philosophers and academic scientists, it nevertheless sums up many of the important currents in Western culture at the end of the century. In the first place he mounted one of his most thoroughgoing attacks on organized religion, pointing out the obstacles to scientific thinking embodied in its anthropocentric point of view: "That powerful and worldwide group of erroneous opinions which opposes the human organism to the whole of the rest of nature and represents it to be the preordained end of the organic creation, an entity essentially distinct from it, a godlike being." It is, of course, this anthropocentrism which has landed us in our present ecological crisis, although Haeckel did not see that far. He did, however, invent the word "ecology." This reduction of man to a modest place in the universe endeared Haeckel to the new wave of freethinkers, mostly sprung from the Socialist wing of the labor movement. He also stated unequivocally that the gap between the organic and the inorganic could be bridged by chemistry and physics, a position which the progress of these two sciences has been steadily substantiating. Haeckel, however, aspired to go further and tried to unite the organic and what are generally called psychic and emotional aspects of life, which had been divided by Descartes, whom Haeckel chided for not going far enough. All organic cells, said Haeckel, were also psychic cells; as the organism grew, so did the psychic component. "Consciousness is a vital property of every cell." Thus he believed he had arrived at "monism," or unity of the world, and a valid pantheistic sort of religion. Actually, this was a mere verbal disguise of dualism, for the organic cells were really being endowed with psychic doubles and the nature of consciousness was in no way clarified.

If Haeckel was not a success as a philosopher, he did, however, make some further statements concerning ethics, insisting that the moral sense was an outgrowth of animal instincts and a working arrangement between the ego and society. This is a point of view which has been sustained by anthropology, but needless to say, it made his clerical opponents foam at the mouth.

Haeckel's theorizing allied him to the nature philosophers, but his sturdy materialism put him in the same physiological camp as Du Bois-Reymond (who could not endure his "monism"). Amusingly enough, because of his attacks on religion and his materialist attitudes, clergymen were accustomed to refer to him as "a man of licentious life." Actually he was happily married to a second wife, who had borne him several children, when he began a correspondence with a young girl, who was fascinated by him. Haeckel in his mid-sixties was handsome, charming, endowed with an impressive Jovian beard. They met, took walks, spoke tenderly of poetry, admired the scenery, discussed his ideas, and found themselves in love. After some years of Platonic, romantic friendship, sex raised its head. They agonized in true nineteenth-century fashion and made the great renunciation. Shortly after the separation, Haeckel received a letter from the girl's sister, saying she was dead. This strange, touching autumnal romance once more dramatizes Haeckel's character. Poetic, a brilliant if erratic thinker, and an apostle of a new scientific humanism, he sums up the cultural position of natural science before the twentieth-century era of genes and biochemistry.

23

The Deep Sea Floor

☙ IN the year 1866 a submarine telegraph cable was laid from Newfoundland to Valentia, Ireland. In performing this engineering feat many difficulties were encountered. Thus it was brought home to Western man that although his knowledge of his land environment had been steadily increasing, his ignorance of the seas, which covered two-thirds of the earth, was immense. The animals that inhabited the profound depths, the shape of the ocean bed, the temperatures and the areas and directions of the currents were among the problems that needed investigation.

The British Navy took the lead in marine study with six weeks of dredging at 650 fathoms by the *Lightning* in 1868 and periods of dredging by the *Porcupine* at 2,400 fathoms in 1869 and 1870. Then in 1872, a corvette (a three-masted sailing ship with a steam engine) of 2,000 tons was fitted up for a long intensive scientific voyage around the world. The cruise of HMS *Challenger* is, therefore, of great historical importance; with it begins the serious study of oceanography.

Even the captain, Sir George S. Nares, was a member of the Royal Society, while the scientists were chosen by a committee of that same body, the leader being plump, good-natured Dr. Wyville Thomson, professor of natural history at Edinburgh University (who had also been in charge of the work on the *Lightning* and the *Porcupine*).

All the corvette's guns, except two, were removed, while the interior of the vessel was remodeled to include a chemical laboratory, a photographic studio and workrooms and storerooms for the analysis and preservation of organic material.

The technique of exploration consisted of dropping the dredge and dragging it along the bottom, sometimes at a depth of 2,435 fathoms (or 14,610 feet). It brought up mostly mud, but many organic objects were buried in it. The scientists attached to it swabs of hemp yarn like beards, and these, dragging along the bottom, entangled objects or animals of a spiny nature. Aside from this type of bottom sampling, short tubes attached to a sounding weight were dropped. Such tubes could bring up a core, showing a section of the bottom ooze, but at this time no one suspected the story that stratification could tell. Besides these instruments, a thin gauze trawl was also dropped, the mesh being fine enough to capture very small marine life, generically known as plankton. The use of a small portable steam engine, called a donkey engine, to run the winch, which pulled up the heavy dredge and trawl, and the substitution of piano wire for hemp cord testified to technology's role in making oceanography a more exact science.

The samples obtained by the *Challenger* created the nomenclature for bottom sediment which has only been recently modified. The scientists labeled the different types: red clay, blue clay, diatomic ooze and globigerina ooze.

Diatoms are tiny one-celled plants of the algae family which secrete silicate shells of two pieces fitting one over the other like the lid of a pillbox. They make up much of the bottom ooze, along with *Radiolaria,* protozoa with starlike silicate skeletons which we have already described in connection with the career of Ernst Haeckel. Since he was the acknowledged expert, he was invited to write the volume on these microscopic creatures which became a part of the final monumental thirty-six-volume report. Globigerina are protozoa with calcium shells, and indeed countless deposits of these tiny shells in ages past created the chalk cliffs of Dover.

Huxley wrote enthusiastically when the ship had been voyaging for a year and a few preliminary reports had been received:

> It is obvious that the *Challenger* has the privilege of opening a new chapter in the history of the living world. She cannot send down her dredges and trawls into these virgin depths of the great oceans without bringing up a discovery.

For some insight into the human side of the expedition we have diaries. One of the senior officers, W. J. J. Spry, wrote a solid account, but a more uninhibited and amusing narrative was jotted down by Sublieutenant Herbert Swire for his own family and was not published until 1938. Young Swire came of a good family and never forgot it. The scientists, who wore the wrong kind of clothes, often unpressed, and were unfamiliar with nautical terminology, aroused his condescending amusement. When, during the early part of the voyage, Thomson lectured the officers on the aims of the expedition, Swire decided he was "a jolly old boy."

The ship carried a garden from which the crew were fed mustard and cress while the livestock consisted of twenty ducks, fifty chickens, some geese and fifteen sheep. Swire, who had a flair for drawing, made amusing sketches. He noted that the ship's band performed "excruciatingly." As far as he was concerned, the dredge brought up nothing but mud; the sailors called dredging "drudging."

When the *Challenger* reached Lisbon, the personnel were honored by their first royal visit. Luiz I of Portugal, who professed an interest in science, probably because some of the more cultured European royalty had made a fad of it, came on board in a naval uniform, surrounded by naval officers. They all were happily photographed together with the notables of the *Challenger*.

On the early part of the voyage, ending at Tenerife, deep-sea fish were brought up by the trawl. Adapted to great pressures, their air bladders expanded when they were brought to the surface. They floated helpless, belly up, mouth agape, eyes popping from their heads; some even burst.

When the ship left the Virgin Islands, where Swire had an eye for good-looking black girls, a line parted and they lost a trawl, probably because it was overfilled with sea slugs. At this time, they suffered their first and only tragedy: A seaman was killed by a flying iron block.

As far as their explorative operations went, the scientists found that they had to be performed with steam up because the ship rolled too much under sail. They continued to take soundings and samples of the bottom regularly and temperature readings every hundred fathoms. There had been much speculation as to whether life could exist at great depths in the sea; the *Challenger*'s efforts were to prove that at no depth they were able to reach was the water empty of organic creatures.

The ship zigzagged over to Brazil, but no landing was made at Bahia because of the ever-present yellow fever. A sloth, however, was brought on board, an animal which amazed young Swire, who quoted a native saying that God made it when he was drunk.

Among the Tristan de Cunha island group, where Swire saw cape pigeons and albatross, while stormy petrels fought over the ship's garbage, members of the expedition visited one rocky pinpoint, appropriately called Inaccessible Island, and removed a couple of German Robinson Crusoes. The men had settled there in the hope of making a fortune from sealing. The first year they captured some seals but incurred the hostility of the inhabitants of Tristan de Cunha who were also sealers. The latter would land surreptitiously and steal their belongings. They also shot the goats which inhabited the rocky heights and were a source of food for the Germans. Finally, a grass fire destroyed the only means the Crusoes had of climbing up the cliffs to hunt goats and wild pigs. For a time, they subsisted on fish in summer and on penguins in winter. By the time the *Challenger* arrived they were ready to give up and be taken off. Swire quaintly noted that they "were not exactly gentlemen."

The sailors from the *Challenger,* in defiance of discipline, ranged about the island, destroying penguins and burning down the hut built by the Germans. Swire was outraged by the lack of discipline, but a larger issue was involved. On many islands there and later in the Antarctic, the seamen encountered tame penguins and nesting boobies (another inoffensive seafowl) which they proceeded to slaughter with sticks for the fun of killing. Many of these birds nest only on certain small islands, and just this wanton type of destruction has, over the years, made them endangered species. The roots of our ecological indifference can now be noted in the nineteenth century, but even the scientists of the *Challenger* were not alert to the implications.

Continuing its dredging and trawling, the ship went on to the Cape of Good Hope, where the officers and scientists were wined and dined and given a ball. Young Swire visited an ostrich farm, where he saw the female sitting on her eggs, her long neck stretched out on the ground. The male and female took turns on the nest during the forty-two-day incubation.

From South Africa, the *Challenger* sailed southwest. Once more, it was a question of the southern continent, which Cook had failed to find but of which there had been mutterings since; indeed, it had been named Wilkes Termination Land. Some time was spent in Rhodes Harbor of Kerguelen Island, one of the Falkland group, which consisted of rock and a little grass, the resort of countless penguins. There, there were whalers, since in the past the area had swarmed with sea elephants, fur seals and whales. In an incredibly short space of time all three species had been reduced to "a mere remnant, and in a few years extinction is sure to follow," Spry wrote with philosophical indifference. Protection did not occur to him, for with true Victorian commercialism, he commented, "In such arduous vocations, men must be expected to make all they can." For the scientists, however, the area yielded a rich harvest of echinoderms.

Sailing ever southward, the ship encountered many icebergs which made navigation difficult, and finally, on March 23, 1874, the expedition gave up the search for a southern continent and set out for Australia.

On the way they stopped at Tonga, where Swire, reminding us of Sir Joseph Banks, romped with the pretty Tonga girls and noted the musical sound of mallets beating mulberry bark into tapa cloth. It was there that Spry and Swire were amusingly at odds in their reactions. The first congratulated himself on the fact that formerly "cannibalism was indulged in to an incredible extent; and this not from mere satisfaction of revenge but to satisfy appetite; friend, relation or foe equally affording food to the most powerful," but was now rapidly passing away under the influence of English missionaries. The sportive Swire considered the missionaries narrow-minded bores and felt that the native advisers should be "gentlemen." At any rate, civilities were exchanged with King George Tuben, who boasted the high civilization of a four-poster bed in his residence, kept for show while he slept on a mat.

In the Fiji Islands, the officers and scientists were treated to a dance which Spry said consisted of "bounding, flourishing clubs and shrieking." Swire, who was on the edge of turning into an anthropologist, was, on the other hand, delighted. He said that the wild and beautiful exactitude of the rhythm was unequaled by anything, except "a ballet on a first-rate London stage." The war dance was followed by a fan dance and a nautical dance with paddles. The hula-like women's dances had been suppressed by the missionaries as "indecent."

Zoologically, an interesting and important discovery was a live nautilus. Shells and dead animals had been found before, but no one had observed the creature in action. Related to but unlike the squids, it lives in a whorled shell, continually moving from chamber to chamber as it grows and hence inspiring Oliver Wendell Holmes to write a moralizing poem. Holmes subscribed to the myth of "purpled wings" and "webs of living gauze," the notion that the animal held up its mantle to the breeze and sailed along the surface of the sea. Swire noted that it was jet propelled like the squid, water being shot from "a small tube," the siphon.

The *Challenger* sailed on to the Philippines where more dancing took place and Swire was charmed by the graceful arm movements of the daughter of a Moro chief. Ports of call were Hong Kong and China, Japan, the Sandwich Islands, the southern areas of South America, Hawaii. At only one point, the coast of New Guinea, were the explorers menaced with spears and refused a landing. Later the incident was smoothed over by trading.

Young Swire, who drew pictures of everything and everyone, remained a gentleman, even if he never became a zoologist, although he was, from the beginning, a convinced Darwinist. Thoroughly enjoying the expedition, he received his small measure of immortality. Near the Carolinas, the *Challenger* reached the limit of her soundings, 4,500 fathoms. The area was named the Swire Deep.

24

Paleontology and the Wild West

ఆ§ THE United States entered the field of zoology late in the nineteenth century, shooting from the hip as it were, for the first important contributions came from two paleontologists who operated on the frontier and carried on one of the great feuds of science, a battle which lasted twenty-five years. Their buccaneering attitudes exactly parallel that of the robber barons of the post-Civil War period, and their swashbuckling careers put the United States on the map paleontologically. Then, too, their contributions were highly significant for the evolutionary controversy in the post-Darwinian era.

Before Edward Drinker Cope and Othniel Charles Marsh, American zoology was sporadic and little systematized. Interestingly enough, there is a quaintly spelled letter in English on record from Lamarck and Geoffroy Saint-Hilaire to Charles Willson Peale, who set up a museum in Philadelphia in 1785, which consisted of shells, rocks, a few stuffed animals and birds and a mounted mastodon incorrectly labeled "Mammoth, a nondescript carnivorous animal." The Frenchmen were interested in the large fossil bones found in America, and they also wanted to exchange specimens, for they wished to prove that no specimens found in the New World existed in the Old.

Zoology, as a science, was brought to the attention of Americans in 1839 with the arrival in the United States of Louis Agassiz, a Swiss expert on fossil fish.

Born in Motier in 1807, he had the familiar struggle with a well-to-do bourgeois father who wanted him to become a doctor. He took his medical degree in Munich, but managed to study zoology on the side. He then went to Paris in 1831, where he met the aging Cuvier, who gave him his notes on ichthyology and set him to cataloging fish in the Jardin des Plantes. At Cuvier's death, it looked as though he might be obliged to go back to medicine, but Humboldt sent him one thousand francs and got him a professorship at the University of Neufchâtel. There he worked on fossil fish, bringing out an important book in 1833. He became interested in the movements of glaciers, a matter of controversy, and spent much time in the mountains, measuring and observing those rivers of ice. His observations led him to the conclusion that glaciers had advanced and receded in the past and that they were capable of carrying boulders and pebbles with them. He also identified glacial scars on beds of rock. This was pioneer work which led to the later geological mapping of glacial ages, and Agassiz's conclusions were accepted by Lyell in 1846.

In 1839 Agassiz came to lecture in the United States and never went back to Europe, joining the Harvard faculty the following year. He left his wife and children in Europe for ten years. When his wife died, he brought his children to America and married Elizabeth Cary of Boston.

Agassiz was a member of the Saturday Club where he mingled with such New England intellectuals as Emerson, Oliver Wendell Holmes and Longfellow. A popular lecturer with an engaging French accent, he did much to make the public conscious of natural science. He also founded the Museum of Comparative Zoology at Harvard. Asa Gray, the Harvard botanist who for years kept his correspondence with Darwin secret (at the latter's request), came out in the open when the great book was published in 1859. Agassiz immediately took the side of the opposition, calling the theory "mischievous" and comparing it to the philosophy of Oken. The conflict in Harvard became so acute that the two naturalists' students cut each other in the corridors while their preceptors carried on public debates in which Gray generally came off best.

Such was the scientific atmosphere in America when O. C. Marsh and E. D. Cope were growing up. The first, the son of a New England farmer, had the good fortune to be the nephew of George Peabody, the founder of the banking firm which was to become J. P. Morgan & Co. A philanthropist influenced by the philosophy of Robert Owen, Peabody was a pioneer in the great American custom of giving away large amounts of money. He took care of all his fourteen nephews but with New England firmness saw to it that they did not marry until they were established in their careers.

Born in 1831, Marsh was put through school, including Andover and Yale, by his uncle and never lacked for funds from then on. Europe and graduate work were both underwritten by the generous uncle. Encouraged by Professor Benjamin Silliman, young Marsh went into paleontology, with the understanding that there would be a place for him at Yale. The Yale academics were worldly enough to hope that some of Peabody's money, which was being siphoned off to Harvard, might be diverted to Yale. Besides, Marsh, after some waste of time at Andover, turned into a model student (and something of an academic politician). Everything worked out well: Marsh got his professorship (without salary) and Yale got $1,500,000 for a museum which would house zoological, archaeological and paleontological collections.

Cope's background paralleled Marsh's to the extent that he, too, was born into a moneyed family. His wealthy Quaker father, who owned steamship lines, was able to finance his son himself. Cope was born in Fairfield, near Philadelphia, in 1840. After primary schooling in a Quaker institution, he was put on a farm because of his delicate health. For some reason, his father wanted to make a farmer of him, and as a result, Cope had little formal higher education.

In his youth, he saw the famous Peale mastodon and collected snakes and amphibians on which he exercised his talent for drawing. His father gave him a farm when he was eighteen. After toying with farming for a while, he got permission in 1860 to attend the lectures on anatomy of Dr. Joseph Leidy at the University of Pennsylvania.

Leidy was really the first native zoologist and particularly concerned with the field of paleontology. In the 1840's, large bones began to be picked up in the Middle West, some of which were sent to Agassiz. Leidy, however, took a more active interest, and soon most of the surface finds were sent to him. By 1847 he had enough material to set him thinking along evolutionary lines. Although he never went collecting in the field, Leidy had a precise, orderly scientific mind and, from the material sent him, was able to write the first great American book in the field of zoology, *The Extinct Mammalian Fauna of Dakota and Nebraska,* which appeared in 1869. Although Leidy did not lecture on paleontology, Cope got to know him and was undoubtedly influenced. The young man frequented the Philadelphia Academy of Natural Sciences and involved himself with cataloging its collection of fish. He also visited the Smithsonian Institution. In 1863, his father was perhaps anxious to get him out of the way of the Civil War, and indeed Cope, as a Quaker, had worried about the issue. At any rate, young Cope was sent to Europe with a pocketful of letters to professors of natural science and directors of zoos and museums. Needless to say, he looked at all the collections of fossil bones available and was shown live hippopotamuses in the Amsterdam zoo.

Young Cope spoke German, picked up some Italian, but had trouble with French. In his letters, he described scenery, criticized the looks of the women, and admitted that he was not impressed by Hiram Powers' prettified sculpture. He met Wallace and the latter's former associate, Bates, and saw Sir John Lubbock at a meeting of the Linnaean Society.

Returning home in 1864, he managed somehow, without degrees, to obtain a professorship in zoology at Haverford College, and the following year he married a distant cousin, Annie Pim, by whom he was to have a daughter, Julia. Still friendly with Leidy, he began collecting, made a house visit to Agassiz in 1867, and began publishing on extinct batrachians. In 1868, he met O. C. Marsh and went collecting peaceably with him in New Jersey, where the former had found an eighteen-foot carnivorous dinosaur.

That same year, Cope left Haverford, apparently disgusted with the administration. He had a home in Haddonfield where he settled down, after selling his farm, to devote himself to paleontology.

It was Marsh, however, who first ventured into the Wild West in search of the great fossils of which so many rumors had percolated to the East. In 1868 he was made secretary of the American Association for the Advancement of Science and went on an expedition, organized by a group of members, to Wyoming. The old seabeds of the western plains excited him. In the far geologic past, the whole central area of the continent had been under water. The land rose, bringing with it skeletons and organic remains preserved in sedimentary rocks. Subsequent risings and fallings had stratified the rock accruing from sediment, and after the latest upsurge, chalky buttes were weathered away and here and there bones began to appear. The Indians said they were animals destroyed by the god of lightning. Marsh heard of a skeleton in Antelope Springs, Nebraska. He managed to persuade a conductor to hold the train while he looked through the earth brought up in digging a well. He found the skeleton of a tiny horse and bones of other animals. The horse, which he named *Equus parvulus,* stood three feet high and had three toes. He read a paper on the beast, now known as *Protohippus,* before the National Academy. Up to this time, nothing systematic was known of Western paleontology. Even Leidy's book had not yet appeared. The field was wide open, and Marsh sensed that there was a possibility of digging up the record of vast areas of the past.

Despite threats of Indian wars, in 1870 he organized an expedition of Yale graduate students, most of whom were romantic about Indians and the Wild West. His base was a depot of the Union Pacific Railroad, and he managed to obtain a guard of soldiers, because the Sioux and Pawnee were hostile to each other and there was danger of being caught between them. Buffalo Bill spent a day with his party and later accompanied him on some of his expeditions. Traveling through a parched country at 110° in the shade, he found six more early horses. Ranging from Nebraska to Fort Russel in Wyoming to Colorado, he discovered, among other things, many remains of the titanothere, an animal which, as it evolved, grew huge, developed horns on its nose, and then became extinct. Finally, Fort Bridger, Wyoming, became his most important base. There were many alarms concerning horse thieves, but once when the horses disappeared, supposedly driven off by Indians, it turned out they had merely been frightened by coyotes. The troops from Fort Wallace came thundering out to save the party, but found no Indians. Marsh became a good horseman, took part in a buffalo hunt, and had buffalo steak for Christmas.

In Kansas he found the remains of 35-foot sea lizards, called mosasaurs, which were characteristic of the Cretaceous, a 70,000,000-year-old formation. Marsh and his students came home triumphantly with thirty boxes of bones. He organized another expedition in 1871, which resulted in the discovery of pterodactyls, the flying dinosaurs, with a 20- to 25-foot wingspread. These had no teeth, while the birds of the same period had teeth.

Marsh developed techniques for handling his material, packing his more delicate bones in cotton and then sewing them up in burlap. Later techniques used by Cope's helpers involved sending East the smaller and more delicate skeletons embedded in the stone matrix so that they could be carefully removed in the laboratory.

Everywhere there were buffalo. Western Union placed a stuffed head as a sign over all its offices. Once, stalking a herd of fifty thousand, Marsh shot one and then found himself in the middle of a stampede. There was nothing to do but gallop in their midst, hoping that his pony would not give out. Finally, he succeeded in swinging out toward the edge of the mass and circling a butte, which he was able to put between him and the herd.

Amusingly enough, Marsh was invited to a meeting with Brigham Young. The Book of Mormon contained a "revelation" of horses in America, and Marsh's discoveries apparently made him a defender of the faith. The expedition, which came home with rich spoils, cost $15,000, the expenses being defrayed by its members.

Meanwhile, Cope was not idle. In 1871 he, too, went to Kansas. Marsh seemed to consider the whole West his monopoly, and from then on, a kind of cold war began between the two men. It consisted mainly of refusing to disclose information and buying off each other's local helpers. Supposedly Professor B. F. Mudge of the University of Kansas had discovered a fossil bird with teeth which he had planned to ship to Cope, but Marsh nipped in and got it for the Peabody Museum. On the 1871 expedition, one of Cope's finest finds was a twenty-five-foot mosasaur whose long tail went through the limestone bluff in which it was embedded and came out on the other side.

Cope was an enthusiastic letter writer, and those he sent to his wife and daughter are full of lively descriptions of the country, the flora and fauna and the people he met. He was particularly successful in collecting fossils from the Cretaceous. He wrote of the dinosaurs:

> We find that they lived in a period called the Cretaceous, at the time when the chalk of England and the green sand marl of New Jersey were being deposited and when many other huge reptiles and fish peopled both sea and land in those quarters of the globe. The thirty-seven species of reptiles found in Kansas up to the present time varied from ten to eighty feet in length. . . . Far out on the expanse of this ancient sea might have been seen a huge snakelike form, which rose above the surface, and stood erect, with tapering throat and arrow-shaped head, or swayed about describing a circle of twenty-five feet radius above the water. . . .

This was *Elasmosaurus platyurus,* a swimming lizard. Unfortunately, when Cope came to mount this forty-five-foot reptile's skeleton, he put the head on the wrong end. Marsh pounced on this mistake and used it against him many years later.

The competition between Marsh and Cope, in which each struggled to establish and name new species first, resulted in the use of the telegraph to establish priority. In 1872, the operator scrambled the name *Loxolophodon,* sent in by Cope to the American Philosophical Society, to "Lefalophodon." But Marsh had gotten in two days before with the name *Tinoceras* for the same species.

Cope got into difficulties with the American Philosophical Society for printing papers before they were read. He eventually became disgusted with the organization's slow procedure and bought a publication, *The American Naturalist,* in which he could publish whenever he pleased.

The rivalry between Marsh and Cope so disgusted Leidy, who was still working in paleontology, that he withdrew from the field. He had only a professor's salary, while the other two were wealthy. Cope inherited from his uncle, Marsh from his father, and neither hesitated to fling their private fortunes into the struggle to corner the fossil bone market. Cope was associated with one of the three Government Surveys that was headed by Lieutenant George Montague Wheeler. This government body organized expeditions on which he worked as paleontologist, and Cope's connection resulted in his monographs being published by the government.

In 1874, Marsh received a good deal of publicity because he became embroiled in a controversy over the Indian question. He had had some dealings with Chief Red Cloud, who was suspicious of him because there had been talk of a gold rush, and the Indians feared more land would be taken from them. Marsh pacified Red Cloud by giving him a banquet. He also took the list of the Indians' complaints to Washington to the Commissioner of Indian Affairs. When this proved useless, he went to President Ulysses S. Grant. That unfortunate executive's administration was a stew of corruption. The Indian issue was fought with charges and countercharges. Marsh's integrity was attacked; he fought back and the press had a holiday. A commission to investigate was appointed, and when the smoke cleared away, Marsh's charges were upheld and Secretary Columbus Delano of the Interior Department was fired. Since the Indians obtained some temporary improvements, Red Cloud became the paleontologist's firm friend.

After the seventies, Marsh gave up active fieldwork and depended mostly on well-trained assistants, directed by Professor Mudge. Through these he obtained many more great dinosaurs, among them the fierce carnivore *Allosaurus,* which ran erect on its hind feet, and the sixty-seven-foot *Brontosaurus,* which was mounted for the Peabody Museum. The Sinclair Oil Company adopted the dinosaur as its trademark for the Chicago World's Fair of 1893 and in recent years has even contributed money for dinosaur-spotting from the air. Indeed, the well-publicized activities of Marsh and Cope made the great lizards of the past come to life for many young Americans, who were eventually able to gaze at mounted skeletons in the American Museum of Natural History.

Cope, with the aid of the Survey, continued to work in the field until 1878, when he returned to Europe and was able to purchase in Paris an important collection of Argentinian fossils, which he sent home to his residence in Philadelphia, two houses on Pine Street which he had turned into a museum.

At this time, Marsh had the better reputation in Europe. He had published *Birds with Teeth* and *The Titanotheres*. He had regular access to the *American Journal of Science* and an able assistant, Oscar Harger, who was later credited with doing much of the writing of his concise, clear monographs. Cope's papers were scattered all about in various publications, and in his competitive haste he often introduced new names that had to be canceled or withdrawn. It is also probably true that Yale gave Marsh a certain standing which Cope lacked. Then, too, when Huxley visited America in 1876, he became friendly with Marsh and was delighted with his series of fossils, which represented the evolution of the horse from a small three-toed animal to the modern species. Huxley revised his own lecture on the horse on this basis, for he said there was the first complete evidence of the evolution of an existing animal. Huxley prophesied that a four-toed horse with a rudimentary fifth toe would be found, and two months after he made this statement, the prophecy came true. Darwin, to whose theories Marsh subscribed, wrote that the American's work was the most important evidence for evolution he had seen in the past twenty years.

Cope was well received, however, by almost everyone in Europe, except Huxley who remained cool. The pampas collection was shipped to America in many boxes, most of which Cope never had time to open before his death. (It was sold to the American Museum of Natural History for $20,000 in 1902.) His assistants were still collecting, and on his return from abroad he revisited the field in 1879. That same year, however, the three existing Government Surveys were combined in the interest of greater efficiency and the director appointed was Clarence King, a schoolmate of Marsh, whose successor the next year, John Powell, promptly appointed Marsh government paleontologist. This was a serious blow to Cope. At the same time, having spent a great deal of his own money, especially on the pampas collection, he endeavored to recoup by investing in mining stock. The result was that he lost most of his fortune and for the next ten years was in serious financial difficulties.

The decade of the eighties saw the intensification and climax of the battle between Cope and Marsh. The Geological Survey had published the first part of Cope's book on the Tertiary and Permian vertebrates. Now the Survey, under the direction of Major John Powell, informed him that the second part was too expensive and that the Survey did not have the funds to publish. Cope blamed this setback on Marsh and spent the years from 1886 to 1888 in and out of Washington, lobbying, trying to get bills through which would include appropriations for his book. He never succeeded. At the same time, he was having trouble keeping *The American Naturalist* going. He had difficulties with publishers who lost him subscriptions, and at one point, his friend, Henry Fairfield Osborn of Princeton, put up $200 to help get out an issue. In spite of Cope's efforts to prevent it, Marsh was reelected president of the National Academy for the Advancement of Science. On top of all these blows came an order from the Secretary of the Interior to deposit all of Cope's Cretaceous and Tertiary collections in the National Museum, the Secretary maintaining that they belonged to the Survey. Cope had spent $75,000 of his own money collecting, and since he never accepted a salary, he felt the bones belonged to him. In these days, Cope was in such a state of frustration that he was probably somewhat paranoid, believing as he did that everyone was intriguing against him with Marsh pulling the strings. Having become friendly with an ambitious young New York *Herald* reporter, William Ballou, he allowed himself to be talked into a public attack on Marsh. Cope had a drawer in his desk, in which he kept every scrap of evidence which he felt proved Marsh was an incompetent. He turned this over to Ballou and furnished the young man with a set of charges against Marsh, Powell and the National Academy.

Circumstances had to some extent played into Cope's hands. Marsh had collected more material than he could handle and being of reserved, dictatorial character, had never given any of his assistants credit for their contributions. Moreover, in the eighties, he seems to have gotten lazy and spent more time hobnobbing with the rich and famous in the Century Club than in his workrooms at Yale. As a result, his assistants revolted in 1890. It was at this point that Cope attacked, claiming that Marsh was a plagiarist of his work, that he had unlawfully appropriated government fossils for Yale, that he used his assistants' work without credit, and listing his mistakes, branded him incompetent. He also attacked Powell and the National Academy as united to conspire against him. All this Ballou wrote up in his article of January 12 for the New York *Herald,* spicing the story with flamboyant headlines. By now Cope had become professor of geology at the University of Pennsylvania, so it was Yale versus Penn, with Marsh and Cope on the opposing teams. Marsh published a cool and learned rejoinder. Cope could point to a horned lizard that Marsh mistook for a bison. Marsh came back with the *Elasmosaurus* whose head had been placed on its tail. Letters, telegrams, interviews, statements of support and denials came from other scientists. Nearly everybody took sides. Provost William Pepper of Pennsylvania got frightened and seemed about to fire Cope, but Osborn came up to Philadelphia and spoke for his friend. Old Professor Leidy came out for Marsh. A newspaper poet summed up the controversy in some humorous lines which went in part, Cope speaking to Marsh:

> You stole your evoluted horse from Kovalesky's brain
> And previous peoples' fossils smashed from Mexico to Maine

Marsh rejoined:

> Your reference to a horncone on an ichium sends a chill.
> Professor Huxley is my friend and so is Buffalo Bill.

While Powell attacked Cope:

> Through natural selection's law you are, as I'm alive,
> Of all survivals found, the least fit to survive.

Although Marsh made a better public impression, Cope seems to have partially triumphed, for Marsh was dismissed from the Survey. Marsh, in his turn, suffered financial reverses, for something went wrong with the Peabody trust fund, and he had to accept a salary from Yale. He had expended a fortune of $200,000 on fossils. On the other hand, Cope was able to sell his North American Mammal Collection to the American Museum of Natural History for $32,000, which made life easier for him up to his death in 1897.

Marsh died in 1899, a reticent, lonely man, who never married. Cope, in contrast, was, on the whole, a genial character who enjoyed a happy relationship with his wife and daughter.

The feud between the two resulted in many mistakes and confusions because of their breathless competition and often put them in a rather comic light, but the public perhaps learned something about paleontology.

Marsh's contribution was concrete. As we have pointed out, his series of fossil horses made the facts of evolution undeniable; his toothed birds established the relationship between the birds and the reptiles.

Cope was more of a theoretician. He inclined toward Lamarck, perhaps because of his devout Quaker background (he used the quaint "thee" in all of his letters). He simply restated Lamarck's point of view with the phrase "use and effort" and went on to reiterate the standard life force position for which he coined some rather empty terms.

As a herpetologist, he had no peer in his time. Important books are *Batrachia of North America,* 1889, *Crocodiles, Lizards and Snakes of North America,* U.S. "National Museum Report of 1898," pp. 155-1270, Washington, D.C. 1900. In his countless monographs, he roamed over the whole field of zoology, fossil and living, adding to our knowledge of the Eocene, Triassic, Jurassic and Cretaceous. His first volume of the Survey report on the Tertiary and Permian vertebrates was referred to as his bible. The second volume was never completed although some of the plates were published in 1915 by the American Museum of Natural History. Perhaps the greatest contribution of all to taxonomy was his establishment that the hooved animals were descended from five-clawed beasts. He also established, as a classification diagnostic, the three-tubercled tooth as the ancestral form which evolutionary variations could be traced. All in all, the work of both Cope and Marsh added a third dimension to taxonomy. Relationships between living groups, which had not been clear, could now be studied in terms of their common ancestors, embedded in the rocks of the past.

The splendid evolutionary series mounted in the American Museum of Natural History, showing the lineage of the horse, the titanotheres, the elephants and many other mammals, are based on the work of these pioneer paleontologists. Again and again, they drove home the indisputable facts of evolution. The how of evolution, however, was to be investigated by another group of specialists, the geneticists, and, eventually, the biochemists.

25

The Reign of the Fruit Fly

☙ MOST of us, as children, have searched for four-leaf clovers and never stopped to wonder about the significance of the extra leaf. Darwin and his immediate followers had thought that new species came about by means of many slight changes, like darkening of color or thickening of hair. But many of these small changes turned out to be temporary; they were not inherited. On the other hand, the fourth leaf, and sometimes even the fifth leaf, of the clover is a significant change whose importance was not appreciated until two botanists, Gregor Mendel and Hugo De Vries, had worked on the problem of inheritance.

In this case advances in botany led the way for similar research in zoology, and therefore it is necessary to sketch the contributions of these two men before approaching Thomas Hunt Morgan, the real father of modern genetics.

The story of the monk Gregor Mendel (1822-84), whose work was forgotten for thirty-four years, has often been told. Like Purkinje he joined a religious order more as a matter of survival than from profound piety. His work in the local school in Troppau, Moravia, was excellent, but in order to earn enough money to gain a higher education, he overworked at tutoring and made himself ill. Eventually a self-sacrificing sister turned her share of the family estate over to him. After two years of university work, the money again ran out. Once more he worked himself, or worried himself, sick and at last solved his problem by joining the Augustinian Order at Altbrunner. Mendel was no model monk, but his diligence as a student and his blameless life gained him admission to this order as a priest in 1845.

He took to teaching with pleasure and had no trouble gaining the affection of his pupils. He is described as a stout man, with a benevolent expression, who wore gold-rimmed glasses and side whiskers, and, strangely enough, dressed not in a soutane but in a top hat, frock coat and trousers generally too big for him.

He soon proved he could not work as a parish priest, for he grew sick at the sight of suffering and could not endure deathbeds. His liberal-minded abbot or prelate, as he was called, let him follow his bent. Mendel made the mistake of taking the examination to qualify as a high school teacher and failed. Since the examiner suggested that he be sent to the University of Vienna for more training, this was done by his order. All the while Mendel had been working on botanical experiments in the garden of the monastery. In Vienna he studied zoology, botany, paleontology and entomology. He took the teacher's examination once more and seems to have failed again. There is a rumor that he disagreed with the botany examiner. At any rate (despite lack of credentials) he never ceased to be a successful teacher and delighted his pupils with a collection of animals: a fox, a hedgehog, mice, and cages of birds. When he caught his boys shooting birds, he took away their slingshots.

The presence of both white and gray mice in his collection suggests that he had thought of using them in heredity experiments, but it is probable that the authorities would not have approved. Although he was loved in his own monastery, he had once satirized a fat bishop which had not made him popular with his superiors.

He began his important experiments in crossbreeding peas in 1856. Mendel crossed round and wrinkled peas, fertilizing them manually. All the peas produced were round. He crossed two hybrids and got three out of four round and one wrinkled. He decided roundness was a "dominant" characteristic and wrinkledness "recessive." When he went ahead and mated the two wrinkled peas, he could again produce wrinkled stock that bred true. Two of the round peas produced from the cross also bred true. Since the same pattern always emerged, there were evidently mathematical laws. He then crossed round yellow-seed peas with those producing wrinkled green seeds. Out of sixteen peas nine were real hybrids. The ratio of recessive to dominant remained the same.

He went on to complicate his experiments and came to several conclusions. He decided that the individual characteristics remained distinct in the hybrid form. Characteristics, therefore, were controlled by heritable units and the heredity of any individual specimen was determined by the units in the parents. The mysterious nature of the process consisted in the fact that the heredity units did not affect each other. They could be manipulated in all possible combinations.

Mendel lectured on his work before the Brunn Society for the Study of Natural Science, which he had helped found. There were no questions, no discussions. No one bothered to check his work. He published it in 1866 and sent a copy to Karl von Nägeli, a Swiss botanist and a very important man in his field. Nägeli was an evolutionist but also something of an old-fashioned Okenist. He believed ideas were more important than observation. Hence, although he wrote condescendingly to Mendel, he suggested that Mendel's results were "empirical" rather than rational and urged further work. He never checked his fellow scientist's results.

Mendel's monograph remained unnoticed until a Dutch botanist, Hugo De Vries (born in 1848 in Haarlem), undertook the study of evening primroses. Observing a group of these plants, which had grown wild in a fallow field, he found a number of plants with striking dissimilarities. Some were dwarf, some giant; some had red-streaked leaves. De Vries kept a record of the number of new forms which appeared in each generation of his primroses. He came to the conclusion that these new forms had appeared without intermediate small changes. He advanced the theory that "mutations," sudden changes in form (which had been observed before and called sports), gave rise to new species. He accepted natural selection as the process which determined their survival value. His work was done in the late 1880's and early 1890's and was violently opposed by orthodox Darwinists, who had become the establishment. But De Vries performed another service. In researching the literature on heredity he came upon Mendel's long-forgotten paper and recognized its importance. He therefore realized that certain elements in heredity, units, were constant and in the case of mutations there was some change in the relationships of these units. De Vries spent years working with many types of plants, seeking mutations which would breed true and finding many examples, including clover which produced from five to seven leaves on the same plant.

Other scientists checked De Vries' findings. In some cases he seems to have been mistaken and to have worked with hybrids, but his emphasis on mutations proved to have a sound basis and the phenomenon was accepted in genetic circles by the end of the century.

The stage was being set for another leap forward, but one more element had to come into the picture, an element supplied by the cytologists, the specialists who were working on more and more careful analyses of the structure of the cell and in particular the sex cell. In other words, although Mendel had indicated that there were "units" of heredity, the reductive process of science had to go further before more could be learned concerning the identity of these units.

In 1882, a German, Walter Flemming, discovered elements in the cell nucleus which took a deeper stain than the rest of the protoplasm and labeled the elements chromosomes from the Greek word for "color." When cells reproduced by fission, or dividing in two, the spots turned into filaments which split lengthways, pulled apart, and each set moved into one of the new cells. In 1885 Wilhelm von Waldeyer labeled these filaments chromosomes. Then Edouard van Beneden, a Belgian cytologist, discovered in 1887 that the chromosomes in the nuclei of the male and female sex cells behaved differently. Instead of doubling, during fertilization each set was reduced by half so that finally the fertilized egg possessed the normal number of chromosomes for the species in its nucleus. (It was Oscar Hertwig who had first observed that in reproduction the nuclei of the sperm and the egg united.)

The elements of the genetic problem were now ready to be assembled into a systematized theory of heredity. The man who was to devise the standard technique of experiment and to put the separate study of genetics on its feet was an American, Thomas Hunt Morgan, a graduate of the State College of Kentucky. Morgan was born in 1866. His father had been an officer in the Confederate Army, was for a time a manufacturer, and had even done a stint in the American consulate in Sicily. Young Morgan seems to have drifted into natural history. Graduate work at Johns Hopkins left him hesitating between morphology and physiology. Toward the end of the nineteenth century the vitalists were in the ascendant. Although the acceptance of evolution had done away with the crude mythology of special creation by supernatural means, in scientific circles there was still the unexplained difference between life and nonlife. The vitalists stuck to their mysterious force as Cope did. The mechanists, however, were carrying the reduction of living tissues farther and farther in the direction of chemistry.

Morgan was attracted to the mechanists, whose approach he considered practical. In the field of heredity Mendel had found certain laws. Morgan decided to look more closely into mutations. The aim of the mechanists was to reduce biological processes to laboratory analysis. But first a suitable experimental animal had to be found. Morgan tried mice, rats, pigeons, plant lice. Then he heard that a laboratory worker had worked with *Drosophila,* the vinegar or fruit fly. If rotting fruit is left exposed to the outside air, these tiny-winged dots cluster on it. When we brush them away, they hover about rather feebly and return to their meal. If we think about them at all, we include them with gnats. Yet these humble insects proved to be the perfect experimental animal. In the first place, they breed with extreme rapidity, developing from egg to winged adult in ten days. Thus thirty generations can be checked on in a year, and they are so small and easily reared that thousands can live in a few milk bottles. Morgan made *Drosophila* the symbol of genetic research. Indeed a wit said that God had created them especially for Morgan.

At first he tried various stimuli to see if he could hurry along the process of mutation. High and low temperatures, acid and alkali diets were tried in all stages of the insects' lives.

The work was started in 1909. By April, 1910, the investigators had a strain with white eyes which had been breeding true for a year. Ordinarily the fly has red eyes.

Mendel's law was checked by crossing red-eyed and white-eyed strains. The resulting ratios between the two strains in subsequent generations were the same as in the case of two types of peas. Other mutations were found, some with wing specks, some with and some without thorax patterns, some with beaded edges to their wings, some with pink eyes, some with very small wings, some with almost no wings at all.

During all this work Morgan was teaching in the graduate school of Columbia University and thus was able to interest young scientists in his work. His first talented assistant was Calvin Bridges.

Once Morgan's research had produced fifteen mutations, new facts began to be noted. The white-eyed strain was transmitted by females only to males. This was named a "sex-linked" trait. Then it was found that two and three traits could be associated and always transmitted together.

Thus patterns of heredity were appearing. Could they be further elaborated? The key proved to be the chromosome structure. In 1902 an American cytologist, Walter Sutton, had pointed out that chomosomes seemed to be connected with heredity. Each cell had a fixed number of pairs of chromosomes. Then they were reduced in number in fertilization. They seemed to behave like Mendel's units of heredity, for they produced replicas of themselves as long as the species continued to breed.

It was one of Morgan's former students, Nettie Stevens, who identified three large and one small chromosome in *Drosophila* sex cells. The three large ones, Morgan thought, easily accounted for the three-fold linkages. The small one, by many careful experiments, was finally associated with the trait of a bent wing and never took part in linkages.

Chromosomes had now assumed their true meaning as carriers of hereditary traits, both dominant and recessive. But could the process be reduced still further? Was it a case of boxes within boxes within boxes?

Apparently it was, for botanists had discovered that there were fewer chromosomes than characteristics. This indicated that chromosomes could be broken down into still smaller units. In 1909 a Danish botanist, Wilhelm Ludwig Johannsen, gave these as yet undiscovered hypothetical entities the name "genes." Although their existence was to be established by experiment and a whole doctrine erected upon them, a gene was, with the development of tremendously high-powered microscopes, first seen in the sex cell of a bacterium in 1969.

Another disciple of Morgan's, Alfred H. Sturtevant, worked on the problem of genes. A further study had shown that linkages were not stable; the traits could separate and appear in new relationships. From this it was deduced that the whole matter of linkages had to do with the pattern of genes within the same chromosome. At first, these unseen ghosts were thought of as beads on a string. On the hypothesis that the beads could cross over and thus disturb the linkages, all sorts of mathematical experiments were carried out as thousands of fruit flies obligingly produced generation after generation.

It was found that the farther apart the genes were, the more easily they separated. Finally the picture of the ghostly genes changed again and they were visualized as the strands of a twisted cable. Tens of thousands of careful experiments, mathematically correlated and based on frequency of unlinking, resulted in actual maps of *Drosophila* genes in 1911, a process as curious as the growth of atom research while the atom remained a purely intellectual concept.

We begin to note the importance of teamwork in modern research, for the Morgan school of genetic research, requiring as it does a multitude of careful experiments, meant that many assistants and coworkers added something to the development in a particular direction. It was Morgan's wife, Lilian, who bred hermaphrodite fruit flies. How were they to be explained? More generations of fruit flies supported the theory that certain chromosomes carried sexual traits. The female egg possessed two of these X chromosomes, the male had one plus a Y chromosome. The hermaphrodite got two X's plus the Y, and this Mrs. Morgan could verify under the microscope because in some eggs she could see the two X's joined which failed to separate properly in reproduction. The doctrine of X and Y sex chromosomes checked with sex-linked heredity. It could be seen that the white-eye trait in *Drosophila* (and the human trait of hemophilia), which never showed up in females, was nevertheless handed down through the female, for the egg contributed the X chromosome to the male.

In 1933 Morgan was given the Nobel Prize, and by that time his work was accepted and fruit fly research was being carried out all over the world. He remains the founder of what is called classical genetic research. That he was not without a certain dry humor is proved by his remark when he was called to the California Institute of Technology in 1928. He said: "Of course I expected to go to California when I died, but the call to the Institute arrived a few years earlier and I took advantage of the opportunity to see what my future life would be like." He retired from Cal. Tech. as head of the biology department in 1941 and died in 1945.

It was one of Morgan's students, Hermann Joseph Muller (born in 1890), who tried to increase the frequency of mutations. First, he raised the temperature of the environment; later he tried X rays, which definitely stimulated instant changes in the genes. This discovery has unfortunately been brought home to us only too vividly by the damaging effect of the radiation poured forth by the explosion of atom bombs.

The classical period of genetics still dealt with the gene as an indivisible entity. The discoverers of the Morgan school made possible more accurate stock- and plant-breeding procedures and also clarified the genetic nature of certain diseases. The *nature* of the gene still remained unsolved.

From this point on, the problem of heredity involved a further reductive process, and as a result, chemistry and physics emerged to play a major role in the investigation of life.

26

Inside the Giant Molecule

☙ THE animal body is made up of the organic substances fat, carbohydrates and protein. This the nineteenth century knew, for it began to have some inkling of the structure of organic substances and the fact that living organisms functioned through chemical processes. The twentieth century has gradually discovered that the proteins are all-important in metabolism, and finally, a detailed study of the chemical composition of the cell has provided the key to the mechanics of heredity.

The investigation of the nature of the gene has followed a zigzag course. Lines of research were opened up and dropped, then later picked up again. Isolated facts were reported before the time was ripe, but finally, from the 1940's up to the present, the pace of discovery increased, thanks to an important technical advance. The electron microscope, invented in the early thirties, focused a beam of electrons instead of light rays, thus making it possible to see the molecules within the cell. Although the X-ray refraction technique was invented by Lawrence Bragg and his father, William, in 1912, it was not until the thirties that it, too, was effectively used to analyze the structure of protein molecules themselves. An X-ray beam is scattered when it passes through matter. When the material is arranged in a regular order, the ray leaves a pattern of dots on photographic film placed behind the observed material. This is an indirect kind of picture, but by using computers which measure the angle of refraction of the rays indicated by the dots, the data can be processed and some insight gained into molecular structure by checking against the known chemical composition.

This, however, is running ahead of our story. Two concepts were to lead to the nature of the gene: the enzyme and the virus.

The enzyme in yeast which caused fermentation was discovered in

1897 when Eduard Buchner ground yeast cells with sand until they were destroyed. To his surprise he found that yeast juice still caused fermentation. Not the live cells but some chemical substance was present which caused the fermentation to take place. It was soon found that all ferments were of the same nature and the name "enzyme" (meaning "in yeast") was given to them. They were proteins and evidently catalysts—that is, they caused change but did not take part in it themselves. By 1935, when the intensive study of protein was beginning, three enzymes produced by the animal body were isolated: pepsin, which came from the stomach lining and broke down other proteins; trypsin from the pancreatic juice which also broke down proteins; and urease, which changed urea from the kidneys into ammonium carbonate. These three important substances were found to consist of giant protein molecules and could be isolated in crystalline form. So far, so good. But way back in 1908 an English doctor, Sir Archibald Garrod, had suggested that certain diseases were caused by the lack of a specific enzyme. This lack was inherited. Therefore some deviation from the normal gene content in the chromosomes was responsible for the failure to manufacture the needed catalyst. This, of course, suggested an intimate relation between gene and enzyme, but the hint was not taken.

In 1940, which is now considered the end of the period of classical genetics, two scientists who had been working on the genetic control of fruit fly eye pigment decided on a new approach. George Beadle (1903-) and Edward L. Tatum (1909-) proposed to begin with chemical reactions controlled by enzymes and to work back to genetic patterns which might be associated with these reactions. What they discovered won them the Nobel Prize.

They chose the pink bread mold *Neurospora,* for their experimental organism. This mold has since become as famous as that other genetic hero, the fruit fly. Its life cycle is brief, it is easy to grow, and it has only one set of seven chromosomes.

During the study of proteins it had been discovered that these substances were composed of chains of molecules containing nitrogen called amino acids. The number of amino acids was enormous and countless different arrangements were possible. If protein was the basic stuff of protoplasm, then amino acids were extremely important in the study of life.

It was found that *Neurospora,* although it lived on substances containing no amino acids, was able to make its own. Then the geneticists searched for mutations and came up with some *Neurospora* which could not create their own amino acids. Further experiment proved that each gene had a single function: to synthesize only one enzyme and to control the chemical reaction catalyzed by that enzyme. Thus the *Neurospora* containing the right gene could manufacture amino acid; those lacking it could not. The work of these scientists resulted in the "one-gene-one-enzyme" theory.

From now on, the communication metaphor begins to be applied to heredity. Each gene had one message to send, each enzyme was the messenger which delivered the message thus carried in the sex cell which was to affect the protoplasm of the newly developing individual.

Later it was thought that proteins were the direct products of the gene and were the basic material from which all other substances were built. It was even held that genes too might be proteins and contain master models of all the proteins that could be required by the cell.

Protein became the focus of genetic research, and at this point it is necessary to take time out to discuss what was known of the protein molecules. We have already referred to them as giant; indeed they were so large and so complicated, being composed of countless subgroups of molecules, that it was not until the thirties that their three-dimensional structure was appreciated. W. T. Astbury in the late thirties began to use crystallography, the X-ray technique invented by Lawrence Bragg and his father more than twenty years earlier. It began to be clear that the giant molecules were put together in certain patterns. Astbury also coined the phrase "molecular biology" for this new type of research, which attracted physicists as well as chemists. The early stages of the movement are alleged to have provided C. P. Snow with material for his novel, *The Search.*

By 1900 it was known, as we have said, that the protein molecules were largely composed of the amino acid group of atoms. Physicists were to show that all atoms were miniature solar systems giving off clouds of electrons. This made it possible to understand how the submolecular components of these giant molecules could be electrically joined together in chains which, in turn, built three-dimensional structures. All protein molecules were built of twenty different amino acids in different arrangements. The submolecules of these acids were connected in series like freight cars and these were called polypeptide chains. The chains were rolled up in structures or corkscrew shapes known as helices. This three-dimensional structure had to be worked out with the aid of the X-ray technique which recorded the density of the electrons that clustered around the atoms. Some chains contained as many as three hundred linked amino acids. Organic chemistry, with the aid of physics, was now reducing the stuff of life to elements more minute than anything ever dreamed by Leeuwenhoek.

At the same time that the analysis of molecules was going on, a new line of investigation, with a considerable history, finally impinged on biochemistry, throwing still more light upon the nature of the gene. This was the study of viruses.

We now turn to medical history. As a result of the work of Pasteur and Robert Koch in the late nineteenth century, it was thought that most diseases could be attributed to bacteria, one-celled plants. Pasteur looked for them as the cause of rabies and failed to find them. In 1892 Dimitri Ivanovski, searching for the cause of tobacco mosaic disease, discovered that the sap from infected plants was infectious even after passing through a fine filter. In 1898 a Dutch botanist, Martinus Beijerinck, performed the same experiment and went on to show that the juice of the leaf infected by the filtered material was also infectious. In short, there was some toxin which seemed to have a continuing life of its own. He called the infection *contagium vivum fluidum* (contagious living fluid). In 1900, Walter Reed discovered that yellow fever was transmitted by mosquitoes through some type of filtrate, but the substance still remained an abstract concept.

For the first thirty years this type of research consisted of injecting something invisible into a host and recording the results. The amount of the infectious substance was estimated in terms of how much a sample could be diluted before its strength was lost. One valuable discovery was the observation that animals which recovered from an infection produced substances which could neutralize the infecting agent. These substances were named antibodies and were the real explanation of vaccination. By taking a weakened culture of the disease from an animal and injecting it into another creature, that creature could be immunized by stimulating the production of antibodies.

Thomas Rivers of the Rockefeller Institute, when he published *Filterable Viruses* in 1928, brought all the material so far accumulated in this research together and gave the infectious entity a name.

But what was a virus—an animal, a plant or something inorganic? In 1935, when work had been done on the protein molecules, Wendell Stanley of the Rockefeller Institute at Princeton managed to isolate the tobacco mosaic virus. It was not a bacterium or a protozoa, but a crystalline substance composed of molecules. When the crystals were put back in solution, they again became infective. It was as if from being a mere inert chemical these crystals became alive. Rivers himself said the virus was either an "organule" or a "molechism," for its characteristics seemed to place it just on the borderline between the living and the inert.

Now a certain group of viruses which preyed on bacteria had been discovered in 1915 by F. W. Twort, a bacteriologist. In 1917, Felix d'Herelle of the Pasteur Institute rediscovered them. They caused clear glassy patches to appear on the jellylike seaweed product, called agar, on which bacteria were grown. Herelle enthusiastically studied them and came up with a great idea. Why could not all bacterial diseases be eliminated by means of the cannibal viruses which he named bacteriophages? He insisted he had great success with dysentery. Public health workers in India and Asia seeded public drinking water with phages known to be inimical to cholera, dysentery and the plague. The results were saddening. Apparently the phages worked well in the test tube but not in the human gut. Also, a slight mutation in host or predator destroyed the relationship.

Thus the phages were relegated to a supposedly closed chapter in the history of medicine. In their stead, antibiotics took over.

In 1930, Max Delbrück, a German physicist, came to the United States. During the war, he was regarded as an enemy alien and was not allowed to engage in "sensitive" work in physics. He therefore turned to biophysics.

It should be pointed out here that in 1922 Morgan's pupil, H. J. Muller, with a flash of insight, wrote: "If these d'Herelle bodies [viruses] were . . . fundamentally like our chromosome genes, they would give an utterly new angle from which to attack the gene problem."

Delbrück had the same thought when interest in viruses had almost died out. Yet fortunately for him, just at this time bacteriophages were seen under the electron microscope, which could magnify two thousand times. Using the new instrument Delbrück and his associate, Salvador Luria (also an enemy alien), found that the little predators had tails; they looked something like tadpoles and their shape was constant, even though one billion could be enclosed in a single lower-case printed letter.

As work went on, it was found that a group of seven had the power to infect one particular species of colon bacterium, *Escherichia coli,* and once more a new experimental organism took its bow in the laboratory, the humble bacterium of the human gut.

Interestingly enough, as the experimental organism got smaller and smaller, the technical equipment for observing it got larger and larger. An electron microscope, which could peer within the cell, was a large console costing about $100,000, while the X-ray refraction machine, which came close to photographing atoms, was the size of a large automobile, filling an entire room and disgorging its secrets to a computer. Needless to say, the cost was astronomical.

More and more physicists were entering the arena of organic chemistry. Curiously enough, the members of the phage group, which was sparked by Luria, Delbrück and Alfred Hershey, were reluctant to call themselves molecular biologists, although Astbury pushed the term.

Gunther Stent wrote that "two kinds of people were then wont to refer to themselves as 'biophysicists': physiologists who were able to repair their own electronic equipment, and second-rate physicists who sought to convince biologists they were first rate." This was soon changed, however, when such distinguished men as Linus Pauling entered the field.

The phage group named their phages T_1 through T_7. Mutants were observed. A strain of T_2, for example, might turn up that could not infect bacterium which was nevertheless infected by the normal strain. Another strain would appear that destroyed the cell of its host more or less rapidly than its parent. Delbrück and Hershey, working separately, discovered almost at the same time that if *E. coli* was infected simultaneously with T_4r and T_2, T_4, T_2r and T_4r and T_2 would appear. Something similar to mating appeared to occur among these curious minute entities which were neither wholly organic nor inorganic!

It followed then that phages had chromosomes, and at this point virology and genetics also mated. By 1949 many more investigators worked on the problem and new techniques made it possible to introduce radioactive sulfur and radioactive phosphorus into a culture infected with the phages. A new generation of phages was traceable containing the radioactive atoms which had replaced the nonilluminated ones. In a famous experiment in 1952, Hershey and Martha Chase put these traceable phages into a fresh colony of bacteria, which they, of course, attacked.

When the culture was analyzed, it was found that the radioactive sulfur was outside the bacterial cell and the illuminated phosphorus inside. This meant that the protein which contained the first chemical was left outside and the deoxyribonucleic acid containing the second was inside. The empty phage heads of protein were merely containers, and when sheared off from the outside of the bacterium, the work of infection continued.

We now turn to this history of deoxyribonucleic acid, which has become famous under the name DNA.

The nucleic acids, so named because they were first discovered in 1869 in cell nuclei by Johann Friedrich Miescher, were also found elsewhere. By the 1880's they were broken down into phosphoric acid, sugars, and a number of nitrogen bases which differed in structure. There were five of these: adenine and guanine consisted of two rings of atoms containing four nitrogen atoms; the other three were cytocine, thymine and uracil, which contained a single atom ring and only two nitrogen atoms. The nucleic acids themselves were divided into two groups. One type contained only five carbon atoms in its sugars (most sugars have six) and lacked thymine. It was labeled ribonucleic acid, more simply RNA. The second type contained one fewer oxygen atom in its molecule and lacked uracil. This was deoxyribonucleic acid, or DNA.

The submolecule groupings of phosphoric acid, sugar, and nitrogen base were called nucleotides and were parallel to the amino acid groups in proteins called polypeptides.

In 1944, an American bacteriologist, Oswald Theodore Avery (1877-1955), working at the Rockefeller Institute with a group which was studying pneumonia found that the bacterium pneumococcus could be isolated into two strains. S had an outer capsule around the cell, while R was without it. S was infectious and R was not. The scientists extracted material from the S strain, which when added to a cell of the R strain converted R into S and bestowed the ability to form the capsule. They then analyzed the extract. Ten years of careful work led up to the startling conclusion that the transforming substance was DNA!

Thus DNA had converted the benign R bacterium into the toxic S type, and this characteristic was heritable.

Now, of course, the theory that protein was the basic substance of the gene was very much weakened. And eventually Hershey's 1952 experiments with the phages were even more suggestive. But DNA had tended to be a disregarded associate of protein in the chromosome mainly because of misdirection. In attempting to extract its molecules they had been broken up and thought to be small. New methods proved that they were as large as the largest protein molecule.

Before Hershey's work took place, the search for the secret of the gene was moving into its final phase. Like the last chapter of a detective story, the clues were piling up, the false leads were being eliminated, and the brilliant solution was in the making. What remained to be clarified was the three-dimensional structure of the DNA molecules, the actual arrangement of the submolecules.

The chemical analysis had been carried out by an English biochemist, Alexander Robertus Todd, in the 1940's. The kinds of atoms and their distribution can be worked out on the basis of the different wavelengths absorbed by different elements. Such wavelengths are measured by spectrophotometers. Between this type of data and that gained from X-ray refraction pictures the solution had to be found.

In the fall of 1951, a young American (age twenty-two), James D. Watson, joined the Cavendish Laboratory of Cambridge University. Watson was a biologist who had been trained under Luria of the phage group. Luria believed the place to concentrate was in biochemistry and sent Watson to Copenhagen to study. Watson became more and more interested in X-ray refraction and DNA. By changing his field he had lost his fellowship, but he scraped enough money together to live at Cambridge while waiting to obtain a new grant. Through Luria's influence he was appointed assistant to Sir Lawrence Bragg, the grand old man of X-ray refraction technique. Under his direction a group of physicists and chemists were working on three-dimensional protein structures. Among these was Francis Crick (1916-) an imaginative young man with a laugh that rocked the laboratory. He was always ready to explain other people's research to them and his nonstop tirades made the dignified Sir Lawrence shudder.

But Crick agreed with Watson that DNA was the proper point of concentration, not the proteins. Unfortunately, Maurice Wilkins (1916-), a physicist at the University of London, had staked out DNA and was using the X-ray technique to investigate its structure. British sportsmanship dictated that no one should poach on Wilkins. The reward was likely to be the Nobel Prize. Watson makes it clear in his narrative, *The Double Helix,* that Americans were more competitive, and indeed Linus Pauling (1901-) working in the California Institute of Technology was also in a position to hit upon the solution. The situation was complicated by the fact that Wilkins had been assigned an assistant, Rosalind Franklin (1927-58), a brilliant crystallographer, far more expert than Wilkins. She was something of a women's liberation type, did not see herself in the role of an assistant, claimed DNA for her own, and sometimes withheld her pictures from Wilkins for considerable periods. (He, in turn, withheld them from Linus Pauling.) Since Wilkins spent most of his time worrying about his battle with Franklin, he was not getting much done.

The result was that Watson and Crick did poach on Wilkins' territory and worked on the problem in their spare time. Pauling's discovery of the structure of the alpha protein of blood plasma in 1950 was both a stimulus and a threat. (It had shocked the Cambridge laboratory.) The shape was a helical chain of amino acids anchored together at the curves by hydrogen bonds, such bonds operating through the attraction between electrical charges. Pauling was a brilliant chemist, but Watson realized he had not used complicated mathematics. Rather, he had worked from models which looked like tinkertoys and had tried to figure out what atoms liked to sit together.

As we have already noted, chemists had decided that nucleic acids were composed of three-part units (nucleotides) consisting of phosphorus, a sugar, and a base (the nitrogen group), linked in chains, the phosphoric acid of one group adhering to the sugar of the next.

Since the amino acid chain of Pauling's protein was helical, it was possible that a helix might enter into DNA structure.

Watson and Crick began building models, postulating that the sugar-phosphate backbone was regular and the order of the nitrogen bases irregular.

Astbury, five years before, had taken a not very satisfactory picture of DNA. Now the two yearned to see Wilkins' closely guarded photographs. When approached diplomatically, Wilkins revealed the fact that he had not even seen Franklin's pictures, but that she was about to give a seminar to which they might come.

Franklin's seminar revealed sharper pictures, but she had made no real progress in pinning down the structure. She of course had no respect for the tinkertoy approach of Linus Pauling. Watson's narrative illustrates the curiously blinkered attitude of scientific schools which adhere to one point of view. Classical geneticists had no use for the study of DNA, and Watson says of them: "All most of them wanted out of life was to set their students onto uninterpretable details of chromosome behavior or to give elegantly phrased, fuzzy-minded speculations over the wireless on topics like the role of the geneticist in this transitional age of changing values."

And this statement of Watson's is also revealing. Nowhere in his book, *The Double Helix,* does he show any awareness of or interest in the social impact of his work. Molecular biology takes on the tone of an intellectual football game in which the investigator scores goals against the other team.

Watson and Crick felt that the known chemical data could mean from two to four chains of chemical groups. Following the model method, they constructed a three-chain affair with bonds of salts. Wilkins and Franklin were invited to inspect it. Franklin criticized the attempt and proved to them that they had not allowed for enough water. This setback came at an unfortunate time. Crick had just had a set-to with Sir Lawrence over credit for an idea used in one of Bragg's papers. Since Crick shot his mind off in all directions, the whole affair was unclear but the result was indignation on the part of Bragg. The old gentleman could not stand Crick's personality and also felt that his young subordinate was all talk and nothing done. The upshot was that Crick and Watson were told to give up work on DNA. They did not, however, stop thinking about it.

Watson turned to RNA, another nucleic acid, and did some X-ray work which persuaded him that this, too, was helical in structure. At this time, too, Watson attended a meeting of the Society of General Microbiology in Paris for which Pauling and Luria were not granted passports by the U.S. State Department because they were considered subversively interested in peace. Watson received a letter from Hershey describing the now-famous 1952 experiments with phages and *E. coli.* Watson presented the material to the meeting, where it was yawned away. The fixed ideas of the various types of microbiologists were proof against a new breakthrough.

Watson got pictures proving to his satisfaction that RNA was definitely helical. His attention was then directed to some work done by the Columbia University biochemist Erwin Chargaff (1905-) who had results to show that the proportions or molecules of the nitrogen bases in DNA were the same in all cells of a given species but varied from one species to another. He drew no conclusion, but the idea occurred to Watson and Crick that his results might bear some relation to the genetic code—that is, the mechanism for producing exact replicas of a species. When Chargaff came to Cambridge, however, he was not impressed by what he evidently considered to be a pair of amateurs. An evening spent with him was not productive of any new insights. A meeting with Delbrück at the International Biochemical Congress in Paris proved only that this distinguished geneticist was not interested in DNA. Linus Pauling, who had finally fought free of the State Department, also attended the meeting, but he, too, had little to say about DNA.

Watson had taped on his wall a sheet with "DNA → RNA → protein" on it, but he was no nearer being able to prove this insight.

Then Linus Pauling struck and announced that he had solved the DNA problem. When Watson and Crick got hold of the paper through Pauling's son, who was now attached to the lab, the structure turned out to be a three-chain helix. The bonds holding the chains together were hydrogen joining the phosphate, and Watson knew that phosphate groups never contained bound hydrogen atoms.

Checking with various chemists seemed to prove that there was a fundamental and absurd error.

Wilkins, who was now duplicating Franklin's work, showed Watson some pictures which definitely proved helical DNA structure, but how the bases were to be packed in was still a problem. Watson had an idea for a two-chain model which he shared with Sir Lawrence Bragg. He also pointed out that Pauling might soon be expected to retrieve his error. British-American competitive spirit prevailed, and Bragg set him free to work on DNA once more. Fortunately, when Wilkins was asked if he minded whether the Cambridge team built more models, he said no. Franklin was leaving in a few months and he was postponing DNA until she was gone.

Finally, the possibility of two intertwined chains, each with identical sequences of nitrogen bases, occurred to Watson. He finally postulated that the hydrogen bonds occurred between the nitrogen bases. Crick came along and figured out that the backbones of the two chains would run in different directions. While Crick went about announcing to the world that they had found the secret of life, they set about building the model. It came out as a spiral staircase with two balustrades both formed of the sugar-phosphates and the steps across made up of the hydrogen-bonded pairs of nitrogen bases. There was a cross-pairing of the bases which agreed with Chargaff's data. Bragg became excited but wanted a check on the chemistry of the enthusiastic pair. Alexander Todd, the chemical expert on DNA, came, saw, and reassured Bragg. Wilkins and Franklin both came, saw and agreed that the new structure tallied with their work. Delbrück and Pauling were informed; both were excited by the new structure and Pauling generously conceded victory to the two young men whom Chargaff had referred to as "scientific clowns." Everybody was happy, for the Cambridge lab had won the game, as Bragg sent off the descriptive paper to the periodical *Nature*. Subsequently, Watson, Crick and Wilkins all received the Nobel Prize.

What was the significance of their discovery, which has sometimes been likened in importance to Darwin's theory? If the DNA molecule was the carrier of genetic information, the potential of great variety was necessary. Any sequence of the four nitrogen bases was possible, since the number of pairs varied from about 5,000 in a simple virus to 5 billion in man's 46 chromosomes. Thus an enormous amount of information could be transmitted in the base pairs.

Likewise, the requirement that the double helix must reproduce itself in cell division was also met. The two helices broke apart at the hydrogen bonds, each formed its twin, and thus each new cell contained the double helix. Other experiments were able to show this process in action. In the California Institute of Technology, when a colon bacterium was grown in heavy nitrogen, the element penetrated the molecules. The experimenters then placed the bacterium in a light nitrogen medium and allowed it to divide once. The result was that the daughter cells each contained one helical strand with heavy nitrogen atoms and one with light nitrogen atoms. This showed that half the parent chain is conserved in each daughter cell while the new strand is formed from the surrounding medium.

It was now firmly established that the atom groups in the molecules of DNA and RNA are the same for all members of a given species but are different for different species. Thus it was clear that these molecules could act as the template or basic pattern for the fixity of the species. Watson and Crick's carefully understated conclusion to their 1953 paper, "It has not escaped our notice that the specific pairing we have postulated immediately suggests a possible copying mechanism for the genetic material," was triumphantly vindicated.

From 1953 molecular biology pushed on rapidly in the area of ribonucleic acid. It was eventually revealed that RNA was also involved in the genetic code. Sometimes DNA worked in combination with RNA, which served as an intermediate messenger, and sometimes RNA itself controlled the proteins of the cell. By 1961 synthetic RNA was being used to work out the specific codes for creating single amino acids. In other words, we now begin to understand how the DNA and RNA molecules coiled up in the nucleus of genetic cells, and also in body cells, control the nature of protein which makes up the rest of the cell.

In the case of crossing over, observed by the Morgan school, it is one or many atom groups in the DNA molecule of the gene that cause the change.

Through the work on nucleic acids, which continues to pile up new data, scientists are beginning to solve the mechanics of heredity and possibly the creation of living protoplasm. Proteins have already been synthesized from inorganic chemicals. Living cells, it is true, have not yet been produced, but even this may possibly be achieved. Mutations are now known to be dependent on a change in the order of amino acids in the protein produced by some slight change in the DNA. The image of a typographical error has been used.

There are skeptics, among them Barry Commoner, who are less enthusiastic. They point out that the interactions within the complex cell mechanism are still not understood. Commoner states: "The simple sum of separate molecular events is insufficient to represent the living whole." He does not feel, therefore, that DNA is the secret of life but rather that the ultimate problem is the baffling wholeness of the cell "in a system which is alive."

By now old-fashioned naïve vitalism has pretty much gone out the window since it has been proven, once and for all, that protoplasm follows the same chemical laws as inorganic chemistry. The beginnings of life on earth have been suggested by Harold Urey to have taken place in a world without an upper layer of oxygen to absorb the ultraviolet radiation. Radiation could reach the oceans and set up reactions that would produce complex organic molecules which would eventually become reproductive.

This process was mimicked with electric discharges in 1962, and numerous organic compounds, including important phosphates, were produced.

Another result of molecular biology is the prospect of gene tinkering. While this may be a blessing, making possible the elimination of genetic diseases, it is also frightening. Salvador Luria, the phage expert, sees the possibility of a destructive use, for instance, by disseminating a virus which changes the gene structure so that victims are sensitized to and destroyed by some simple element in the environment.

Of course, Aldous Huxley already suggested a dictatorship in which a nation's offspring were produced in laboratories, and it does not take much imagination to see dictatorial genetic engineering producing a worker and soldier class, as in ant culture, and carefully eliminating artists and nonconformists.

On the other hand, the molecular biologist Gunther Stent, one of the first to show appreciation for Watson and Crick's work, has written a book asserting, perhaps naïvely, that the era of discovery is now over in both science and art and a kind of relaxed Golden Age is about to begin.

In contrast, the biochemist Erwin Chargaff, of whose work we have had occasion to speak, wrote:

> We manipulate nature as if we were stuffing an Alsatian goose. We create new forms of energy; we make new elements; we kill the crops; we wash the brains. I can hear them in the dark, sharpening their lasers. Soon hereditary determinants themselves will begin to be manipulated. I am afraid the "dark satanic mills" of which Blake wrote will be no less satanic for being brightly illuminated.

It could well be that the molecular biologists may, in time, find themselves in the same state of remorse as the physicists who gave us the atom bomb.

Perhaps the most amusing extrapolation of molecular biology is a story by the science-fiction writer S. Damon Knight in which he describes the death of contemporary urban technology for lack of metals and the rise of a new agrarianism in which plants are induced to grow plastic knives and forks and all sorts of animal forms created, such as six-legged transport beasts and improved birds capable of functioning as airplanes.

Golden Age or Brave New World, whichever confronts us, there is no stopping the onrush of discovery, and the need for responsibility and controls grows ever more critical.

It must be remembered, however, that although molecular biology has told us much about heredity, the basic problem of evolution still remains. How did it come about?

The firmly embedded dogma of the scientific establishment, upheld since August Weismann rather naïvely cut off the tails of generations of mice, is that random mutations during eons of natural selection have nurtured traits with survival value and given us all the finned, furred, and feathered species and the invertebrates which we have now, plus many fantastic entities which have died out during the millions of years of prehistory.

A Michigan biologist, Carl C. Lindgren, has put together the evidence produced by a few heretic Neo-Lamarckians. In his book, *The Cold War in Biology* (1966), he brings up examples which seem to show heritable changes stimulated by the environment.

Scientists found that high temperatures applied to *Drosophila* larvae resulted in a change in the wing-folding of the adult, which was transmitted for several generations and if reinforced by the same treatment, was permanent. Cecil P. Martin claimed that a change in feeding habits transformed a head louse into a body louse. Michael

Guyer took the pulped serum of the eye lenses of rabbits and injected it into the peritoneum of fowls; then the blood serum of the fowls was injected into pregnant rabbits. Eye defects resulted, which were passed on for eight generations. Peter Klopfer said of the *Drosophila* experiment:

> The point of greatest significance is that a phenomenon mimicking a Lamarckian form of inheritance (that is, an effect of somatic responses on the genotype) can be obtained without recourse to the orthodox assumption that Lamarckian explanations require. In the case above the "new" trait was present from the start, though at a very low and originally nondetectable frequency.

But perhaps any number of new traits are present at very low and nondetectable frequencies in all organisms and need only certain environmental stimuli to bring them out. The establishment will use any verbal quibble to avoid giving credit to Lamarck.

Lindgren points out Guyer's rabbit experiment, performed in 1921, was never checked or repeated but simply ignored. Lindgren claims he, himself, has induced mutations in yeasts by nutriment. He feels that the idea of accidental mutation has been a dogma in reaction to old vitalist ideas of a supernatural purpose. He points out that it is hard to conceive how the multitude of small changes which led from four-toed eohippus to the modern horse all had survival value. Likewise the countless small mutations needed to create so complicated an organ as the mammalian eye are difficult to envisage without some coordinating mechanism. He sums up:

> The autonomous nervous system exercises considerable control of body mechanisms, only a very small fraction of which ever reaches the consciousness. . . . Is it not possible that there are mechanisms in the brain that have something to do with the adjustment and adaptation of the animal toward the environment? Here one returns to Lamarck's original idea that the animal achieves at least some adaptation because he *tries*.

There have been some interesting and controversial experiments in the realm of the inheritance of habit. James V. McConnell, in the late 1950's, reported that when fragments of conditioned flatworms were eaten by unconditioned worms, the second group seemed to acquire some of the habits of the first. Similarly it has been claimed that RNA from the brains of conditioned rats injected into unconditioned animals produced in part the behavior taught the conditioned group.

Although the functioning of the nervous system remains to be explained chemically, Gunther Stent feels that molecular biology will soon solve its mysteries.

In any case, even if, for instance, the formula for the production of a poem could be written, the poem would still remain a many-leveled structure of symbolic meanings which are socially communicative and qualitatively different from the atomic models of the molecular biologists. It seems unlikely that molecular biology will bridge the Cartesian gap between the intangible products of mind and the measurable aspects of matter.

Perhaps Chargaff deserves the last word:

> But can we really believe that if we keep on plodding for another 200 years or so, suddenly microscopic angels will be seen carrying a sign, "Now you know all about nature." Actually knowledge of nature is an expanding universe, continually creating ever greater circumstances of ignorance, a concept that can be expressed in the words: "The more we know, the less we know."

27

Insect Homers

❧ IN his book on insects, *L'insecte,* 1859, Jules Michelet wrote:

> We take an insect, we pierce it with a long pin, we fix it in a cork bottomed box, we place under its feet a label with a Latin name and that is all there is to be said about it. This method of understanding entomological history doesn't satisfy me at all. There is no use telling me that such and such a species has so many joints in its antennae, so many wing veins . . . I do not really know the animal until I comprehend its way of life, its instincts, its habits.

This statement is a good summary of Jean Henri Fabre's attitude.

This obstinate, dedicated Frenchman, whose ten volumes of *Entomological Souvenirs* are a contribution to both literature and science, leads us, temporarily at least, away from the reductive process and into a zoological direction which has recently assumed a new importance, the study of ethology. Defined as the study of habits and behavior, it is the area of investigation which has attracted the naturalist temperament, the type of man who contemplates the natural world and derives a poetic pleasure from it, the type of man who develops an affection for and respect for other species of living things. This, however, does not preclude scientific accuracy and experimental techniques in the field. An important point is the fact that living creatures in their natural environment behave differently than they do under laboratory conditions or when submitted to experiment in captivity. The ethologist with the patience to observe under natural conditions is able to discover facts about animal psychology, social relationships, and relationships between the living being and its environment which are hidden from the scalpel wielder or the chemical analyst. It is true that the latter may subsequently assist in explaining behavior, but relationships and reactions to natural stimuli must first be observed in the field. It is not an accident that ethology and ecology have in recent decades risen in importance together, for the two are interwoven.

As we have seen, the nineteenth and early twentieth century made dramatic and revolutionary advances in exploring the internal mechanisms of the living entity and the minute universe of the cell which was to culminate in molecular biology. A truly scientific interest in psychology, however, was historically late in developing. Nineteenth-century materialism in a sense shied away from the intangibles of psychic life. With the rise of behavioristic techniques, depth psychology and gestalt (the study of forms of perception), the investigation of mental processes, even those of animals, became respectable. And so we come to the problem of instinct.

This rather portmanteau word has always been used to cover every motivation or organic activity believed to be distinct from human intelligence. Yet it is very difficult to draw a hard and fast line between the two. The need to explore this area has stimulated work in ethology. The earlier naturalists such as Gilbert White or Réaumur were minute observers of animal activities, but neither challenged or probed the Cartesian formulation. Fabre, an admirer of Réaumur, is a link between eighteenth-century naturalists and modern ethology. One of his chief concerns was the study of instinct, and his simple, even primitive experiments led to the more sophisticated work being done today. The French naturalist was a curiously independent figure. Ultraconservative in theory, scornful of the scientific establishment, he brought unlimited patience to the daily contemplation of ants, bees, wasps, beetles, and scorpions, and was not daunted by the odor of putrescence when he came to make extensive studies of the behavior of maggots. He was christened by Victor Hugo "The Homer of the Insects."

Fabre was born in a tiny village in central France in 1823. He grew up shaped by peasant tradition (his mother was illiterate), while his early education was obtained in a primitive rural school. He attended normal school at Vaucluse and was subsequently put in charge of a primary school at Carpentras, a grim little prison where the pupils did not even have tablets on which to write. Outdoor surveying lessons pleased both pupils and master. Fabre noticed that his charges tended to stray and he observed them licking straws. He then discovered a big black bee making masonry nests on pebbles and filling them with honey. The small boys, practical naturalists, had learned to rob the nests with a straw. This was Fabre's first contact with an interesting insect upon which he was to write extensively.

The local bookseller had a gorgeous volume entitled *Histoire naturelle des animaux articulés.* It was a luxury out of reach of the struggling young schoolmaster who existed on a pittance. Nevertheless by heroic economies he saved a month's salary and bought it. He wrote:

> The book was devoured: there is no other word for it. In it I learned the name of my black bee; I read for the first time various details of the habits of insects; I found, surrounded in my eyes with a sort of halo, the revered names of Dufour and Réaumur: and, while I turned over the pages for the hundredth time, a voice within me seemed to whisper, "You also shall be of their company."

The voice was right, but Fabre continued to struggle with poverty all his life. During his formative period he developed a passion for the poetry of Lamartine and wrote some passable verses himself. He also became acquainted with the work of other zoologists, and although Geoffroy's son, Isidore Saint-Hilaire, was on the committee that awarded him his doctorate for a thesis on the myriapods in 1856, Fabre stuck fast to the doctrines of Cuvier, special creation, catastrophe theory and all, and defended them all his life.

It will be remembered that evolution was ignored for many decades in France, and although Fabre lived well into the twentieth century, his intense piety caused him to turn a deaf ear to the accumulating evidence for Darwinism.

It is not clear why he never obtained a university post. It is true that during the early part of his life he was probably without influence. He married the year he obtained his doctorate and reared five children. Becoming a practical chemist, he worked with dyes in an attempt to improve his financial status. Just as he was about to put a pink dye from the madder root on the market, the invention of aniline dyes, which were much cheaper, drove him out of business.

While he was teaching in the lycée at Avignon, in Provence, he began his field studies. He was, however, not allowed to work peacefully at his beloved studies in his spare time. Like most intelligent men, Fabre was an unconventional teacher. Although his ideas, which included admitting girls to his science classes, had interested Napoleon III's minister of education, local bureaucrats were outraged when it was discovered that the girls were exposed to descriptions of the fertilization of flowers! In 1870, during the Franco-Prussian War, Fabre was dismissed in disgrace. It is at this point that we wonder why he did not use the influence of his friends (he already had some reputation and even Darwin had praised his papers) and fight back. A stubborn independence led him to reject the school system, however, and to support himself by writing. John Stuart Mill, who wintered in Avignon and with whom he was evidently friendly, came to his aid with a loan of $600. Fabre managed to support his family for the next ten years by hack writing, turning out two or three popular books on science a year for which he was miserably paid. Yet by 1879 he managed to save enough to buy a modest piece of ground at Sérignan on which he built himself a house. It was as if Fabre was compensated for his exhausting struggles by the length of his life. From the age of seventy to his death at ninety-two he was able to devote himself to entomology, and during his last five years he tasted world fame.

Writing for a living had made him a clear and forceful stylist. He shared Réaumur's lack of interest in classification. In a few cases he did not bother to identify specimens he discussed. This, of course, aroused the hostility of the orthodox. He always remained somewhat apart from official scientific circles and was wholly intolerant of people who disagreed with him. No one, however, has lived more intimately with insects than he. As his ten volume *Souvenirs entomologiques* (completed in 1907) appeared, he began to achieve a fame both literary and scientific. As early as 1867, Napoleon III had awarded him the Legion of Honor. Fabre remarked that the red ribbon was sometimes helpful since it prevented rural policemen from arresting him for suspicious conduct when they found him squatting intently by the side of the public highway.

He lived as a recluse with his family in the little house he had built surrounded by what the peasants called *harmas,* worthless stony land covered with scrubby vegetation which became his outdoor laboratory. There he spent hours in the sun, observing the teeming world of insects, which he found more sympathetic than that of men. At one point the Provençal poet Frédéric Mistral created a small furor by announcing that the patriarchal entomologist was starving. As a result, the government granted him a pension. After his second wife died in 1912, he lived quietly with one of his daughters in Sérignan, dying in 1915, just when the German armies were sweeping toward Paris.

His rejection of the official world seems to have been complete, for there is no record of his participating in any learned societies and he seems to have had almost no scientific friends. He was scarcely known in Sérignan, which, it was said, he did not visit more than twenty times in thirty years. Apparently his dismissal from the school system rankled profoundly and together with the fact that no university invited him to join its faculty turned him into a lone wolf. One cannot help feeling that there was something of the self-martyr in his fierce egotism.

Fabre was obliged in the beginning to carry out his work with the simplest equipment. Later, when a friend gave him a powerful microscope, he seldom used it; the asceticism of his daily life applied to scientific technique, for he stuck to a hand lens and for dissection used small scalpels and modified needles. He brought to fieldwork enormous patience and a human point of view. His special literary virtue consisted in making the lives of insects dramatic and fascinating, for when he set out to unravel some small mystery of behavior, he used all the art of the detective story.

His essay on the black-bellied tarantula, the lycosa, is a good example of his method. He begins by noting the peasant superstition that its bite is mortal, that it causes a dancing sickness which can be cured only by music. He then quotes what Léon Dufour, the naturalist he most admired, had to say about the animal and then introduces us to the creature's habitat—arid, uncultivated, sunny areas. There the spider digs a burrow, about a foot deep, bent at an angle about five inches down. The burrow is an inch in diameter and is lined with silk. Posted at the elbow, the lycosa waits for its prey. It has, however, also constructed a kind of well curb at the surface of the ground, about an inch high, composed of dry bits of wood joined by a little clay. Fabre thought that this turret was to prevent water from seeping into the burrow.

After vainly attempting to dig the spider out, the naturalist discovered that it could be caught by waving a grass stalk at the entrance and thrusting it into the tube. When it seized it, he was able to draw it out and imprison it in a paper bag. Spiders in captivity became so tame that they took a live fly from his fingers. He discovered that two males would fight fiercely, the victor eating the vanquished. After copulation, the female also ate its mate.

Fabre experimented with the spider's poison. Bumblebees as big as itself it overpowered and killed in a matter of seconds. He then wanted to know just where the spider inserted its fangs in order to bring about such sudden death (proportionately so much faster than a rattlesnake). By putting a bottle containing a carpenter bee over the nest, he succeeded in observing the murder. The spider rushed out and plunged its fangs into the space between the head and the thorax, accurately striking the cervical ganglia, equivalent to the insect's brain. When he succeeded in having a grasshopper bitten in the abdomen, the insect lasted for a day. Bitten in the neck, it died immediately. The tarantula therefore possessed a mysterious instinctive knowledge of anatomy, allowing it to strike its victim in the exact spot to kill it immediately. Fabre than tried injecting ammonia into the bee or the grasshopper's neck ganglion. He also produced speedy death but not as rapidly as the spider's deadly poison.

His next experiment was on a young sparrow. "A drop of blood flows; the wounded spot is surrounded by a reddish circle, changing to purple." The bird immediately lost the use of its leg but ate well for two days and then died. "There was a certain coolness among us at the evening meal. I read mute reproaches, because of my experiment, in the eyes of my home circle; I read an unspoken accusation of cruelty all around me. . . ."

Although he felt guilty, he persisted and allowed a spider to bite the tip of the snout of a mole which had been caught ravaging his garden. It scratched its nose but ate well for a couple of days, only to die on the third. He stated:

> The bite of the Black-Bellied Tarantula is therefore dangerous to other animals than insects: it is fatal to the sparrow, it is fatal to the mole. Up to what point are we to generalize? I do not know, because my enquiries extend no further. Nevertheless judging from the little I saw, it appears to me that the bite of this spider is not an accident which man can afford to treat lightly.

Fabre went on to discuss the ability to select the right spot to kill the victim or paralyze it (in the case of the *Sphex* wasp). "If the instinct of these scientific murderers is not, in both cases, an inborn predisposition inseparable from the animal, but an acquired habit, then I rack my brain in a vain attempt to understand how the habit can have been acquired. Shroud these facts in theoretic mists as much as you will, you shall never succeed in veiling the glaring evidence which they afford of a pre-established order of things."

Fabre never abandoned an uncomplicated religious attitude. Although he corresponded with Darwin and although during his long life he saw geological fact after fact added to substantiate the developmental point of view, he resolutely shut his eyes to it. Rigorously logical in his experiments, he refused to accept the larger logic of theory.

His experiments with hunting wasps and mason bees are

fascinating, for he carefully traced out the limitation of instinctive behavior. He discovered that in most cases, the animal was programmed to perform a series of actions in an elaborate reproductive cycle but it was unable to turn back or to cope with changed circumstances. When a hole was made in the mason bee's nest, allowing the honey to drip out, the bee went ahead if it had finished its cycle of provisioning, laid its egg, and sealed up the ravaged nest.

He observed that although garden spiders constructed their webs with geometrical accuracy, they were unable to repair them when they were damaged; they had to start all over again from the beginning.

Yet what, for lack of a more descriptive word, we call instinct was capable of some baffling achievements. How was it that one species of hunting wasp inevitably chose a particular species of grasshopper which it stung, paralyzed, and stowed away for its larva? How was it that the same wasp could find its nest after Fabre had carried it a couple of miles away in a paper bag? How was it that another species of wasp could find the entrance to its nest, even though Fabre had changed the landmarks of the locality and covered the entrance with a flat stone? The wasp promptly dug under it. He placed animal dung on the spot to change any possible odor. The wasp dug through it. Even a vial of ether baffled it only momentarily. And then when deprived of its antennae, supposed seat of its sensory organs, the wasp was still capable of finding the doorway to its home through a blanket of moss. The whole subject of instinct was an obsession with Fabre. The young mason bees, he discovered, could pierce the entrance to the cell when the time came for them to emerge, whether it was sealed with the original cement, wood pulp or brown paper. Once they had emerged, if he placed them in a bottle covered with paper, the insects were unable to repeat the act of cutting their way out. "The Mason bee perishes for lack of the smallest gleam of intelligence. And this is the singular intellect in which it is the fashion nowadays to see the germ of human reason! The fashion will pass and the facts remain, bringing us back to the good old notions of the soul and its immortal destinies."

The mystery of how complicated instinctive actions are transmitted from an insect to its offspring which it never meets is profound. To assume, as Fabre did, that the creature was specially created with these habits from the beginning of time is to refuse to take into account the now heavily documented evidence of paleontology concerning the development of new species.

Fabre's own material, however, proves that instinct is not absolutely and rigidly mechanical. The yellow-winged *Sphex* wasp, which catches crickets, habitually left its prey at the entrance of its burrow, entered on a rapid journey of inspection, came out and pulled its cricket in by its antennae. While the wasp was out of sight, Fabre moved the cricket away. The insect came up, uttered a little cry of surprise, searched about, found the victim, brought it back to the edge of the burrow, and repeated its visit of inspection. Continuing the experiment forty times, he found that the *Sphex* wasp always made its tour of inspection and never got the cricket into its burrow. This of course demonstrated the blindness of the instinctive series of actions. All wasps in one colony behaved in the same fashion. Yet the wasps of another colony, some distance away, after three frustrations seized the cricket and went immediately into the hole. There appeared to be, therefore, a glimmer of intelligence, at least the ability to learn, which gave the lie to Fabre's insistence that instinct was totally divorced from mental ability.

Fabre's researches point up the strange limitations of instinct. Homing ability, demonstrated not only by insects but by pigeons and cats, and much of the elaborate social behavior of ants reveal abilities outside the range of human rationality. Yet Fabre's mason bees, which died because they could not cut their way twice out of an imprisoning substance, behaved like Cartesian machines.

Although Fabre is isolated from the mainstream of the ethologists, a more orthodox early ethologist (whose name will always be associated with bees) is Friedrich Karl von Frisch, who was born in 1886 in Vienna. His academic credentials were highly respectable, for he studied under Richard von Hertwig, brother of Haeckel's pupil, Oscar.

Frisch's father was a surgeon and a professor, and Karl grew up in a typically middle-class intellectual environment. He and his three brothers played Haydn quartets, his mother corresponded with a famous poet, his uncle was a professor of anatomy. Karl, as a boy, kept pets, including a woodpecker named Ignatz, which hammered on dishes on the table when it wanted mealworms. His father wanted him to study medicine, which he did, taking his doctorate at the University of Vienna in 1910, and in the same year became Hertwig's assistant at the University of Munich. His early work involved him in a controversy with a Professor C. von Ness who maintained that fish and invertebrates were color-blind. Supported by Hertwig, Frisch opposed this view and from then on, he worked on the sense perceptions of animals. By 1912 he was investigating bees to prove that they could recognize color. His method was to allow them to take sugar water from blue paper, then to put the sugar water on gray paper. It turned out that they were conditioned to color, for they at first went to the blue. Frisch took a leave from Munich and went home to Brunwinkl, where he drafted his relatives to help with his work on bees. A slight struggle developed with Hertwig, who wanted him to return to teaching. Frisch prevailed and triumphantly disproved Ness.

His research was interrupted by World War I, during which he worked in a hospital. After the war he returned to Munich only to be upset by the unsuccessful Socialist revolution, but he stayed on at the university, where he was made a full professor in 1919.

He kept his bees in the garden of the Zoological Institute as he continued to investigate sensory perception. Further studies in color recognition showed that the bees could distinguish blue and violet from orange and yellow, but red was to them the same as black or dark gray. Even more striking was the fact that they could distinguish ultraviolet as a color and preferred it to all others. He pointed out that there were few red flowers in Europe but many in America. These, however, were mostly pollinated by birds. Butterflies were the only insects known not to be red-blind. White flowers reflected ultraviolet rays and hence were also attractive to bees.

For experiments with smell Frisch used different boxes with a small entrance hole containing scents. The bees could distinguish scents as readily as human beings. He localized the organs of smell in the antennae, cone-shaped depressions which alternated with touch organs, small palps or flagella. The insects were also fairly discriminating as to taste. They required a 20 percent solution of sugar for real satisfaction; below 5 percent they rejected the food. Out of thirty-four sweet chemicals they only accepted nine. They also distinguished salt, sour and bitter flavors. His experiments with butterflies proved that they had taste organs in the tips of their legs.

It was in 1919 that Frisch first observed the phenomenon on which he spent the major part of his time. He had begun the practice of marking bees with groups of spots of color by which he could use three-digit numbers and identify 599 bees. One of his marked bees came back to the hive after tasting some sugar water and did a round dance. Another bee became excited and flew to the sugar water.

Frisch performed many experiments, analyzing what came to be called "the dance language of the bees." He distinguished two types of dances. There was the "round dance," a circle to the right and then one to the left, repeated a certain number of times. By careful observation he correlated this with a honey source close to the hive. The "waggle dance" consisted of a short run while the bee waggled its abdomen, a turn to the left, a turn to the right without waggling, then a repeat of the whole pattern. Frisch correlated this with a more distant source of honey. The farther away the food source, the faster the pace in the cycles. The dances took place only when the food source was especially plentiful. Frisch went on with further experiments which coordinated the angle of the sun's rays with the direction of the food source. As the sun moved, the angle of the straight run changed, and this occurred even on cloudy days. It was suggested that in this case ultraviolet rays were the guide. In the dark of the hive gravity seemed to be the guide; 60° left of vertical indicated 60° left of the sun's direct rays. Other experiments showed that the bees recruited to the new food source would fly straight up over mountains and down again to reach their goal.

Frisch also claimed that the dances were used in the case of swarming. Scouts returned to the swarm clustered outside the hive and danced to indicate the direction of a suitable new home. The most enthusiastic dancers gained more and more recruits. In the round dance, he said:

> Those sitting next to the dancer start tripping after her trying to keep their outstretched feelers in close contact with the tip of her abdomen. The maneuverings are such that the dancer herself in her madly whirling movements appears to carry behind her a perpetual comet's tail of bees.

Finally one scout prevailed; the others became converted in a kind of automatically democratic decision.

One rather charming point made by Frisch is that American and German bees have no difficulty understanding each other. Theirs is an international language.

Frisch lectured in a number of American universities in 1949 and convinced many skeptics. One critic of his work, Adrian M. Wenner, sticks to the position that Frisch's correlations are only by inference and that we do not know exactly what guides the new recruits. His experiments conclude that a sound signal produced during the straight run of the waggle dance is also a stimulus. He seems inclined to believe that the scent of the particular flowers (carried in a pouch on the abdomen) is what guides the recruits and also, discovery of the hive scent on the food source. This would be admitted by Frisch as the method of pinpointing the food source, but it does not seem to be an argument against the communication of general direction and distance, particularly over mountains. Wenner cannot accept the bees' imparting of abstract information, but he does admit that the dance acts as a stimulus to food-seeking.

The work of Fabre, admittedly primitive, leads to the more rigorous Frisch, and the latter, in turn, with his repeated experiments, leads to the newest group of ethologists who concern themselves with what is called ritual behavior and its relation to basic psychological drives in human beings.

28

Only Human

☙ THOMAS HUXLEY was not the only member of a remarkable, distinguished family to work in the area of zoology. His grandson, Sir Julian Huxley, has followed in the footsteps of Darwin's champion by both making a pioneering contribution to ethology and continuing to elucidate evolutionary theory. Also, in public life he became, like his grandfather, a kind of "statesman of culture." Both stand out as great men, in addition to being talented scientists, because of their sense of social responsibility.

Born in 1887, Julian has been independent and fearless all his life. Indeed, when he was a child, his grandfather remarked that he liked the little chap because he looked him straight in the eye and immediately disobeyed him. When Julian was about seven, he caught his grandfather in an incautious statement about fish. The old man remarked that they didn't go in for paternal care. "What about the sticklebacks?" inquired the *enfant terrible.*

Leonard, Julian's father, seems to have been an admirable schoolteacher, biographer and eventually an editor, but he never made the same mark in the world as his father and his two sons, Julian and Aldous. Since Leonard had married Matthew Arnold's niece, the third generation had an especially interesting set of genes.

Julian went to Eton, which was followed by Oxford, and he was then lured away to, of all places, Texas where the newly formed Rice Institute was being organized, supported by the fortune of a deceased timber baron. Julian was invited to head the Biology Department in 1911, but he did not leave for America until 1913. In the meantime, while visiting in Hertfordshire, he engaged in a bit of fieldwork that is now considered an important landmark in ethology.

Huxley obtained a permit and the use of a punt to watch birds on the Tring reservoirs. There, for three weeks he patiently observed the antics of *Podiceps cristatus* through field glasses. The grebe in question is a swimming and diving bird with a long neck. Its upper parts are mottled brown, its underside snow white. A dark ruff of erectile feathers surround the face and above the eyes are ear tufts. These feathers can be half opened or slicked down; they can also be fully extended and tilted laterally.

The birds arrived in Hertfordshire in February in flocks. Pairing went on in March, nesting in April. Huxley did not arrive in time for the pairing, but he noticed fascinating behavior on the part of the paired birds, which he decided to call courtship habits. (At the time his observations took place, terms for various types of social behavior had not yet been generally accepted.)

The first activity consisted of the pair, as they approached each other, raising their necks and extending their ruffs, then when face-to-face, shaking their heads, wagging them from side to side four to five times, first slowly then fast. The rhythms of the two were not coordinated and the cycles of wagging were repeated six to seven times. Then the beak was put under the wing feathers near the tail as if to preen or, as Huxley described it, "as of a stereotyped and meaningless relic" of preening. Then, after twelve to fifteen more shakes, both lowered their ruffs and went about their business. Huxley called the action with the bill "habit preening."

The second activity was initiated by the hen, who, just after alighting, would swim toward her mate, her neck arched, her bill pointed down almost to the surface, and her ruff sweeping the water. She gave a short barking call, now and then lowering her ruff and then resuming the attitude with her wings half spread on the water. At 30 or 40 yards away the cock dived, and when he came up near her, he stood up on his tail like a penguin, his ruff spread, his ear tufts horizontal. He then sank down and the pair went into a bout of head shaking. This Huxley called a "discovery" ceremony.

A third action, which took place after a long bout of preening and head shaking, was a double dive after which both came up holding ribbony waterweed in their beaks. They lowered their heads, swam toward each other, and when about a yard apart, both did the penguin stand until they came together, their breasts touching, their feet working madly, weed still dripping from their beaks. They then gradually sank down and resumed the usual head shaking.

Later Huxley was to discover that the cock often mounted the hen in the water in an imitation of coition which could take place only in the nest. What was even stranger was the fact that the cock sometimes assumed a passive position and was symbolically trodden by the hen. Huxley also noted that several nests or platforms of weed were built, and on one of these coition took place. Huxley called these activities of the pair "rituals." He felt that the courtship behavior was "self-exhausting" and compared bird behavior to that of humans. Huxley's attempts at explanation went back to Darwin's notions of sexual selection which tended to credit animals with an esthetic reaction to color display and sound, an idea which is no longer accepted.

The kind of activities observed, however, and the fact of their emotional and social significance in the animal group resulted in a trail-blazing piece of work. It was to set the pattern for extended analysis and wider observation of many different types of social behavior in animals and foreshadowed the new school of comparative ethology which has grown up since.

In Germany, the tradition began with the work of Oskar Heinroth, who in 1910 published a study of ducks and geese which discussed their threat postures and included data on hostility within the species. His pupil, the Viennese Konrad Lorenz (1903–), continued work with geese and jackdaws from 1935 on. Niko Tinbergen, who was born in The Hague in 1907, was steered into animal behavior by his zoology teacher while attending the University of Leiden. His doctoral thesis, on the homing of wasps, was sparked by the work of Frisch. While participating in the Dutch expedition to East Greenland during the International Polar Year, 1932-33, he studied the snow bunting. In 1937, he worked for three months with Lorenz. During World War II he was held by the Nazis as a hostage with twenty-one other professors for protesting the firing of Jewish members of the faculty. Some were shot, but he survived to return to the University of Leiden. In 1949 he was invited to Oxford, where he introduced the new ethology.

Lorenz, who is director of the Max Planck Institute for Behavioral Psychology in Bavaria, earned his doctorate in medicine at the University of Vienna. He also studied medicine in New York. His popular book *On Aggression,* published in German in 1963, has perhaps gained a greater public than any other study which has come out of the new ethology.

What is essentially new and different in contemporary ethology is the interest in social behavior, defined as the way in which animals react to each other, including activities of the pair. The study involves a psycho-physiological approach.

The growth of behaviorist psychology stemming from the work of

Ivan Pavlov in the late nineteenth century often resulted in a highly Cartesian attitude toward the organism. Pavlov sprinkled meat powder on the tongues of dogs, which caused salivation, and at the same time rang a bell. The dogs soon associated the ringing of the bell with food and salivated when they heard the bell, even if no food was provided. The resulting picture was that of a machine which could be made to operate by pushing a button. B. F. Skinner of Harvard carried out further studies in which rats as a result of various stimulants, mostly rewards, could be trained to pull a chain to obtain a marble from a rack, pick up the marble with their forepaws, carry it to a tube projecting two inches above the floor of the cage, lift it to the top of the tube, and drop it inside. While this approach documented facts about the learning process (which most intelligent animal trainers already knew), it treated the organism as an inert object acted upon by outside forces. The laboratory approach told us nothing about what happens in nature and why animals (and perhaps people) do what they do when they live together.

Such investigators as Fabre and Frisch observed actual behavior in the field and performed simple experiments in order to understand its meaning. It remained for the new ethologists to stress the distinction between innate behavior—that is, potential reactions which have been bred into a species by evolution—and conditioned learning acquired by the individual during its lifetime and not passed on. They then went further and tried to understand how innate reactions or instincts came about and what constituted their survival value.

A number of fascinating concepts have emerged. One of the earliest is that of territoriality, first definitely established by the British bird watcher Eliot Howard, who discovered that the singing of male birds in the spring not only served the purpose of courtship but also acted as a warning to keep other males of the same species away from the area the pair had marked out for its own as a place to nest and carry on its domestic life. (As we have seen, both Friedrich II and Gilbert White had some insights in this direction.)

Territoriality has since been studied in many species, especially fish which, in the case of butterfly fish of the coral reef habitat, for instance, attack but do not fight seriously unless they are cooped up in an aquarium. In their coral reef habitat, where there is room to mark out adequate territory, they merely threaten each other when they approach the confines of each other's boundaries. The conclusion drawn from this and many other observations is that aggression of this sort serves to disseminate the species and avoids overcrowding without unnecessary casualties.

A second concept is the pecking order, so-called because it was first established as a characteristic of barnyard fowls. In this case every individual in a flock or herd knows its relation to those animals more dominant and less dominant; in other words, it accepts its position in a hierarchy. Although there is some tyrannizing, in the case of jackdaws, Lorenz points out:

> Those of the higher orders, particularly the despot himself, are not aggressive toward birds that stand far beneath them; it is only in their relationship toward their immediate inferiors that they are constantly irritable. . . . Very high caste jackdaws are the most condescending to those of lowest degree and consider them merely as the dust beneath their feet. . . .

He also points out that the aristocrats interfere in the quarrels of subordinates and more or less prevent the weakest birds from losing their nesting sites. Changes in status can occur, for, in one case, a jackdaw, which had been absent for a summer, returned to Lorenz's colony sophisticated and toughened by his travels. He immediately beat up the number-one bird and took his place. When he mated, he chose a very-low-ranking bird which had been pecked by 80 percent of the colony. She was immediately promoted to his rank, the rule in the jackdaw flock, and happily tyrannized over all her former superiors.

Pecking order (which is remarkably parallel to rank in human society) seems to be a practical adjustment to group living which eliminates unnecessary conflict and thus can be considered to have survival value.

Another phenomenon, observed in waterfowl by Lorenz, is that of "imprinting." Geese, it seems, run after the first moving thing they see and are thus imprinted to consider it their mother. Young mallards, Lorenz found, were most selective. He knew, however, that they were willing to follow a white domestic duck. Suspecting that the stimulus was auditory, he quacked for a newly hatched litter of mallards and was successful, for they followed him when he drew away. He was obliged to squat in order to present the right image, however; they began to peep desolately when he stood up. One day, when he was quacking and waddling, he looked up to see a row of tourists peering over the fence at him in absolute horror. His behavior was all the more insane to them because the ducklings were hidden by the grass.

As more and more patterns in social behavior began to emerge, a body of theory developed. Tinbergen's study, *The Herring Gull's World,* is a detailed picture of this bird's social life which links Tinbergen with the gull as Frisch is now linked with the honeybee. Tinbergen began to study gulls while he was at school in Leiden and developed various observation techniques, chiefly that of banding so that individuals could be identified. Lorenz also banded his jackdaws but said that he could recognize them all by their facial expressions. Tinbergen netted his birds while they sat on their nests in order to ring them and employed powerful binoculars to watch from a canvas shelter. Often the shelter was made use of by the birds; frightened chicks rushed inside for protection and adults plucked at the cords in emotional displays. Two important concepts were arrived at in this series of observations. One involved the feeding behavior of the chicks. In order to stimulate the adult birds to regurgitate food (mostly fish) they pecked at a red spot near the tip of the bill. Tinbergen experimented with head models, using different-shaped heads and different colors for the spot, sometimes omitting it. His statistics showed that the red spot is the "releaser" of the pecking behavior, as this type of stimulus has been named by Lorenz.

The second discovery involved aggression. Most fights resulted from territorial problems. Gulls like to nest in a colony, which probably provides more protection from predators but also means that nesting sites are sometimes too close together. The fights started with the threat posture, neck stretched upward, the wings slightly lifted. This was followed by a charge that generally put the antagonist to flight. Sometimes, after threatening, the opponents tore off tufts of grass with their bills. If the fight went further, they tried to wrench the grass from each other's bills. Finally, they grabbed and pulled at each other's wings and struck with their wings.

The tearing up of grass was the most interesting point, for Tinbergen identified it as a nest-building behavior. He labeled it "displacement activity."

The concept of displacement, or redirected behavior, has become very important in ethological theory. All aggression is believed to carry with it the alternative of fear and flight. When the gull resorted to its displacement grass pulling, it hovered between the two drives. Lorenz compared this behavior to that of a human being in an embarrassing situation lighting a cigarette. Once the mechanism of displacement was understood, it could be seen in operation in various situations. Male birds fed their mates when courting them just as they fed their young. Tinbergen's gulls before and after copulation imitated the begging behavior of the full-grown young bird. Huxley's grebes, seen in this light, displaced the nest-building activity and used it in courtship. Their ritual preening was a similar redirection.

Interestingly enough, there is a parallel in this concept to Freud's idea of sublimation, the difference being that, etymologically, sublimation carries an implied value judgment while displacement, according to the ethologists, is the result of the interaction of innate drives shaped by evolutionary selection.

The picture of animal behavior, in the words of Lorenz, involves "a great parliament of instincts" which interact, affecting social behavior, the results being bred into the species by the test of survival value.

One of the most interesting mechanisms described by Lorenz is that in which, by ritualization, aggressive movements turn into their opposite. A female duck is accustomed to threaten strange drakes by stretching its neck toward them and running or swimming toward its own drake whom it is inciting to attack. These movements in various species take various ritual forms until, in the mallard, this behavior on the part of the female is an invitation to pair and the male responds by turning its head away, quacking "Rabrab, rabrab," taking ceremonial sips of water and indulging in sham preening. All this, which replaces an attack since there is no intruder, means he is willing.

Elaborate courtship rituals are interpreted as having survival value because they serve as identification, acting as releasers which prevent interspecies breeding and the production of hybrids.

In the case of animals which tend to be overaggressive, such as female empid flies, which are larger than the males, a ceremony has evolved to protect the male from being eaten. The male presents the female with a dead insect of the proper size which she eats while he copulates. Jan van Iersel, studying the courtship behavior of the male stickleback, an aggressive little fish which builds a nest and tends the eggs and young, noted a balanced situation. The male would rush toward the female and rush back toward the nest he had prepared. The procedure becomes a zigzag dance ritual. Through a number of experiments Iersel was able to show that the zig toward the female represented territorial aggression while the zag toward the nest was an invitation to pair.

Another area in which interaction is interesting and important is the relation between adults and young. Tinbergen found that his gulls respected eggs only when they were broody; otherwise they ate them. One-day-old strange chicks were accepted when put into the nest. After four days, however, strange young were pecked and rejected. This showed that a learning process had taken place; the birds had come to recognize their own young.

On the other hand, Lorenz points out that a turkey hen will attack anything moving near her nest, even her own chicks, unless she hears them cheeping, sound causing the inhibition.

Canines possess strong inhibitions against attacking puppies under the age of seven or eight months. Tinbergen found, in the case of Greenland Eskimo dogs, that this extended only to the young of the same pack. Male canines also are not aggressive toward females and show complete passivity when bitches attack them. In the case of bitches the inhibiting sign is probably chemical, smell, while puppies turn over on their backs and produce urine as an appeasement gesture.

Throughout the animal kingdom appeasement gestures serve as a signal to inhibit aggression in the stronger or more dominant individual. Wolves turn their heads away and present their vulnerable jugular vein to the enemy, who is unable to bite it. Monkeys present themselves sexually (males in a homosexual gesture). Fish dim their provocative bright colors. Tinbergen found that kittiwake gulls, who lived crowded together on a lofty ledge, turned the head away as a peaceful gesture, and indeed this is common, for such weapons as the bill of a bird or the teeth of a mammal are thus removed from a threat position. There is probably an analogy in shaking hands among human beings which indicates that the hands are empty of weapons.

Lorenz's book *On Aggression* attempted to draw some parallels and to relate animal and human hostile behavior. He wrote:

> If, in the greylag goose and in man, highly complex norms of behavior, such as falling in love, strife for ranking order, jealousy, grieving, etc., are not only similar but down to the most absurd details the same, we can be sure that every one of these instincts has a very special survival value, in each case almost or quite the same in the greylag and in man.

At the same time man's culture, which is built up of intellectual concepts, has disturbed the automatic functioning and conditioning of his behavior by biological evolution. Culture is passed on by learning, and thus man is able to inherit acquired traits which have not been tested by natural selection. Animals, as has been shown, are motivated by built-in inhibitions which prevent useless killing of their own species. Natural selection works for its own survival morality. Man has no such automatic inhibitions against killing, particularly when he can do it anonymously by pressing a button and does not have to employ his hands or teeth. In addition, Lorenz believes that the basic aggressive drives, without which an organism does not function to its fullest capacity, are present in man but do not find useful outlets, being transformed into power drives, or into the amassing of material possessions. In Lorenz's view, the instincts, if properly balanced, make for healthy community relations, while the intellect, which can rationalize anything, is an untrustworthy guide.

Some of Lorenz's speculations concerning the conditioning of Paleolithic man and the expression of his aggressions are, however, rather questionable in light of the fact that bands of such men were widely scattered and need not often have come into conflict.

It was such rather naïve forays into anthropology, including some remarks about the Ute Indians, which brought down much of the criticism with which the book was assailed in anthropological circles. Actually Lorenz was caught in a kind of crossfire. Robert Ardrey's unsound book on territoriality happened to be published at the same time. Ardrey, an uncritical popularizer, took Lorenz for his hero but did not even interpret him correctly. His book was a reversion to a dated "struggle for existence" concept positing all of man's hostilities as based on a territorial imperative with evidence drawn largely from birds. The fact that most of the apes and monkeys that live in groups do not exhibit this characteristic was waved away.

At any rate, Lorenz's rather literary speculations, which treated too many of man's activities too dogmatically as inherited innate drives, served as a releaser for considerable aggression. Ashley Montagu edited a volume containing a group of essays criticizing Lorenz. Montagu, writing from the point of view of an anthropologist, stated: "Man is man because he has no instincts, because everything he is and has become, he has acquired from his culture, from the manmade environment, from other human beings."

This statement is supported by no evidence, and no experimental proof is cited. In its dogmatism it is more extreme than any written by Lorenz. Actually some experiments are being conducted on babies which seem to show that certain activities are innate. At any rate the area remains controversial.

Other criticisms, such as that by John Paul Scott, who does not believe there is an innate aggressive instinct in man, are more temperate:

> In any practical situation this [control of aggressive behavior] may be quite difficult because a vertebrate animal has a behavioral organization to produce fighting behavior in response to certain kinds of external stimulation and the complete elimination of such stimulation may be impossible.

We begin to perceive that there is a basic disagreement which arises from the Cartesian attitude toward the organism inherent in the work of psychologists and experimental behaviorists of the B. F. Skinner school. The scholar of this type is impressed by motor organization, neural organization and external stimuli. When his experimental animal runs mazes, he is concerned with what and how much the creature can learn, and from the results he constructs curves. He is frankly at a loss when confronted with individual differences.

The ethologist, working from behavior observed in the natural habitat, is impressed by the dynamism of the organism, the way in which its actions are reinforced by emotion, the seeming capacity for spontaneity. He believes that behavioral gestalts, complexes of activity, have been bred into it by natural selection and that these are analogical to the physical organs evolved by natural selection. As W. H. Thorpe wrote in *Current Problems in Animal Behavior*:

> To the ethologist, the psychologist, obsessed with the rat in the maze, was apt to appear so oblivious of the realities of animal behavior as to be almost a figure of fun . . . on his side the psychologist insofar as he knows anything at all about the work of the ethologist—was apt to regard it as anecdotal and lacking in scientific method.

However, more and more experiments have been done in both the field and the laboratory which help bring the two points of view closer together. In the case of imprinting, a dummy duck wired for sound on a turntable made it possible to analyze the duckling's behavior.

Lorenz's account of instinctive mechanisms involves IRM's, innate releasing mechanisms, by which inherited behavior gestalts are allowed to react to the releasers. T. C. Schneirla of the American Museum of Natural History denies any such inherited tendencies, quoting the studies of Zing Yang Kuo. As its heart beats, a chick automatically brings its bill forward inside the shell. Gradually it begins to open and shut its beak and swallow as this happens and thus is conditioned toward pecking. Lorenz sees no reason why this semiautomatic action should be transformed into the pattern of pecking at food, which requires a visual stimulus. Daniel Lehrman of Rutgers answers this a little vaguely by saying that the visual and tactile centers of the brain are connected.

On the whole Lehrman's extended critique is closely reasoned. He objects to the lumping of a series of actions together under the word "instinctive," or innate. Taking the example of the rat which retrieves its young and puts them back in a nest made in the corner of the cage, he points out that this is labeled innate behavior because rats, reared in isolation from other rats, perform this action. Experiments, however, have linked self-licking to licking of the young and carrying of food pellets to carrying the young. When rats were kept in a mesh-bottomed cage which did not allow them to pick up and carry food pellets and when they were also equipped with collars which prevented them from licking their genitals, they failed to retrieve their young.

Lehrman concludes that rat behavior "develops in certain situations through a developmental process in which at each stage there is an identifiable interaction between the environment and organic processes within the animal." Thus in Lehrman's view, the test of "isolation" used by the ethologists to prove innateness of behavior is unsound and in many cases the "gestalt" can be broken down into a step-by-step process.

A second weakness of the Lorenz theory is the tendency to make broad generalizations in which homology and analogy were confused—that is, types of behavior which arose in different ways but served the same purpose. In the area of anatomy the parallel is lumping together the bird's and the bat's wing. For instance, the amoeba and the young infant both react toward weak stimuli and away from strong ones. But in the case of the amoeba, this is the result of different chemical relationships occurring in the protoplasm, while in the child there is probably a different arousal threshold in the flexor and extensor muscles. Both cases reveal survival value because the movement toward results in obtaining food and the movement away helps avoid harmful elements in the environment. Confusing the two types of behavior because of their results, Lehrman contends, prevents any real study of evolutionary development.

On the whole Lehrman criticizes the Lorenz-Tinbergen position for lack of close analysis and careful differentiation and for unprecise use of terms.

The ethologists, who by and large write with charm, are inclined to use a good deal of nonacademic vocabulary. It is true that "instinct" has historically been used to mean almost anything from sexual appetite to maternal love. "Aggression," too, is a blanket term involving a complex pattern of activity. What often happens is that both ethologists and their critics move in and out of various areas of discourse. Ethologists should probably stay within a sociological frame of reference or work in teams with psychologists and physiologists. In fact, Tinbergen has very sensibly called for interdisciplinary cooperation. After all, *something* is inherited and *something* learned; the disagreement is over how much of what.

Laboratory scientists sometimes cannot forgive ethologists for revealing their liking for and emotional involvements with the animals they study, for being "anecdotal." For example, the following rather endearing bit, quoted by Lorenz, can be condemned as anthropomorphizing by the cold logic of the experimentalist. Geese, according to the statistics, pair for life. While working with his assistant, Helen Fisher, Lorenz remarked that their records showed far too many broken marriages. "What do you expect?" Fisher said. "After all, geese are only human."

One more criticism raised by Lehrman involves an important problem. Lorenz and Tinbergen see the innate releasing mechanism as a specific coordinating center in the central nervous system. Psychic energy builds up like water behind a dam. Another center prevents the activity from taking place continually until the releaser triggers the action and drains away a certain amount of the energy. Lorenz points to "self-exhausting" behavior which can be more easily stimulated to take place again after a pause during which, supposedly, a fresh buildup of energy occurs.

Lehrman contends that there is no proof of any buildup of psychic energy in the central nervous system. On the other hand, Donald K. Adams, professor of psychology at Duke University, points out that the psychologist William McDougall, in 1905, arrived at the same damming-of-water image. He remarked:

> Such independent convergence upon a hypothesis would seem, incidentally, to be a kind of evidence for the reality of something like

> the process envisaged, no matter how its particulars may have to be qualified or revised.

As we have said, many ethologists are impressed by the dynamism of the living organism, while even such terms as "innate drive" upset some scholars. For instance, J. B. Crook, a Cambridge zoologist, in discussing the impulse to flock, which he feels is innate in some species, will only go so far as to use the term "tendency," defined as "the readiness to show a particular behavior under natural conditions." "Readiness" and "tendency" are fairly inert words, while "drive" is dynamic. Behind this controversy the horns of vitalism and mechanism can be heard faintly blowing.

Perhaps the most important contribution of the ethologist to the study of behavior in general is the concept of ritualization. Regardless of whether it is learned or inherited, its form and meaning in social groups are extremely interesting. In 1966, Julian Huxley, who has never lost interest in the study he helped found, edited a volume of transactions of the Royal Society of London in which papers by a group of psychologists and zoologists discussed the ritualization of behavior in animals and man. Huxley pointed out that the ritual approach could develop into a new angle for the study of human behavior, for it was present throughout human culture from education to diplomatic maneuvering to the theater. As psychiatrist George Morrison Carstairs put it: "What we are doing is so often at variance with what we think we are doing." He analyzed the healing ritual performed by sorcerers in India. After it was carried out, the kinfolk felt better, even if the patient did not. Thus the ritual had the social function of relieving anxiety in a crisis.

Erik Erikson, with a broad background of competence in both depth psychology and anthropology, pointed out that one might be able to show the equivalent of animal ritualization in human behavior. He suggested that ritualization could be a kind of compromise between the selfish drives of the id and the inhibitions of the superego or social norm.

To sum up in the words of Eugene L. Bliss of the Department of Psychiatry of the University of Utah concerning the study of animal behavior:

> This is clearly only one of the many ways through which man's nature will eventually be revealed, but it appears to be a potentially

productive one—one that should be made better known to many working in related areas.

So far we have said nothing of the primates. Since ethology has recently investigated their behavior, we shall now concern ourselves with the zoological order to which man, himself, belongs.

29

Man's Nearest Relatives

⁂ THE primates have always occupied a peculiar position in zoology. From the confusions of the eighteenth century, in which scholars were not sure whether they were men or animals, through the nineteenth century in which, to the dismay of many, they were established as man's relatives, our knowledge of them consisted more of myth and inaccurate travelers' tales than of careful observation. Somehow monkeys and apes have always held up a kind of distorting mirror. Their remarkably human qualities could be dismissed, but the lord of creation could not help feeling that they were a malicious parody of himself.

In the 1930's, however, pioneering work was done by Sir Solly Zuckerman, who published his studies on the social life of captive animals, *The Social Life of Monkeys and Apes,* in 1932, and by Robert M. Yerkes of Yale, who published *The Great Apes: A Study of Anthropoid Life* in 1929. While areas of study were marked out and individual psychology and intelligence investigated, neither scholar worked in the natural habitat. It remained for R. C. Carpenter, also of Yale, to do the classic introductory work on field behavior in the early thirties during expeditions to Panama, where he observed howling monkeys. Later he gathered data on spider monkeys of Central America, gibbons of Thailand, and rhesus monkeys in Puerto Rico.

Carpenter studied such matters as dominance (gibbons showed no sex difference in dominance) and aggression; the same species had a pecking order which eliminated fighting within the band and vocalization took place when different bands met. He took the number 18 and pointed out that in a group of this size 156 paired relationships were possible. Of these he considered 11 to be most significant. These were interactions between adult males of organized groups, adult females of organized groups, adult males to adult females, adult males to young, adult females to young, relations of solitary animals to groups of the same species, relations of one species to other species (not only primates but also all animals and plants), relations to the whole ecology and climate, and finally relations of the individual to the whole group.

Carpenter did much to set the pattern for primate study. Advancing in the primate series toward man, S. L. Washburn and Irven DeVore's studies of baboons added more interesting data concerning social behavior. In a park near Nairobi, Kenya, they observed twelve troops of baboons containing 450 members. They then proceeded to the foot of the famous Mount Kilimanjaro where the troops of baboons were twice as large, sometimes containing as many as 80 members. Altogether they observed as many as 1,200.

As was to be expected, there was a pecking order. The dominant males, however, achieved social importance not only because of superior strength but, it appeared, through strength of personality. Females, young, and less dominant males liked to sit near them and engaged in grooming them, an act of social affability. The most interesting observation consisted in the recognition of "friendships" between animals. The Victorian notion of both primate social structure and of primitive man was a sexually dominant "old man" raping all the females and cowing or exiling the less dominant males. The two observers found that sex had little to do with social structure; if anything, it was disruptive. Estrous females solicited the males and mated with many of them, finally spending a pairing period with a single male, which might last from an hour to several days. On the other hand, there was no family or specific harem. In friendships two females or two males seemed to enjoy pairing off, and among the young play groups persisted for several years. Moreover, important males formed groups which stuck together and outranked individuals. Such dominant males stopped fights within the group and also formed the nucleus of the defense of the band. When danger arose, the males stayed behind and could drive off cheetahs and other predators with the exception of lions from which the whole band fled.

Infants were always of great interest, for friends and relatives surrounded them, always ready to handle them and groom them. Later the young joined juvenile groups where the social bond was also strong.

While baboon bands occupied a definite territory which sometimes overlapped, when different groups came to water holes at the same time, they did not fight but neither did they mix.

Finally, the ethologists observed that although their diet consisted mostly of grass, fruit and plant shoots, they occasionally ate meat such as newborn Thomson's gazelles. They did not, however, actively hunt for meat.

Meanwhile, the Japanese concerned themselves with acculturation studies of their native species, the macaque. In one experiment, candy was left for the troop to discover. The young monkeys discovered it first and relished the sweet taste. From them the innovation in food habit spread to the mothers and from the mothers to young males and the rest of the troop. This episode seems to show less rigidity in the primate establishment than in the parallel organization among humans. In the light of the many recent studies describing the quenching of curiosity and creativity in contemporary school systems, it is interesting that a subhuman society was able to learn from its children.

It was not until 1960 that extended studies were made of the social life of the two great apes, the chimpanzees and the gorillas. Although the chimpanzees had been used in psychological experiments, they had not been observed in their natural habitat until a young Englishwoman, Jane Goodall, went to Nairobi and became the secretary of the famous paleontologist and anthropologist Dr. Louis B. Leakey. She worked with him in the Olduvai Gorge dig and convinced him of her serious interest in animals. He suggested the chimpanzees as a subject for study and finally obtained a grant for her from the Willkie Foundation of Illinois. Since Jane was only in her very early twenties, the authorities refused to allow her to live alone in the bush, but her mother gamely volunteered to accompany her.

The rain forest of the northern shore of Lake Tanganyika in central Africa was the site chosen. Kigoma, a few hours away by outboard canoe, was the nearest link with the capital to which it was joined by a railroad which gave up in the rainy season. The two women were accompanied by an African cook, Dominic, who proved loyal and invaluable and, indeed, stayed with the project for several years. The first obstacle that had to be overcome by the women was a malarial fever which kept them in bed, in a delirium, for ten days while the solicitous Dominic watched over them. After two months, once more able to walk but still weak, Jane encountered a large group of chimpanzees feeding on figs. They became used to her, for she sat still, always in the same place near one of their feeding areas. Later, when Baron Hugo van Lawick, a photographer sent by the National Geographic Society, arrived in 1962, feeding stations were set up which brought the apes regularly into camp. Finally they became so tame that the tents had to be abandoned for huts which could be locked, since the chimpanzees stole bananas and loved to chew clothing, cardboard boxes and even the tent canvas.

Jane's mother, after successfully curing a native's infected ankle, achieved fame as a doctor and set up an amateur clinic at which she treated the Tanganyikans with common-sense remedies. Five months later, Mrs. Goodall went home and Jane carried on by herself.

She was soon able to move around the fifteen miles of forest by using animal trails and presently her team was joined by Dominic's wife and daughter and Dr. Leakey's boatman, Hasan, who took over the trips to town. She settled down into a routine which often involved rising at five thirty in order to watch the apes when they got out of their treetop beds, which were made by bending together leaves and branches and used only for one night.

One of Jane Goodall's most thrilling discoveries was the preparation and use of tools by the big apes. They would carefully strip a plant stem of twigs and plunge it into a termite's nest to draw it up covered with termites, which they enjoyed eating. Since man is defined as a toolmaker, the chimpanzees' behavior rather blurred the borderline between man and animal.

Goodall found that the bands varied and fluctuated in composition. The band would split up, old males roaming separately or groups of mothers and children sticking together. A plentiful food supply or a sexually receptive female brought a larger group together, sometimes numbering as many as forty. On the whole, they did not associate for defense as the baboons did.

The pecking order had a certain variety. While a male, named Goliath by the investigator, was the most dominant at the beginning of the study, it appeared that he kept his position through threat display waving of a broken branch, yelling, and beating the ground and charging. However, another chimpanzee, who started out in a low social position, discovered he could make a more impressive row by banging three empty kerosene cans, purloined from the camp, batting them around in front of him as he charged. Goliath was so impressed that he ran to submit, bowing to the ground, kissing, touching, and grooming him.

Interestingly enough, touching had great emotional importance to chimpanzees. When excited or afraid, they embraced and kissed each other on the neck. Friends also embraced when they met after an absence. Friendships between males paralleled those we have mentioned among the baboons. When one of two friendship pairs outranked the other, the person of lower status was able to dominate those dominated by his friend when the latter was present. The chimpanzee named Mr. Worzle was bullied by everybody except when he was accompanied by the high-ranking Mr. Leakey; then he threatened his superiors. When Mr. Leakey was away, he sank down to his usual humble station.

Goodall observed the treatment of the young, the relationship between brothers and sisters, and the gradual emergence into their sometimes hereditary place in the hierarchy. Females all were dominated by the males. A female in heat mated with numerous males; there seemed to be no pairing and the father of the infant was unknown. At first the females protected the infants from males and even other females, but when the little ones were a couple of years old, they played with their peers and sometimes with grown-ups.

Relationships between chimpanzees and baboons varied. At times they threatened each other, the baboons giving way when they met, but Goodall noted a friendship between a young female chimpanzee and a female baboon which lasted for months until the chimpanzee grew too heavy and strong to play with the smaller primate.

The first observation of meat eating occurred when a large male chimpanzee called Rudolph killed a small baboon by smashing it against the ground while three other males shrieked and embraced each other hysterically. Rudolph clung to his kill jealously and was pursued by the others, begging for a share (which they finally got), most of the day. While meat eating was an exception, it seemed to release a great deal of emotion. Their kills extended to young bushpig and bushbuck and the small *Colobus* monkey. They did not seem to hunt meat actively, but once the investigator saw three of them combine to stalk a young baboon.

Jane Goodall's grant had been supplemented by the National Geographic Society. She returned to London, obtained her PhD, and married Baron Hugo van Lawick, who had taken hundreds of photographs of the chimpanzees and also prepared a film on primate behavior. In 1965, the couple returned to the Gombe reserve and to the laboratory which had been kept going by an assistant in their absence. The value of continuing observation consisted in being able to watch relationships form as the young grew up and also to observe the effect of maturity and age upon individual positions in the social network. The study of great apes comes close to being anthropology, for, as Hugo van Lawick remarked, the chimpanzees were not exactly animals but somewhere between an animal and a human being. The scientific report on the work has not yet been published, but Jane Goodall's popular books describe it in intimate detail and are significantly entitled *My Friends the Wild Chimpanzees* and *In the Shadow of Man.*

The same friendly emotional identification pervades George B. Schaller's account of his gorilla studies, *The Year of the Gorilla,* which took place from 1960 to 1962 in the heart of the rain forest of the Congo, in and around Mount Mikeno, an area not too far from Goodall's chimpanzee observatory. Before Schaller, the gorilla had been given such a buildup for ferocity that no one had dared get close enough to study it. A few pioneers went on expeditions but got few results. Carl Akeley, the sculptor and naturalist who prepared the gorilla exhibit for the American Museum of Natural History, was so impressed by the animals that he urged the establishment of a sanctuary, which was eventually set up by the Belgian government and today includes the chain of Virunga volcanoes.

Schaller, fascinated by gorillas, wanted to study the mountain species, *Gorilla gorilla beringei,* which is on the verge of extinction. He talked with various people who knew the habitat of the species and was told that his project was impossible. Finally, the pioneer of primate study, R. C. Carpenter, remarked: "When I went to Asia to study the gibbon, everyone said it could not be done. But if you develop your own techniques and find the right place to work, I'm sure you will be successful."

Schaller persevered until John T. Emlen, professor of zoology at the University of Wisconsin, finally obtained a grant from the National Science Foundation for the projected study. He was to lead the expedition for six months and then Schaller and his wife were to continue the work for another eighteen months. They left New York in February, 1959.

Some months later, the two scientists climbed Mount Mikeno in the Virunga range until they reached the 10,000-foot level, where they found a rough hewn cabin, "the place of rest." In a corner of the meadow lay a flat stone with the inscription "Carl Akeley, Nov. 17, 1926." Akeley had died at the very beginning of his attempt to study gorillas. Now Schaller and Emlen had inherited his project. The beautiful meadow was surrounded by 50-foot forest trees. At one end there was a pond muddied by buffalo. They made their camp and prepared to look for gorillas the following morning.

For three days they had no luck. Then they both stumbled on a feeding group, a four-hundred-pound old male, a juvenile, three females, and in the fork of a tree a female with a newborn infant on her shoulders. The old male stood up and slapped his chest (described by chroniclers as the prelude to a ferocious charge). Since Schaller did nothing, the old man stopped roaring and lay down on the ground. Schaller was filled with excitement. For two hours he and Emlen watched the group while they fed unconcernedly. At their very first meeting with the animals, the myth of gorilla ferocity was exploded.

Later they discovered that it was necessary to learn to follow jungle trails, recognize the age of the spoor of the traveling groups, and sketch out their territory, for they were not too numerous. So far, they had seen twenty-two gorillas and they wondered about the average size of the band.

Later the party moved on to Kisoro where Walter Baumgartel, the owner of the Traveler's Rest Hotel, was an old friend of the gorilla and promised expert guides to help with their work. There, too, the investigators worked at an altitude of 10,000 feet, living in bamboo huts. Unfortunately, it was the rainy season which made the trails slippery, and at the same time low-hanging clouds obscured the view.

Little progress was made in finding gorillas until they encountered the Batwa, nomadic pygmies about five feet high or less, who were truly hunting people. Guided by their Batwa scouts, the two zoologists set up a camp deep in the forest, temporarily leaving their wives behind. When the scholars questioned the Batwa about gorilla habits, the forest people avoided answering. They warned the Europeans that if they called an animal by its name, they would never find it. No gorillas were seen at close range, although they found plenty of signs and occasionally glimpsed the animals in the distance.

Tramping the steaming forest in the rain, the two zoologists scouted various areas in Albert Park and mapped the range of the gorillas. They skirted the northern shore of Lake Tanganyika, and moved on to the Congo basin, considerably extending the habitat of the animals.

Six months after their arrival Schaller and his wife returned to Kabara and the Virunga mountains to set up camp on Mount Karisimbi in the depths of the rain forest. They had supplies of canned meat, fruits and vegetables, chicken and maize and a young African to do chores. They draped grass mats and native prints on the walls of the bamboo huts to keep the wind out and settled down to their detailed observation of gorilla life. Schaller soon located eleven gorilla bands numbering from eight to twenty-seven members. He learned how to follow their trails, located their nests, which were much like those of chimpanzees, and developed his techniques of study. Like Jane Goodall he made it a rule never to follow the gorillas, for that would seem like pursuit. When they discovered him, he schooled himself to assume an attitude of "humility" and he was careful not to stare at them when he met them face-to-face, for in the case of a number of intelligent animals, including gorillas, an unwavering stare is considered a threatening gesture.

The male leaders of the bands were always ready to roar and beat their chests, but after this threat gesture, when Schaller neither retreated nor made an aggressive movement, they lost interest and sometimes calmly went to sleep. Finally they became so used to him that the leaders did no more than grunt when he appeared. He noticed that one of the animals, when it approached him, shook its head from side to side. Since it seemed an appeasement gesture, he took to using it himself when he came upon the animals suddenly.

He observed two copulations during his year of study. In one case the female was the aggressor, clutching the male around the waist and pressing herself against him. The animal was not the dominant male, and although he retreated when the latter reacted slightly, he returned and completed the act undisturbed. On the whole, gorillas did not seem highly sexed:

> Since most females are either pregnant or lactating, the silver-backed male or males in the group may, on occasion, spend as much as a year without sexual intercourse, for they seem to make no overtures to the females unless they indicate their receptivity.

Schaller, too, felt that sex had little to do with group cohesion and stability. "Gorillas always gave me the impression that they stay together because they like and know each other." His observation of the peaceful acceptance of an outsider male by the group also tended to upset the Oedipal notions of Freud, which were even accepted by the anthropologist Carleton Coon, that the dominant male must drive out his young rivals.

The general conclusion on sex relations among primates as summed up by the expert Carpenter is rather suggestive:

> The females in natural groupings of primates are usually the aggressors and initiators of sexual responses. This active striving for copulation is, as is well known, motivated by the physiological process underlying estrus. During estrus, a female's capacity for copulation generally exceeds that of any one male, therefore, frequency of copulation during a typical male-female association is limited by the strength of the available males' sex drives.

If females were the sexual aggressors among man's early ancestors, and through the devious myths of male superiority which grew up in a male-dominated culture, males convinced themselves that they must be sexually aggressive and always prodigies of lust, it is easy to see how this, together with the passive role forced upon the females, could have created a situation contrary to normal primate biology.

To return to the gorillas, there was, of course, a pecking order, which seemed stable among males, somewhat less so among the females. The offspring did not, as had been often asserted by armchair naturalists, abandon their mothers as soon as they were weaned. Weaning generally took place from the age of eight to eighteen months, but the juveniles often remained in friendly association with their mothers up to four years of age.

The Schallers came to love their mountain fastness, despite its lack of modern technological comforts. They enjoyed the antics of a pair of white-necked ravens who speedily became tame and even jumped on the kitchen table in search of a handout. Although rain and humidity were a trial, they never tired of the cloud-topped mountain vistas and the green darkness of the forest.

The store of observations on gorillas grew. Schaller discovered that unlike monkeys or chimpanzees, they had no manual curiosity. A tin can, left by the trail, was never picked up. Except for one lone black-backed male, they never attempted to inspect the cabin. The exception hung around the end of the meadow and occasionally roared when he saw human activity. In comparison with the emotional chimpanzees, gorillas turned out to be rather lazy, sluggish creatures whose greatest joy was to lie on their backs in the sun.

Just as Jane Goodall's animals had been humanized by naming them Huxley and Leakey after famous scientists, or more descriptively by such names as David Graybeard, so Schaller christened his gorilla friends whom he could easily tell apart. There was Big Daddy, for instance, the boss, and Mr. Dillon and Mrs. Patch and Callosity Jane. Thus, when Schaller came home chilled from the rain at four o'clock and his wife made him hot cocoa, they could discuss the activities of various individuals he had observed during the day.

Gorillas ate about twenty-nine plants, most of which Schaller discovered had a rather bitter taste. He had no way of learning why some types of vegetation were enjoyed and others not. He found, in contrast to Jane Goodall's observation of the chimpanzee to which he referred, that there was absolutely no evidence of meat eating. Then, too, the mothers never handed food to their young as chimpanzees have been observed to do.

Schaller was convinced that the gorilla bands shared their ranges and did not defend specific territories. The relation between bands varied. Some avoided each other; some fed peacefully side by side. In one case there was antagonism; the dominant males bluffed each other, put on exhibitions of roaring, and threw weeds into the air until finally one group retreated. Meanwhile the other members of the bands had shown no antagonism or interest in the contest.

The display activity of the male gorilla was used as a threat but also as a kind of emotional discharge. The complete sequence included a series of soft hoots that were built up to a crescendo (Schaller compared it to drumbeat climax); then the displaying male would sometimes pluck a leaf and place it between his lips. He then rose on his hind legs and ripped off vegetation (displaced feeding, Schaller suggested) and finally he slapped his chest, or belly, or a tree trunk and generally ended by slapping the ground and roaring. Schaller pointed out that orangutan display was comparable to that of the gorilla and that of the chimpanzee parallel, although less organized. He then suggested that the leaping, shouting and stamping crowd at a baseball game was displaying similar tactics in releasing emotional tension. In the area of threat he might have compared gorilla behavior to the war dance of the Maoris, which included shouting, stamping and face-making, both as a threat and emotional self-stimulus before beginning a battle.

Schaller, commenting on a statement of Ashley Montagu's that man has no instincts, disagreed sharply. The smile which can be evoked even in blind babies is well established, as was the tendency to throw and beat things when excited and to crouch down in submission when threatened, the last two activities being common to both men and apes.

The simple life that the gorilla leads, in which there is no competition for food and no pressure to do more than eat it, Schaller suggests, accounts for the animal's failure to evolve further toward a more human type of intelligence. Without enemies except for man, he remains content in his jungle fastness. But man is steadily shredding away the forest and endangering all beautiful and interesting creatures that inhabit it.

This Schaller was to experience before he left because in January, 1960, after the agitation led by Patrice Lumumba, the Congo was granted independence from Belgium. This was to take place in June, but the Schallers decided to leave before that date, for it appeared anything might happen. Europeans began to leave; refugees streamed over the border.

After a last return to his camp in the jungle, where Schaller found the newly appointed native guards battling invading Watusi, who had taken advantage of the anarchical interregnum to move their cattle into the park, he left full of misgivings for the wild creatures under the new, disorganized regime.

Returning three years later, he was happy to discover that Albert Park had been preserved. But elsewhere, with the coming of independence in Africa wildlife did not fare so well.

And this brings us to the last chapter in the history of our fellow inhabitants of the planet. What is their future, in the light of man's growing technology and social irresponsibility? This question also pertains to man's own future as well.

30

Fellow Citizens of the Planet

"THERE is no risk in making the flat statement that in a world devoid of other living creatures, man himself would die. This fact—call it a theory if you will—is far more provable than the accepted theory of relativity. Involved in it is, in truth, another kind if principle of relativity—the relatedness of all living things."

So wrote Fairfield Osborn, director of the New York Zoological Society, in 1948. This is one of the earliest serious warnings issued to the human race, warnings which have become ever more imperative.

Osborn was writing of ecology, the relationship between various types of life and the world environment. The word was coined by Ernst Haeckel in 1869 from the Greek word for "home," hence environment. Today, ecology is on everyone's lips and it is equated with survival. In the past, it has been treated as an academic subject. Animal ecology, for example, has dealt with such elements in the relationship of the organism to its environment as the effect of temperature or light, diet and metabolism, hibernation, distribution versus types of habitat, the checks and balances between predators and prey, the sharing of food and space between members of the same and different species, the integration of more than one species into a community.

Now ecology has assumed a wider meaning. The crisis belatedly recognized as worldwide involves the chemical balance needful to sustain life of any kind. Man is forced to survey his own role in ecological change.

The fact that man is the most dangerous animal on the planet is well known, dangerous to all living things—and now dangerous to himself. Self-awareness and insight began with conservation after the havoc that improved weapons, improved transportation and a selfish delight in killing had inflicted upon other live species. Since the beginning of the Christian Era 250 species have completely vanished, two-thirds during the past fifty years. Today there are another 250 which are on the verge of extinction.

The story of the mindless slaughter of the American bison is well known. In 1913, there was one solitary female passenger pigeon left in the Cincinnati Zoo, the sole survivor of immense flocks which, when they alighted, were so numerous that they bent the branches of trees. Between 1850 and 1910 the Eskimo curlew, the Carolina parakeet and the great auk were exterminated by man.

It is hard to realize today that such songbirds as the robin, the meadowlark, the goldfinch and the Baltimore oriole were slaughtered for the table and were in grave danger of becoming extinct by the early twentieth century.

In the course of this book we have had occasion to mention the reckless extermination of seabirds, seals and whales.

It is true that something has been done. The rise of the conservation movement in the twentieth century was motivated partly by a sense of ethics, partly by esthetics. It has been pointed out that the extermination of a species can be equated with the burning of the "Mona Lisa." Both contribute to human culture, both add interest and emotional color to life. And from an objective, ethical point of view animals also have rights as fellow citizens of the planet.

These are, of course, fairly sophisticated points of view, but they have carried some weight. Gradually in the privileged areas of the world laws have been passed, wildlife parks have been set up, game wardens have been given police power, young people have been educated in conservation ideas. Even a few politicians have seen the light.

Such measures have helped check plain ordinary killing.

Osborn's book, *Our Plundered Planet,* however, sounded a new note. He was primarily concerned with what is happening to the earth's topsoil, the relatively thin layer of humus which contains organic elements, particularly bacteria and protozoa. When it is lost, there is no more vegetation, and without vegetation water evaporates and runs off, and of course in the ensuing desert no animal life can subsist.

We have paid little heed to the lessons of antiquity. Once the Tigris-Euphrates Valley was a thickly populated area, full of vegetation and animal life. Today, mounds of sun-dried brick emerge from the desert sand. In classical times Greece and other lands in the Mediterranean area were rich in forests and wildlife and highly productive. Today, brush covers the former forest land and the exhausted soil produces one-fifth of a normal yield.

Coming nearer home, in 1934 great dust storms dramatized the fact that the once fertile plains of five states—Kansas, Texas, Oklahoma, Colorado and New Mexico—had become a desolate dust bowl.

The main causes of erosion are well known. First, there is deforestation. The shade of the forest and the cohesive power of the roots plus all the minor vegetation and humus from dead leaves preserve the life-giving moisture. Man has blindly destroyed the wooded environment and still does. Even our national parks are under constant attack from the lumber interests. Land that has been deforested loses its topsoil in heavy rainfalls which wash the organic materials into the rivers where the powerful currents of the waterways bear them out to sea.

The second form of destruction arises from the misuse of animals, the overgrazing of grasslands by cattle. The pressure of profit-taking results in the maintenance of the largest number of animals possible on a limited range. As a result the grass and bushes are so closely cropped that the ground can be gullied by rains and eventually eroded. George Schaller reacted to the herdsmen who were infiltrating and destroying the jungle, and indeed tropical areas such as Africa are suffering acutely from the too-rapid spread of pastoralism.

Finally, in the past, before the use of fertilizers and crop rotation, the slash and burn technique meant that the soil was cleared by burning trees and cutting bushes, the land being exploited until it was exhausted. It was then abandoned and a new area denuded in the same way. This exhaustion of the productive soil has even been suggested as the reason for the desertion of the cities of the old Mayan Empire. It has been estimated that the processes of nature require five hundred years to build an inch of topsoil.

The fact that Asia, India and China have suffered for decades from famines and are still caught in the squeeze between unchecked population increase and spoliation of the land does not seem to have been taken seriously until recent years.

Of course, as man uses up or destroys the productive surface of the earth, organic life is steadily diminished. The jungle dwellers cannot exist without the jungle, and the savannas, overgrazed by cattle, cannot support wild antelopes or buffalo. It has been shown that in east African areas which have no agricultural value, careful cropping of wild game yields more food value per acre than that which could be garnered from domestic cattle supported by the same terrain.

In 1948, Fairfield Osborn was actually raising his voice against the inherited ethic, which we have pointed out stemmed directly from the Christian heritage (with the honorable exception of Saint Francis), the belief that all animals and natural products of the earth were put there expressly for man's use and it was his privilege to do as he liked with them. Today, it seems almost unbelievable that no one in the past realized that natural resources are finite and that so little heed was paid to the possible results of blind interference with the stability and balance produced automatically by the forces of evolution. Very often, for instance, a species was introduced into a new area with some special aim in mind, nearly always with harmful effects. A good example of such tampering took place in Tahiti. Vanilla does not pollinate itself and so the vanilla growers decided to introduce wasps to perform this task. In the words of Philip Kingsland Crowe:

> The wasps did not comply and spent their time stinging people. Myna birds from India were then brought in to devour the wasps but soon developed a liking for less prickly bugs. Finally hawks were imported to keep down the myna birds, but the hawks found chickens easier to kill and better to eat. Result is that there are now lots of wasps, lots of myna birds, lots of hawks and almost no indigenous birds. Rats and cats, brought by the early whalers, undoubtedly started the decline of the island's bird population but the few survivors have certainly been helped toward the status of the dodo by the voracious newcomers.

As we have said, Osborn sounded an early grave warning that men were ignoring the effect of their culture upon world ecology. A second shock to civilized complacency was the controversy over fallout from atomic bomb tests, which started in 1953. Despite loud propagandistic disclaimers from government scientists, such pioneer ecologists as Barry Commoner proved that strontium 90 was being rained all over the world (from both Soviet and American tests), drifting down from clouds of fallout to turn up via grass and cow's milk in babies' teeth. The resulting dismay launched a movement that resulted in greater public awareness of what man was doing to the environment. In this case the agitation became so intense that it even penetrated government circles and resulted in the 1963 test ban treaty which drove nuclear experiments underground.

But only a year before this temporary victory for humanity, *Silent Spring,* a book by a quiet, gentle nature student, Rachel Carson, who had written beautifully about the sea, rocked modern man's complacency once again.

Back in 1948, Osborn had written in passing:

> More recently a powerful chemical known as DDT seems the cure-all. Some of the initial experiments with this insect killer have been withering to bird life as a result of birds eating the insects which have been impregnated with the chemical. The careless use of DDT can result in destroying fishes, frogs and toads, all of which live on insects. This new chemical is deadly to many kinds of insects—no doubt of that. But what of the ultimate and net result to the life scheme of the earth?

DDT, which was developed in World War II, was sprayed from airplanes by the Department of Agriculture and individual farmers, carried to far corners of the earth to destroy tropical insect pests, and brought into houses in compression bombs to eliminate cockroaches and ants.

Rachel Carson's book, heavily documented and ably written, answered the question Osborn had raised fifteen years before. She cited the widespread destruction of songbirds and small animals, the poisoning of fish, the death and sterility of livestock as a result of thoughtless spraying programs which scarcely ever eradicated the pest.

As it goes up the food chain, the concentration of the poison in animal bodies increases; it particularly attacks the liver, the nervous system and the sex organs. The vulnerability of birds is especially evident in the status of the brown pelican, the peregrine falcon, the osprey and the bald eagle, all of which prey on marine life. By the time the DDT reaches fish it is already a concentrate. It has been established that the shells of eggs laid by these birds are so thin that they no longer survive the brooding period.

In cow's milk and other foods DDT enters the bodies of human beings until all of us who have been around since 1945 are affected. Although at low levels it is not lethal, we still do not know what effect it will have on human beings in the long run. We do know what it does to wild animals. Paul Ehrlich points out that the load of DDT stored in the tissue of every child born since 1945 may seriously curtail its life-span.

This miserable poison is persistent. Body chemistry fails to break it down: instead it is stored in the animal's fat. It may play a role in causing cancer and mutations. It is now so widespread throughout the biosphere (the inhabited area of the earth) that it has even been detected in the bodies of caribou, Arctic Eskimo and penguins!

Although the facts are well known, thanks to Rachel Carson and others since, the difficulty encountered in combating this ecological evil is alarming. Ms. Carson, who bravely finished her book when she was dying of cancer, was attacked and vilified by the pesticide interests and their allies. Even the Department of Agriculture, which should be concerned with protecting the population, played the same role as the government scientists who defended the bomb tests.

Although at last, in 1969, the federal government put an end to the use of DDT in the United States, it still allows the chemical to be sold elsewhere, and other poisons, heralded as but not proved to be non-toxic to animals, are still in use. If the countries with advanced technologies are just waking up to the perils of pesticides, the poison pedlers still have plenty of customers in less-developed nations which still regard pesticides as one of the blessings of modern civilization.

During the last decade biologists have realized that the ecological problem is not one of winning a few battles but that man's technological culture is upsetting the natural balance of the whole biosphere.

The four classical elements of earth, air, fire and water so important in Greek thought, engage our serious attention once again. Prometheus, when he conferred combustion on the human race, may have granted a two-edged benefit, for man has used fire to pollute the other three. Not content with using it to destroy woodlands, it has become one of his chief sources of energy as he burns various fuels. But it also produces various toxic wastes. Interestingly enough, in 1306 a citizen of London was tried and executed for burning coal within the city. It did no good, for today our cities are blanketed with smog.

Simple village communities do little damage to the environment. With the evolution of large cities, men huddled together like ants and began concentrating their wastes, dumping them indiscriminately, excreting them into the air, or pouring them into streams and bodies of water which they also drank. Even this, despite the ensuing epidemics, did not create a crisis situation until the Industrial Revolution sparked advances in applied chemistry which produced factory wastes containing chemicals in forms and quantities which never occur in nature.

Another problem that is leading to a real emergency is human reproduction. Medicine has cut down mortality and lengthened life while technology has made it possible for weaker individuals to survive. As the animal ecologists point out, in nature the predators act as a natural check on the overbreeding of their prey, while diminishing prey controls the number of predators. Man, except for unpredictable wars, is not subject to natural checks upon his reproductive efforts. This is a case in which his vaunted intelligence must be used if there is eventually to be even standing room on the planet. So far as population control goes, he is a prey to apathy, ignorance and myth, often indulging in the false optimism that science will take care of everything.

We have spoken of natural cycles. Back in 1832, Boussingault discovered and publicized the nitrogen cycle. Bacteria and algae take nitrogen from the atmosphere and convert it into ammonia which is toxic and would poison us all. Other organisms in soil and water turn ammonia into nitrate—green plants absorb both nitrate and ammonia and use the nitrate to build plant protein. Microorganisms and animals get nitrogen for their proteins directly or indirectly from proteins of plants. When animals and plants die, microorganisms break down protoplasm and release ammonia. Still more microorganisms convert some of this to molecular nitrogen, which returns again to the atmosphere.

Now we are adding vast quantities of chemicals to earth, air and water without knowing or caring whether one of these or some unforeseen combination of them might disrupt one of the steps in the nitrogen cycle and put an end to all life.

Another question of balance occurs in the case of the oxygen-carbon dioxide cycle. The only reason that oxygen is available in our atmosphere is that plants keep putting it there. Plants take in carbon dioxide and give off oxygen. Animals and microorganisms take in oxygen and give off carbon dioxide. Aside from the fact that we seem intent on blacktopping the earth and removing every speck of green, we are also endangering the 70 percent of the free oxygen in the air put there by planktonic diatoms (tiny plants in the sea), for pollution in waterways finally reaches the sea to destroy the life within it.

In addition, such mechanical sources as furnaces, factories and automobiles pour carbon dioxide into the air. The amount in the earth's atmosphere has increased 25 percent during the past hundred years and will increase another 25 percent by the year 2000. One possible result is an overheating of the earth because carbon dioxide tends to trap heat. On the other hand, smoke poured into the atmosphere has increased the turbidity of the air because small particles flying overhead not only shade the earth but produce cloud formations (aided by the contrails of jet planes). This would reduce the amount of solar energy reaching the earth. Whatever the precise effect of combustion, the natural state of affairs is once more being disturbed and we cannot be sure of the result.

So far we have been dealing mainly with damage to earth and air. Water damage is even more obvious. By now the death of Lake Erie has been well publicized. Practically every aspect of our technology contributes to water pollution. An exhibit of flounders in the American Museum of Natural History shows the normal fish and those taken from New York Harbor, appalling specimens with fins eaten away, peeling scales and unsightly blotches. The Cuyahoga River at Cleveland, oily and chocolate brown from industrial waste, is a fire hazard and actually burst into flames in 1969. In this sink of corruption even leeches and sludge worms cannot live.

The number one hazard is raw sewage, and it seems incredible that this is still poured into rivers, lakes and harbors. Aside from disease germs introduced into the water, bacteria feeding on the pollutants multiply and consume the oxygen. Fish can no longer survive. Industrial wastes such as those from paper factories also have the property of killing fish, and even the heated water returned to the river or lake by factories using it in cooling processes (and this includes atomic energy plants) can so raise the temperature as to destroy fish and aquatic life.

One-crop farmers, to compensate for the exhaustion of the soil, use chemical fertilizers which pour nitrates and phosphorus into streams in the runoff of heavy rains. This stimulates the growth of algae until the waterway or lake becomes choked and animal life dies out. Phosphorus from detergents used in the everpresent washing machine does the same thing when sewage systems overflow or enter the watertable.

Present-day water pollution is therefore highly complex, for the life-giving liquid may now contain bacteria and viruses from sewage, acid from mine drainage, poison from pesticides and weed killers, and chemicals so new that their effort upon organic life is not known.

This alarming environmental destruction is not limited to the United States. The Swiss lakes of Geneva, Constance and Neuchâtel are so murky with pollution that the trout and perch are nearly gone. The Rhine is known as Europe's sewer and even the hardy eel has difficulty surviving. The famous Aswan Dam, which was going to do so much for Egypt's economy, is filling with weeds which may evaporate so much water that the reservoir can no longer drive generators. The dam has also prevented the nutriment-filled silt from moving down into the Mediterranean, with the result that the sardine catch has declined from 18,000 tons to 500 tons. The disastrous effects of leakage from oil tankers and offshore drilling has been painfully dramatized by dying seals and seabirds. Even the Russians who do not need to cater to private profit have built a polluting paperplant on the shore of beautiful Lake Baikal.

In the face of these threats to our environment a group of biologists and zoologists have awakened to their social responsibilities. Their number includes Paul Ehrlich, Eugene P. Odum, LaMont C. Cole, René Dubos and Barry Commoner. Some, however, are highly pessimistic as to whether the world's problems will be solved in time.

Barry Commoner, born in 1917, and dubbed by *Time* magazine the Paul Revere of the movement, is moderately hopeful. His father was a Russian emigrant tailor who lived in Brooklyn, where young Commoner spent his early years mingling with street gangs. By the time he entered high school he was enamored of biology and spent his spare time in Prospect Park seeking specimens to study under his microscope. He put himself through Columbia, graduating with honors in zoology in 1937. A liberal, he supported the Spanish loyalists and other causes of the period. Three years at Harvard earned him his doctorate in biology. After serving in the Navy during World War II, he went to Washington University in St. Louis where he headed the Biology Department. His research took him into the study of viruses and the field of biochemistry and biophysics.

As we have already noted, he became an activist with the fallout crisis in 1953. He was quick to see that fallout was only a part of the vast conspiracy against nature launched by technology. He investigated the death of Lake Erie and began to publicize his conviction that pollution could spell the end of life on earth. Thus he joins the honorable ranks of the Huxleys and Ernst Haeckel in bringing information to the public. In an attempt to unite physical and social scientists into a cooperative whole focused on the environment, he founded Washington University's Center for the Biology of Natural Systems in 1966, the first of its kind.

Energetic and sometimes "abrasive," an able speaker and brilliant teacher, Commoner is an advocate of relevance. His center undertook a study of the ecology of ghetto rats which helped the St. Louis Health Department more efficiently rid the city of rats. Commoner said: "We could just as well do a study of the fence lizard, but that wouldn't be relevant to human problems." Such scientists as Commoner, who travel, lecture and write diligently, are challenging the old attitude of research for research's sake which has dominated academic science (and alienated the young). Activists like Commoner are creating an attitude that in a threatened world there are such things as priorities.

While the struggle to deal with the population explosion and to save the environment goes on, efforts to protect game animals continue with varying success. As early as 1943, Julian Huxley was in Africa making a survey of the condition of wildlife on that continent and pointing out that many species were in danger.

Huxley has continued to lead a life of public service and statesmanship. From 1934 through World War II he was secretary of the London Zoological Society. He made every effort to publicize the zoo and turn it into an educational institution, giving it a new and more popular image, much to the disgust of the stuffy, academic fellows. In 1945, he was chairman of a Special Committee on Wildlife Conservation. In 1946, he was made head of the newly formed UNESCO and served as director general for two years. He outlined a humanist program for the organization which annoyed both religionists and Marxists. Like his grandfather he has attempted to forge a philosophy of humanism, expressed in his book, *Religion Without Revelation,* which is based on an evolutionary point of view. His philosophy calls for population control and a partnership of man with nature. His 1960 survey of Africa, after the game had suffered severely from unchecked depredations during the formation of the independent nations, resulted in a report to UNESCO in which he wrote:

> With the alarming increase in organized poaching [in parks], the equally alarming increase in population, the new methods in controlling the tsetse fly and the diseases of livestock, the fashionable urge toward technological and agricultural development, the spread of money values among the African governments, the situation is critical. The future of African wildlife is bound up with that of the conservation of natural resources. . . . The ecological problem is fundamentally one of balancing resources against human needs, both in the short and the long term.

A survey by Philip Kingsland Crowe of various undeveloped areas of the map, which was undertaken from 1963 to 1966, brings out the fact that weak governments and uneducated populations have done little more than, in some cases, put laws on their books for which they have no machinery of enforcement. In Egypt and other Arab countries practically nothing is done to preserve endangered species. In Latin America there are some spotty efforts, and a few nature preserves are being established.

George Schaller, in a 1967 study of the deer and tiger in India, points out that originally the major mammals were decimated by sporting civil servants and native rajas. After independence a new orgy of killing went on because controls were rejected as "colonial." Other enemies of Indian wildlife are the countless, useless but sacred cattle which are allowed to graze everywhere. The cheetah is extinct, the Indian rhinoceros is almost gone, wild buffalo are scarce, and the blackbuck is an endangered species.

Even more sophisticated governments have a poor record when it comes to the largest of all mammals, the whale. Only Russia, Norway and Japan are still in the whaling business, but the world population of blue whales has dropped in thirty years from 100,000 to 1,000. Other species are also rapidly declining because of that lethal weapon, the bomb lance. The annual take in the Arctic is about 15,000 blue whales. To save the species it must be cut to 2,000. But international agreements are necessary, and so far, none has been implemented. A United Nations proposal for an International Whaling Authority to sét quotas, administered by the UN and involving inspection, seems to be the answer. This proposal was brought up in 1966, but it has still not been accepted.

It is a pity to end with a discouraging picture of planetary zoology, but facts have to be faced before conditions grow better. Man's study of his animal relatives has proved intellectually stimulating and exciting as new attitudes and insights have solved problem after problem and laid bare the fascinating patterns and mechanisms of nature of which we are a part. We now know that earth is an island, to use Margaret Mead's image, and we are all, from protozoan to President, marooned on it together. As Osborn stated, over twenty years ago, we cannot live without animal life and animal life is dependent on us. We have now been very clearly warned that the resources of our global island are being both used and destroyed at a dangerous rate. Whether the science of zoology (and anthropology) has a future depends upon how rapidly we move into action.

We have proved that a concerted effort involving science, government and private industry could put a man on the moon. The cost was astronomical and the human relevance small. Surely when the stakes are so overwhelming we can summon up the self-discipline necessary to make the world safe for life.

Bibliography

GENERAL

ASIMOV, ISAAC, *A Short History of Biology*. New York, 1964.

BURCKHARDT, GABRIEL, *Geschichte der Zoologie*. Berlin, 1921.

GABRIEL, MORDECAI, and FOGEL, SEYMOUR, eds., *Great Experiments in Biology*. Englewood Cliffs, 1955.

HALL, RUPERT A., *The Scientific Revolution, 1500-1800*. London, 1962.

KELLER, GOTTFRIED, *Daten zur Geschichte der Zoologie*. Bonn, 1949.

NORDENSKÖLD, ERIC, *A History of Biology*, trans. by Leonard, Buchnall, Eyre. New York, 1928.

SINGER, CHARLES, *A History of Biology*. New York, 1950.

Chapter 1
The Invention of Science

ARISTOTLE, *Works: The Oxford Translation*, J. A. Smith and W. D. Ross, eds., 11 vols. Oxford, 1962.

CORNFORD, FRANCIS, *Before and After Socrates*. New York, 1962.

EDELSTEIN, LUDWIG, "Motives and Incentives for Science in Antiquity," in Alistair C. Crombie, ed., *Scientific Change, a Symposium on the History of Science, University of Oxford, 9-15, July, 1961*. New York, 1963.

Chapter 2
A World of Wonder

LOISEL, LOUIS, *Histoire des Ménageries*, 3 vols. Paris, 1912.

LUCRETIUS, *De Rerum Natura*, trans. by Rolfe Humphries. Bloomington, Indiana, 1969.

PLINY THE ELDER, *The History of the World*, commonly called *The Natural History of G. Plinius Secundus*, trans. by Philemon Holland, selected and introduced by Paul Turner. New York, 1963.

———, *Natural History*, with an English translation in 10 volumes by H. Rackham. Cambridge, 1956.

Chapter 3
Emperor and Two Saints

AIKEN, PAULINE, "The Animal History of Albertus Magnus," *Speculum,* XXII. Cambridge, 1947.

ALBERTUS MAGNUS, *De animalibus,* H. Stadler, ed., *Beiträge z. Geschichte des Mittelaltertums,* XV-XVI, 2 vols. Munich, 1916-20.

Bestiary, The, trans. from a Latin bestiary of the twelfth century by T. H. White. New York, 1960.

EGAN, MAURICE FRANCIS, *Everybody's St. Francis of Assisi.* New York, 1912.

FRANCIS, SAINT, *The Little Flowers of Saint Francis,* trans. and edited by Cardinal Manning. Boston, 1915.

FRIEDRICH II, *The Art of Falconry,* trans. and edited by Casey A. Wood. London, 1943.

KANTOROWICZ, ERNST, *Frederick the Second 1194-1250,* trans. by E. O. Lorimer. New York, 1957.

Penguin Book of Italian Verse, George K. Kay, ed. Harmondsworth, 1958.

THEORIDES, JEAN, *La zoologie au moyen âge. La conférence faite au Palais de la Découverte, Série D, Histoire des Sciences,* No. 55. Paris, 1958.

WHITE, LYNN, JR., "The Historical Roots of Our Ecological Crisis." *Science,* Vol. CLV, No. 3767, pp. 1203-7.

Chapter 4
The Assault on the Establishment

ANDERSON, F. H., *The Philosophy of Francis Bacon.* Chicago, 1948.

BACON, SIR FRANCIS, *Essays,* trans. from *De augmentis scientiarum,* by William Willymott, 2 vols. London, 1742.

CROWTHER, J. G., *Francis Bacon, The First Statesman of Science.* London, 1960.

TILLYARD, E. M. W., *The Elizabethan World Picture.* New York, n. d.

WHITE, LYNN, JR., "What Accelerated Technical Progress in the Western Middle Ages," in Alistair C. Crombie, ed., *Scientific Change, a Symposium on the History of Science, University of Oxford, 9-15, July, 1961.* New York 1963.

Chapter 5
Doctors and Picture Books

ALDROVANDI, ULISSE, *The Ornithology of Ulisse Aldrovandi,* Vol. II, Book XIV, trans. L. R. Lind. Norman, Oklahoma, 1966.

———, *La vita, scritta per lui medisimo.* Bologna, 1907.

BELON, PIERRE, *L'Histoire de la nature des oyseaux avec leurs descriptions et naïfs pourtraicts.* Paris, 1555.

———, *La Nature et diversité des poissons.* Paris, 1555.

Crombie, A. C., "Cybo d'Hyre, a 14th-century Zoological Artist." *Endeavour,* Vol. XL, No. 44. London, 1952.

Delaunay, Pierre, *L'aventureuse existence de Pierre Belon.* Paris, 1926.

Gesner, Konrad von, *Historia animalium.* Zurich, 1554.

———, *Thierbuch.* Zurich, 1563.

Gudger, E. W., "The Five Great Naturalists of the 16th Century." *Isis,* Vol. XXIII. Brussels, 1936.

Ley, Willy, *Konrad Gesner, Leben und Werke.* Munich, 1929.

Rondelet, Guillaume, *Libride piscibus marinis.* Lyons, 1554.

White, Lynn, Jr., "Natural Science and Naturalistic Art in the Middle Ages." *American Historical Review,* Vol. LII, No. 3 (1947), pp. 21-35.

Chapter 6
Inside the Animal Body

Chauvois, Louis, *William Harvey.* New York, 1957.

Harvey, William, *Movement of the Heart and Blood in Animals,* trans. by Kenneth J. Franklin. Oxford, 1957.

Morley, Henry, *Clement Marot & Other Studies.* London, 1870.

O'Malley, Charles D., *Andreas Vesalius of Brussels.* Los Angeles, 1964.

Vesalius, Andreas, The last chapter of the *De humani corporis fabrica,* trans. by Benjamin Farrington. Reprint, transactions of the Royal Society of South Africa, Vol. XX, Pt. I. Johannesburg, 1831.

Chapter 7
The Animal as Machine

Buchdahl, Gerd, "Descartes' Anticipation of a Logic of Scientific Discovery." *Scientific Change,* a Symposium on the History of Science, University of Oxford, 9-15, July, 1961. New York, 1963.

Keeling, S. V., *Descartes.* Oxford, 1968.

Chapter 8
Who Sees What No One Saw . . .

Adelmann, Howard B., *Marcello Malpighi and the Evolution of Embryology,* 5 vols. Ithaca, 1966.

Dobell, Clifford, *Antony van Leeuwenhoek and His "Little Animals."* London, 1932.

Leeuwenhoek, Antony van, *Brieven.* Amsterdam, 1939.

Meyer, A. W., "Leeuwenhoek as Experimental Zoologist." *Osiris,* Vol. III. Bruges, 1937.

Redi, Francesco, *De animaculis vivis quae in corporibus animalium vivorum reperiuntur.* Florence, 1708.

———, *Experiments on the Generation of Insects* (1688), trans. by Mab Bigelow. Chicago, 1908.
SCHIERBEEK, A., *Antoni van Leeuwenhoek, zyn leven en vornammste ondekkingen*. Den Haag, 1963.
———, *Jan Swammerdam*. Lochem, 1947.
SWAMMERDAM, JAN, *The Book of Nature and the History of Insects*, trans. by Thomas Floyd. London, 1758.
VIVIANI, HUGO, *Vita, opera, iconographia, etc. Francesco Redi*. Arezzo, 1924.

Chapter 9
The Book of Nature

LOCKSLEY, R. M., *Gilbert White*. London, 1954.
RAVEN, CHARLES E., *John Ray, Naturalist*. Cambridge, 1942.
RAY, JOHN, *Further Letters*, Robert Gunther, ed., Royal Society Publications, No. 114. London, 1928.
———, *Synopsis methodica animalium quadrupedum*. London, 1693.
———, *Three Physio-Theological Discourses*. London, 1713.
WHITE, GILBERT, *The Natural History of Selborne*, James Fisher, ed. London, 1960.
WILLUGHBY, FRANCIS, *The Ornithology*, John Ray, ed. London, 1626.

Chapter 10
Method Is the Soul of Science

GOURLIE, NORAH, *The Prince of Botanists, Carl von Linné*. London, 1953.
LINNAEUS, CARL, *Systema Naturae*, Tome I, Facsimile 10th ed. London, 1939.
———, *Juvenile Works*. Stockholm, 1888.

Chapter 11
Gentleman Amateur

BANKS, SIR JOSEPH, *The Endeavour Journal*, J. C. Beaglehole, ed. Sydney, 1963.
CAMERON, HECTOR CHARLES, *Sir Joseph Banks*. London, 1952.

Chapter 12
Giraffe in the Hallway

CAMPER, PETRUS, *Oeuvres*, 3 vols. Paris, 1805.
FOOT, JESSE, *The Life of John Hunter*. London, 1794.
GRAY, ERNEST A., *Portrait of a Surgeon*. London, 1952.

Chapter 13
The Elegant Popularized

BIREMBAUT, A., *et al., Réaumur, la vie et l'oeuvre.* Paris, 1962.
BOURDIER, FRANK, *et al., Buffon.* Paris, 1952.
BRUNET, PIERRE, *Maupertuis.* Paris, 1929.
BUFFON, GEORGES, *La Nature, l'homme, les animaux,* Roger Hein, ed. Paris, 1957.
———, *Oeuvres complètes.* Paris, 1778.
RÉAUMUR, RENÉ ANTOINE, *Mémoires pour servir à l'histoire naturelle des insectes.* Paris, 1734.
———, *The Natural History of Ants,* tr. and annotated by William Morris Wheeler. New York, 1926.

Chapter 14
The Great Debate

BEER, RUDIGER ROBERT, *Der Grosse Haller.* Sachingen, 1947.
BONNET, CHARLES, *Contemplation de la nature.* Hamburg, 1732.
COLE, FRANCIS JOSEPH, *Early Theories of Sexual Generation.* Oxford, 1930.
GUMPERT, MARTIN, *Das Leben für die Idee.* Berlin, 1935.
HOCHDOERFER, MARGARETE, "The Conflict Between the Religious and Scientific Views of Albrecht von Haller." *University of Nebraska Studies in Language, Literature and Criticism,* No. 12. Lincoln, Nebraska, 1932.
LEMOINE, ALBERT, *Charles Bonnet.* Paris, 1850.
ROSTAND, JEAN, *Les origines de la biologie expérimentale. Les conférences du Palais de la Découverte, Série D,* No. 63. Paris, 1951.
SPALLANZANI, LAZZARO, *Dissertations Relative to the Natural History of Animals and Vegetables,* trans. by Thomas Beddoes. London, 1859.

Chapter 15
The Breath of Life

COHEN, BERNARD, *Lives in Science.* New York, 1957.
GABRIEL, MORDECAI L., and FOGEL, SEYMOUR, *Great Experiments in Biology.* Englewood Cliffs, 1959.
JAFFE, BERNARD, *Crucibles: The Story of Chemistry.* New York, 1936.

Chapter 16
Revolution and Evolution

CANNON, H. GRAHAM, *Lamarck and Modern Genetics.* London, 1959.
CÚVIER, BARON GEORGES, *A Discourse on the Revolution of the Surface of the Globe.* Philadelphia, 1831.

———, *Recherches sur les ossements de quadrupèdes,* Paris, 1821.
———, *Le règne animal.* Paris, 1817.
FLOURENS, M., "Memoir of Geoffroy Saint-Hilaire." *Smithsonian Report.* Washington, 1892.
GEOFFROY SAINT-HILAIRE, ÉTIENNE, *Principes de philosophie zoologique.* Paris, 1830.
GEOFFROY SAINT-HILAIRE, ISIDORE, *Vie, traveaux, et doctrine scientifique de Geoffroy Saint-Hilaire.* Paris, 1847.
LAMARCK, JEAN BAPTISTE DE, *Histoire naturelle des animaux sans vertèbres.* Paris, 1815.
———, *Hydrogeology,* trans. by Albert Carozzi. Urbana, Illinois, 1964.
———, *Recherches sur l'organisation des corps vivants.* Paris, 1802.
LANDRIEU, MARCEL, *Lamarck, le fondateur du transformisme.* Paris, 1909.
LEE, MRS. R. (BOWDITCH), *Memoirs of Baron Cuvier.* New York, 1837.
PEATIE, DONALD CULROSS, *Green Laurels.* New York, 1936.
TSCHULOK, S., *Lamarck.* Zurich, 1937.

Chapter 17
Contemplation of the Universe

ANCELET-HUSTACHE, JEANNE, *Goethe,* trans. by Cecily Hastings. New York, 1960.
BRAUNING-OKTAVIO, HERMANN, *Oken und Goethe im Lichte neuer Quellen.* Weimar, 1959.
FAIRLEY, BAKER, *A Study of Goethe.* London, 1961.
HUMBOLDT, ALEXANDER VON, *Kosmos,* 3 vols., trans. by E. E. Otte. New York, 1851.
———, BONPLAND, AIMÉ, *Personal Narrative of Travels to the Equinoctial Regions of America During the Years 1799-1803,* trans. by Thomasina Ross. New York, 1881.
KLENKE, HERMANN, *Alexandre Humboldt,* trans. by Juliette Bauer. London. 1852.
OKEN, LORENZ, *Lehrbuch der Natur philosophie.* Jena, 1809.
PFANNENSTIEL, MAX, *Lorenz Oken, sein Leben und Werke.* Freiberger Universitätsreden, *Neue Folge, Heft* 14. Freiburg, 1953.
SCHELLING, F. W. J. VON, *Ideen zu einer Philosophie der Natur.* Jena, 1803.

Chapter 18
Everything Alive Has Cellular Origin

Allgemeine Deutsche Biographie, Vol. 31. Schleiden, Leipzig, 1890.
BAER, K. E. VON, *De ovi mammalium et hominis genesi.* Leipzig, 1827.
FLORKIN, MARCEL, *Theodor Schwann et les débuts de la médicine scientifique.* Paris, 1956.
JOHN, HENRY, *Jan Evangelista Purkinje.* Philadelphia, 1959.

KOLLER, GOTTFRIED, *Das Leben des Biologen, Johannes Müller*. Stuttgart, 1958.

MEYER, ARTHUR WILLIAM, *Human Generation*. Stanford, 1956.

MÜLLER, JOHANNES, *Handbuch der Physiologie des Menschen*. Coblenz, 1837-40.

PURKINJE, JAN EVANGELISTA, *Beobachtungen und Versuche zur Physiologie der Sinne*. Berlin, 1825.

SCHLEIDEN, MATTHIAS JAKOB, *Beitrag zur Physiogenesis, Archiv für Anatomie und Physiologie*. Berlin, 1838.

SCHWANN, THEODOR, *Mikroskopische Untersuchungen über die Übereinstimmung in der Struktur und dem Wachstum der Thiere und Pflanzen*. Berlin, 1839.

Chapter 19
The Triumph of the Atom

BERZELIUS, JÖNS JACOB, *Autobiographical Notes*, trans. by Olaf Larswell. Baltimore, 1934.

———, *A View of the Progress and Present State of Animal Chemistry*, trans. by Gustavus Brunmark. London, 1813.

JAFFE, BERNARD, *Crucibles*. New York, 1936.

PATTERSON, ELIZABETH C., *Dalton*. New York, 1970.

VALENTIN, JOHANNES, *Friedrich Wohler*. Stuttgart, 1949.

Chapter 20
A Living Laboratory

BERNARD, CLAUDE, *Leçons de physiologie expérimentale*. Paris, 1855.

———, *La Science Expérimentale*. Paris, 1878.

BOUSSINGAULT, JEAN BAPTISTE, *Rural Economy*, trans. by George Law. Philadelphia, 1850.

CHARDON, CARLO E., *Boussingault*. Ciudad Trujillo, 1953.

DU BOIS-REYMOND, EMIL, *Über die Grenzen des Naturkennens*. Leipzig, 1884.

GABRIEL, MORDECAI, and FOGEL, SEYMOUR, *Great Experiments*. Englewood Cliffs, 1959.

OLMSTEAD, J. M. D., *Claude Bernard, Physiologist*. New York, 1938.

TASHIS, JEROME, *Claude Bernard*. New York, 1968.

VOIT, C., *Emil Du Bois-Reymond*. Königliche Bayerische Academie der Wissenschaft, *Sitzungsbericht der Mathematischen, Physikalischen Klasse, Band* XXVII. Munich, 1897.

Chapter 21
"It's Dogged as Does It"

DARWIN, CHARLES, *The Autobiography of Charles Darwin*, Francis Darwin, ed. New York, Dover, 1958.

———, *The Descent of Man.* London, 1871.
———, *The Origin of Species,* 5th ed. New York, 1959.
———, *The Voyage of the Beagle,* Leonard Engel, ed. New York, 1962.
EISELY, LOREN, *Darwin's Century.* New York, 1958.
WALLACE, ALFRED RUSSEL, *Contributions to the Theory of Natural Selection.* London, 1870.
———, *Island Life.* London, 1880.
———, *My Life,* 2 vols. New York, 1905.
———, *The Malay Archipelago.* London, 1869.

Chapter 22
Darwin's Champions

ACKERKNECHT, ERWIN H., *Rudolf Virchow: Doctor, Statesman, Anthropologist.* Madison, 1953.
BIBBY, CYRIL, *Thomas H. Huxley.* London, 1959.
BOELSCHE, WILHELM, *Ernst Haeckel.* Berlin-Leipzig, 1900.
HAECKEL, ERNST, *Anthropogenie.* Leipzig, 1874.
———, *Freedom & Science in Teaching.* London, 1892.
———, *Love Letters,* trans. by Ida Zeitlin. New York, 1930.
———, *Die Radiolarien.* Berlin, 1862.
———, *The Riddle of the Universe.* New York, 1901.
HERTWIG, OSCAR, *The Embryology of Man and Mammals,* trans. by Edward Mark. New York, 1899.
HUXLEY, THOMAS H., *Evidence as to Man's Place in Nature.* Ann Arbor, 1953.
———, *Lectures on the Elements of Comparative Anatomy.* London, 1864.
———, *The Oceanic Hydrozoa: A Description of the Caliphoridae and Physophorodae Observed During the Voyage of the H.M.S. Rattlesnake in the Years 1846-1850.* London, Royal Society Publications, 1859.
VIRCHOW, RUDOLF, *The Freedom of Science in the Modern State.* London, 1878.

Chapter 23
The Deep Sea Floor

HUXLEY, THOMAS H., "Some of the Results of the Expedition of H.M.S. *Challenger.*" *Popular Science Monthly,* Vol VII, May 1875 pp. 26-45.
SPRY, W. J. J., *The Cruise of the H.M.S. Challenger.* Toronto, 1877.
SWIRE, HERBERT, *The Voyage of the H.M.S. Challenger 1872-76.* London, 1938.

THOMSON, SIR C. WYVILLE, *The Atlantic: A Preliminary Account of the General Results of the Exploring Voyage of H.S.M. Challenger,* 2 vols. New York, 1898.

Chapter 24
Paleontology and the Wild West

COPE, EDWARD DRINKER, *The Vertebrata of the Tertiary Formations of the West.* Book I, Report, the U.S. Geological Survey of the Territories. Washington, D.C., 1884.

———, "*Batrachia* of North America." *Bulletin, American Natural History Museum,* No. 34. New York, 1889.

JAFFE, BERNARD, *Men of Science in America.* New York, 1944.

MARSH, OTHNIEL CHARLES, "The Dinosaurs of North America." *United States Geological Survey Annual Report.* Washington, D.C., 1894.

———, *Fossil Horses in America.* Salem, 1874.

———, "Odontornithes: A Monograph on the Extinct Toothed Birds of North America." *United States Geological Exploration,* No. 4 Washington, D.C., 1880.

OSBORN, HENRY FAIRFIELD, *Cope: Master Naturalist.* Princeton, 1931.

SCHUCHERT, CHARLES, LEVENE, CLARA MAY, *O. C. Marsh, Pioneer in Paleontology.* New Haven, 1940.

Chapter 25
The Reign of the Fruit Fly

CUNY, HILAIRE, *Morgan et la génétique.* Paris, 1969.

JAFFE, BERNARD, *Men of Science in America.* New York, 1944.

———, *Outposts of Science.* New York, 1935.

———, *Embryology & Genetics.* New York, 1934.

MORGAN, THOMAS HUNT, *The Mechanism of Mendelian Heredity.* New York, 1915.

———, *The Physical Basis of Heredity.* Philadelphia, 1919.

Chapter 26
Inside the Giant Molecule

BARRY, J. M., *Molecular Biology, Genes & the Chemical Control of Living Cells.* Englewood Cliffs, 1964.

CHARGAFF, ERWIN, "The Paradox of Biochemistry." *Columbia Forum,* Vol. XIII, No. 2 (summer, 1969).

COMMONER, BARRY, *Science and Survival.* New York, 1966.

CURTIS, HELEN, *Biology: The Science of Life.* New York, 1963.

———, *The Viruses.* New York, 1968.

KENDREW, JOHN, *The Thread of Life.* Cambridge, 1966.

LINDGREN, CARL C., *The Cold War in Biology.* Ann Arbor, 1966.

STENT, GUNTHER, *The Coming of the Golden Age.* New York, 1969.

TAYLOR, GORDON RATTRAY, *The Biological Time Bomb.* New York, 1968.

WATSON, JAMES D., *The Double Helix.* New York, 1968.

Chapter 27
Insect Homers

BOUVIER, EUGENE, "Jean Henri Fabre, Life and Work." *Smithsonian Institution Annual Report.* Washington, D.C., 1916.

FABRE, JEAN HENRI, *The Mason Bees.* New York, 1914.

———, *The Life of the Fly.* New York, 1913.

———, *The Life of the Spider.* New York, 1915.

———, *Souvenirs entomologiques.* Paris, 1914-24.

———, *The Hunting Wasps.* New York, 1916.

FRISCH, KARL VON, *Bees: Their Vision, Chemical Senses and Language.* Ithaca, 1950.

———, *The Dancing Bees.* London, 1954.

———, *Erinnerungen eines Biologen.* Berlin, 1957.

REVEL, E., *J. H. Fabre, l'Homère des Insectes.* Paris, 1957.

WENNER, ADRIAN M., "Honey Bees," in Thomas Sebeok, ed., *Animal Communication.* Bloomington, Indiana, 1968.

Chapter 28
Only Human

BLISS, EUGENE L., *Roots of Behavior.* New York, 1962.

CLARK, RONALD, *The Huxleys.* New York, 1968.

HUXLEY, JULIAN, "The Courtship Habits of the Great Crested Grebe." *Proceedings the Zoological Society, Sept. -Dec.* London, 1914.

———, ed., *A Discussion of the Ritualization of Behavior in Animals and Man,* Philosophical Transactions, Royal Society of London, Vol. CCLI, No. 722. London, 1966.

LEHRMAN, DANIEL, "A Critique of Konrad Lorenz's Theory of Instinctive Behavior." *The Quarterly Review of Biology,* Vol. XXVIII, No. 4 (December, 1953).

LORENZ, KONRAD, Z., *Evolution and the Modification of Behavior.* Chicago, 1965.

———, *King Solomon's Ring.* New York, 1961.

———, *On Aggression.* New York, 1963.

MONTAGU, ASHLEY, *Man and Aggression.* New York, 1968.

SCOTT, JOHN PAUL, *Aggression.* Chicago, 1958.

THORPE, W. H., and LANGWILL, O. L., *Current Problems in Animal Behavior.* Cambridge, 1961

TINBERGEN, NIKO, *Curious Naturalists.* New York, 1969.

———, *The Herring Gull's World.* New York, 1961.

——, *Social Behavior in Animals*. New York, 1954.
——, *The Study of Instinct*. New York, 1952.
"Nikolaas Tinbergen." *The New Scientist*, Vol. III, No. 66. London, 1958.

Chapter 29
Man's Nearest Relatives

ALTMAN, STEWART A., "Social Behavior of Anthropoid Primates," in E. L. Bliss, ed., *Roots of Behavior*. New York, 1962.
CARPENTER, R. C., "A Field Study of the Behavior and Social Relations of Howling Monkeys. *Comparative Psychology Monographs*, Vol. X, No. 2 (1934).
DE VORE, IRVING, ed., *Primate Behavior*. New York, 1965.
GOODALL (LAWICK), JANE, *My Friends the Wild Chimpanzees*. Washington, D.C., 1967.
——, "New Discoveries Among African Chimpanzees." *National Geographic*. Washington, D.C., 1965.
——, "Tool Using and Aimed Throwing in a Community of Free-Living Chimpanzees." *Nature*, 1964.
SCHALLER, GEORGE B., *The Mountain Gorilla*. Chicago, 1963.
——, *The Year of the Gorilla*. Chicago, 1964.
SOUTHWICK, CHARLES H., ed., *Primate Social Behavior*. Princeton, 1963.
STOLLNITZ, FRED, "Behavior of Nonhuman Primates," in E. L. Bliss, ed., *Roots of Behavior*. New York, 1962.
YERKES, R. M. and A. W., *The Great Apes*. New Haven, 1929.
ZUCKERMAN, SIR SOLLY, *The Social Life of Monkeys and Apes*. London, 1932.

Chapter 30
Fellow Citizens of the Planet

CALDWELL, KEITH, *Environment: A Challenge for Modern Society*. New York, 1970.
CARSON, RACHEL, *Silent Spring*. Greenwich, 1970.
COMMONER, BARRY, *Science and Survival*. New York, 1966.
CROWE, PHILIP KINGSLAND, *The Empty Ark*. New York, 1967.
DUBOS, RENÉ, *The Human Animal*. New York, 1968.
——, *Reason Awake*. New York, 1970.
EHRLICH, PAUL and ANN, *Population, Resources, Environment*. San Francisco, 1970.
FISHER, JAMES; SIMON, NOEL; VINCENT, JACK; *et al*, *Wildlife in Danger*. New York, 1969.
GRAHAM, FRANK, *Since Silent Spring*. Greenwich, 1970.
HELFRICH, HAROLD W., ed., *The Environmental Crisis*. New Haven, 1970.
HUXLEY, SIR JULIAN, *The Conservation of Wildlife*. London, 1961.
——, *Religion Without Revelation*. New York, Mentor, 1957.

KLOPFER, PETER H., *Behavioral Aspects of Ecology*. Englewood Cliffs, 1962.

ODUM, EUGENE P., *Fundamentals of Ecology*. Philadelphia, 1953.

OSBORN, HENRY FAIRFIELD, *Our Plundered Planet*. Boston, 1948.

"Paul Revere of Ecology." *Life* (February 2, 1970), pp. 56-63.

PEARSE, A. S., *Animal Ecology*. New York, 1939.

SCHALLER, GEORGE, *The Deer and the Tiger*. Chicago, 1967.

SCHEFFER, VICTOR B., *The Year of the Whale*. New York, 1969.

Index